Günther Ludwig

# An Axiomatic Basis for Quantum Mechanics

## Volume 2
## Quantum Mechanics and Macrosystems

With 4 Figures

Springer-Verlag
Berlin Heidelberg NewYork Tokyo

**Günther Ludwig**
Fachbereich Physik der Philipps-Universität
Arbeitsgruppe Grundlagen der Physik
Renthof 7, D-3550 Marburg
Federal Republic of Germany

Translated from the German manuscript by Kurt Just

ISBN-13:978-3-642-71899-1 e-ISBN-13:978-3-642-71897-7
DOI: 10.1007/978-3-642-71897-7

Library of Congress Cataloging-in-Publication Data
(Revised for volume 2)
Ludwig, Günther, 1918– . An axiomatic basis for quantum mechanics. Bibliography: v. 1, p. [235]–238; v. 2, p. Includes indexes. Contents: v. 1. Derivation of Hilbert Space structure – v. 2. Quantum mechanics and macrosystems. 1. Quantum theory. 2. Axioms. I. Title. QC174.12.L82 1985 530.1 85-2711
ISBN-13:978-3-642-71899-1


Softcover reprint of the hardcover 1st edition 1987



2153/3020-543210

# Preface

In the first volume we based quantum mechanics on the objective description of macroscopic devices. The further development of the quantum mechanics of atoms, molecules, and collision processes has been described in [2]. In this context also the usual description of composite systems by tensor products of Hilbert spaces has been introduced.

This method can be formally extrapolated to systems composed of "many" elementary systems, even arbitrarily many. One formerly had the opinion that this "extrapolated quantum mechanics" is a more comprehensive theory than the objective description of macrosystems, an opinion which generated unsurmountable difficulties for explaining the measuring process. With respect to our foundation of quantum mechanics on macroscopic objectivity, this opinion would mean that our foundation is no foundation at all.

The task of this second volume is to attain a compatibility between the objective description of macrosystems and an extrapolated quantum mechanics. Thus in X we establish the "statistical mechanics" of macrosystems as a theory more comprehensive than an extrapolated quantum mechanics.

On this basis we solve the problem of the measuring process in quantum mechanics, in XI developing a theory which describes the measuring process as an interaction between microsystems and a macroscopic device. This theory also allows to calculate "in principle" the observable measured by a device. Neither an incorporation of consciousness nor a mysterious imagination such as "collapsing" wave packets are necessary.

In XII fundamental problems such as the EPR-paradox are clarified on the basis laid down in the previous chapters.

Chapter XIII is devoted to general problems of any physical theory concerning the "desired" form of a theory, the physical significance of the "laws" and the concepts of "real" and "possible". All this is demonstrated with the example of quantum mechanics, whereby the reality of the atoms is examined.

References in the text are made as follows: For references to other sections of the same chapter, we only list the section number of the reference; for example § 2.3. For references to other chapters, the chapter is also given; for example III 4.2 refers to section 4.2 of chapter III. The formulas are numbered as follows: (3.2.7) refers to the 7th formula in section 3.2 of the current chapter. References to formulas of other chapters are given, for example, by III (2.1.8).

The numbers in the bibliography are the same as in the first volume. Thus the bibliography of this second volume is only an extension of that of the first volume.

I would like to express my gratitude to *Professor Leo F. Boron* for the translation of the first version [47] of the manuscript. A considerable extension has been initiated by additional questions of *Professor Kurt Just*. The significance of the improvements by these additional parts can only be estimated by a comparison with the first version. To *Professor Just* I owe not only the English version of the additional parts but also many improvements of the whole text, achieved by many discussions. Without his help many items would not have been clarified. Therefore I thank him very much for his extensive collaboration.

I also thank *Dr. Reinhard Werner* for reading the text of X § 3 and for proposing many condensations to make the text easier to understand.

The developments in XIII §§ 2.2 to 2.5, going beyond those of [3], are the result of many discussions with *Professor Erhard Scheibe* for which I thank him very much.

Marburg, March 1987 G. Ludwig

# Contents

**Contents of Vol. 1:**

**Derivation of Hilbert Space Structure**

# IX Further Structures of Preparation and Registration

As already mentioned in Chapter VIII §6, the quantum mechanics of atoms and molecules is not yet closed by the base sets and structure terms introduced until now. We must yet go over to standard extensions (in the sense of XIII §3 and [3] §8). Perhaps we then can obtain a g.G.-closed theory (in the sense of XIII §4.3 and [3] §10.3) for the fundamental domain of "atoms and molecules". This extension problem was explained in detail in [2]. Readers who already know quantum mechanics, however, should neither study [2] nor refer to it again and again. Hence let us here, without mathematical axioms and proofs, summarize the essential further structures. This will enable us, in X and XI to invoke these structures by referring to this chapter.

## §1 Transformations of Registration Procedures Relative to Preparation Procedures

In III §7 we have already introduced a further base term $\Delta$ and a further structure term III (7.1), with physical interpretations and several axioms. We must now inquire, what these further structures imply in connection with the preparation and registration axioms introduced in VI. In III §7, the elements $\delta$ of $\Delta$ could already be defined as mappings of $\mathscr{R}_0$, $\mathscr{R}$ into themselves. This leads immediately (in a canonical way) to a mapping $\mathscr{F} \xrightarrow{\delta} \mathscr{F}$, with $\mathscr{F}$ as in III D4.3. The mapping $\delta$, representing a displacement of the registration by the time $\tau$, will be important in X §2 and XI §6. This $\delta$ has been denoted by $\delta_\tau$. Thus $\delta_\tau b_0$ is the registration method $b_0$ displaced by the time $\tau$, while $\delta_\tau b$ are the registration procedures of the method $\delta_\tau b_0$.

In V §2 we defined the mixture-morphisms of $K$ into itself and also the mixture-isomorphisms of $K$ into itself, i.e. the mixture-automorphisms of $K$. Mixture-automorphisms are just automorphisms of $K$ into itself relative to the convex structure of $K$. If $S$ is a mixture-automorphism, the dual mapping $S'$ maps the set $L=[\mathbf{0}, \mathbf{1}]$ (bijectively and $\sigma(\mathscr{B}', \mathscr{B})$-continuously) into itself (see V §2). We call the mapping $S'$ dual to a mixture-automorphism a $\mathscr{B}$-continuous effect-automorphism. Let $\mathscr{A}$ be the group of all $\mathscr{B}$-continuous effect automorphisms.

Under physically natural assumptions for the mappings $\mathscr{F} \xrightarrow{\delta} \mathscr{F}$ with $\delta \in \Delta$, one concludes that the Galileo group $\Delta_{\mathscr{G}}$ from III §7 is represented in $\mathscr{A}$ ($\Delta_{\mathscr{G}} \to \mathscr{A}$, see [2] V, VI). With $\delta \in \Delta_{\mathscr{G}}$ and $g \in L$, for $g' = \delta g$ we therefore find the physical interpretation:

The effect $g'=\delta g$ arises from $g$ if one subjects $g$ to the Galileo transformation $\delta$ relative to the preparation (i.e. if $\delta$ maps a device that registers $g$ on another device, as described in III §7).

According to Wigner's theorem (see [2] V §5), the part of $\mathscr{A}$ connected with the unit consists of all transformations $g'=U g U^+$, where $U$ in each of the irreducible parts $\mathscr{B}'_\nu$ (determined by a Hilbert space $\mathscr{H}_\nu$; see VII §5.4 and VIII) is a unitary operator in $\mathscr{H}_\nu$. For $g=\sum_\nu g_\nu$, this says $U g U^+ = \sum_\nu U_\nu g U_\nu^+$ with unitary operators $U_\nu$.

Therefore let us as $\Delta_{\mathscr{g}}$ consider only those Galileo transformations which are connected to the unit (only these found a physical interpretation in III §7; the interpretation of the reflections is *not* so self-evident; also see [2] VII). Then to each representation $\Delta_{\mathscr{g}} \to \mathscr{A}$ there corresponds a "series" of representations (of $\Delta_{\mathscr{g}}$ in each $\mathscr{H}_\nu$) via unitary transformations up to a factor. Thus it is customary to consider the representations of $\Delta_{\mathscr{g}}$ up to a factor separately for each system type (VII §6).

The well known meaning of a "representation $U(\delta)$ up to a factor" is that $\delta_1 \delta_2 = \delta_3$ only implies $U(\delta_1)\, U(\delta_2) = e^{i\lambda}\, U(\delta_1 \delta_2)$ with a "factor" $e^{i\lambda}$. Two representations up to a factor may briefly be called factor equivalent if they are identical as representations in $\mathscr{A}$. Two representations $U(\delta)$ and $\tilde{U}(\delta)$ up to a factor are factor equivalent if and only if

$$\tilde{U}(\delta) = e^{i\alpha(\delta)}\, U(\delta).$$

A system type is called "elementary" if the representation of $\Delta_{\mathscr{g}}$ for this system type is irreducible (also see [2] VII §2). The system types which are not elementary are called "composite".

According to AV 4a, the Hilbert space corresponding to each system type is either infinite-dimensional or one-dimensional. We now make the further assumption that there is only a single "one-dimensional" system type, in which $\Delta_{\mathscr{g}}$ then must experience the identity representation(!). This single system type is called the "vacuum". Moreover, we assume that $\Delta_{\mathscr{g}}$ does *not* experience that identity representation for any other system type. Since the identity is the only finite-dimensional unitary representation of $\Delta_{\mathscr{g}}$ up to a factor, AV 4a requires that no system type (except the vacuum) has a finite-dimensional Hilbert space.

The irreducible representations up to a factor of the Galileo group are well known (see [2] VII). One naturally combines factor equivalent representations up to a factor into "the same" representation, since they yield the same representation in $\mathscr{A}$. Thus, for each elementary system type, two parameters $m$ and $s$ uniquely characterize the irreducible representation. While $m$ can be an arbitrary positive number, $s$ can only take the values 0, 1/2, 1, 3/2, .... From among the factor equivalent representations up to a factor, one can be choosen in such a way that the $U(\delta)$ have the following form:

For $\delta$ a translation in space by a vector $\boldsymbol{a}$, we have

$$U(\boldsymbol{a}) = e^{i\boldsymbol{K}\cdot\boldsymbol{a}},$$

where the components $K_\nu$ of $\boldsymbol{K}$ are self-adjoint operators in the Hilbert space $\mathscr{H}$ of the irreducible system type considered. The $K_\nu$ commute.

For $\delta$ a proper Galileo transformation, i.e imparting a velocity $\boldsymbol{v}$, we have

$$U(\boldsymbol{v})=e^{i\boldsymbol{X}\cdot\boldsymbol{v}}, \tag{1.2}$$

where the $\boldsymbol{X}$ are self-adjoint operators in $\mathscr{H}$. While the $X_\nu$ commute, the $K_\nu$ and $X_\nu$ have the "commutators"

$$K_\mu X_\nu - X_\nu K_\mu = im\delta_{\nu\mu}\mathbf{1} \tag{1.3}$$

with the parameter $m$ mentioned above.

For $\delta$ a time translation by $\tau$ (i.e. for $\delta=\delta_\tau$) we have

$$U(\tau)=e^{iH\tau}, \tag{1.4}$$

where $H$ is the self-adjoint operator

$$H=\frac{1}{2m}\boldsymbol{K}^2. \tag{1.5}$$

The Hilbert space $\mathscr{H}$ can be written as a product space

$$\mathscr{H}=\mathscr{H}_b\times\mathscr{H}_s \tag{1.6}$$

so that the $K_\mu$ and $X_\nu$ become

$$K_\mu : K_\mu\times\mathbf{1}; \qquad X_\nu : X_\nu\times\mathbf{1} \tag{1.7}$$

and are irreducible in $\mathscr{H}_b$.

For $\delta$ a rotation through the angle $\alpha$ about the $\nu$-axis, we have

$$U_\nu(\alpha)=e^{iJ_\nu\alpha}. \tag{1.8}$$

The $J_\nu$ are self-adjoint operators, which relative to (1.6) become

$$J_\nu=L_\nu\times\mathbf{1}+\mathbf{1}\times S_\nu \tag{1.9}$$

with

$$\boldsymbol{L}=-\frac{1}{m}\boldsymbol{X}\otimes\boldsymbol{K} \tag{1.10}$$

($\otimes$ denotes the exterior vector product). The $iS_\nu$ are the well-known infinitesimal rotation operators for an irreducible unitary representation of the rotation group up to a factor in $\mathscr{H}_s$. The possible irreducible representations are specified by an $s=0, 1/2, 1 \ldots$ so that $\mathscr{H}_s$ has the dimension $(2s+1)$ and each $S_\nu$ has the eigenvalues $-s, -s+1, \ldots, s$.

Since the $L_\nu$, $S_\nu$, $J_\nu$ are self-adjoint operators, to each there corresponds a scale observable (see VIII T 4.3.4). The observable corresponding to $\boldsymbol{L}$ is called the orbital (angular) momentum, that corresponding to $\boldsymbol{S}$ is called the spin (angular) momentum, and the observable for $\boldsymbol{J}$ is called the total angular momentum (of the elementary system type).

The characteristic parameter $m$ is called the mass of the elementary system and the characteristic parameter $s$ is called the spin of the system.

For an elementary system type, the scale observables "position" and "momentum" are defined uniquely (see [2] VII §4).

## §2 Composite Systems and Scattering Experiments

We assume that the reader knows the usual description of composite systems in quantum mechanics. The most significant results of quantum mechanics rest precisely on the possibility of completely describing not only elementary system types but also composites such as atoms, molecules, and colliding systems.

In §1 we have already distinguished elementary and composite systems. It is typical for the elementary systems that their structure is completely specified by the representation of the Galileo group, i.e that they have no "inner" structure. For composite systems the description by the Galileo group does not suffice. The inner structure of composite systems remains a problem.

Concerning the representation of the Galileo group for a composite system (see [2] VIII), first of all we can represent the Hilbert space $\mathcal{H}$ in the form

$$\mathcal{H} = \mathcal{H}_b \times \mathcal{H}_i. \tag{2.1}$$

Here the representing operators (for the notation see §1) have a form

$$U(\boldsymbol{a}) = e^{i\boldsymbol{K}\cdot\boldsymbol{a}} \times \mathbf{1}, \tag{2.2}$$

$$U(\boldsymbol{v}) = e^{i\boldsymbol{X}\cdot\boldsymbol{v}} \times \mathbf{1}, \tag{2.3}$$

$$U(A) = V(A) \times R(A). \tag{2.4}$$

$A$ is a rotation in space, $\mathcal{H}_b$ is irreducible relative to the operators $\boldsymbol{K}$ and $\boldsymbol{X}$.

One frequently calls $\mathcal{H}_b$ the "orbit space" and $\mathcal{H}_i$ the space of the inner structure. This terminology is not entirely logical since $\mathcal{H}_i$ is identical with the spin space $\mathcal{H}_s$ for elementary systems without inner structure. The decisive difference between elementary and composite systems just lies in the representation of the rotation group, which in $\mathcal{H}_i$ is *not* irreducible. This implies that for a time translation $\delta_\tau$ (see §1) we have

$$U(\tau) = e^{iH\tau} \tag{2.5}$$

with

$$H = \frac{1}{2M} \boldsymbol{K}^2 \times \mathbf{1} + \mathbf{1} \times H_i. \tag{2.6}$$

Though $H_i$ commutes with the $R(A)$, it is not a multiple of the $\mathbf{1}$-operator. The determination of $H_i$ (i.e. of $H$) is one of the essential problems in a theory of composite systems. For this problem there still is no mathematically exact theory. Either one can guess $H$ by the correspondence principle, or obtain it as an approximation from quantum electrodynamics (using a renormalization for eliminating mathematically incorrect expressions). In this book we need not elaborate the $H$ so obtained (see [2] VIII (5.8)).

Yet we must more precisely say what it really means that "atoms and molecules are composed of electrons and atomic nulcei", what after all is really meant by the word "electrons" and "atomic nuclei".

Up to now, we only have defined composite systems but not yet said what it means that every composite system type is composed of a *definite* number of elementary systems. Naturally, the structures introduced until now in the mathematical picture do not suffice to describe this composition.

As a further structure, we must introduce the composition of finitely many preparation procedures into a new preparation procedure. This composition is of fundamental importance for scattering theory (see Fig. 1). We shall consider it mathematically only for the case of two preparation procedures (one will immediately recognize how to extend it to finitely many procedures).

We introduce the new structure via a subset $\Pi \subset \mathcal{Q}' \times \mathcal{Q}'$ and a mapping $\Pi \xrightarrow{\gamma} \mathcal{Q}'$ (also see [2] XVI §1). In the physical interpretation, $(a_1, a_2) \in \Pi$ means that the two preparation devices corresponding to $a_1$ and $a_2$ may be combined as described in III for a preparation and a registration device. The two combined preparation devices provide a new preparation procedure, for which we simply write $a = \gamma(a_1, a_2)$.

If the $a_1$ and $a_2$ represent precisely such preparation procedures which produce *only* systems of an elementary system type, then $\gamma(a_1, a_2)$ represents systems of a composite system type. One can formulate this axiomatically:

If $\varphi(a_1) \in K_{\nu_1}$ and $\varphi(a_2) \in K_{\nu_2}$ (where $\nu_1$ and $\nu_2$ denote irreducible parts and their system types are elementary), we have $\varphi(\gamma(a_1, a_2)) \in K_\mu$, where $\mu$ characterizes a composite system type.

If one composes a finite number of preparation procedures, $\gamma$ generates mappings of the following form: Let $\nu_1, \ldots, \nu_r$ be elementary system types and $n_1, \ldots, n_r$ positive integers. Then $\gamma$ generates mappings

$$(n_1, \nu_1), (n_2, \nu_2), \ldots, (n_r, \nu_r) \to \mu, \tag{2.7}$$

where $\mu$ is a composite system type.

In principle, there could be many such mappings (2.7), which could further depend on the preparation procedures $a_1, a_2, \ldots$ used for the composition. For non-relativistic(!) quantum mechanics, the following axiom (natural law) turns out as useable:

The mappings (2.7) are bijective. Here, bijective means that for each composite system type there is precisely one sequence $(n_1, \nu_1), (n_2, \nu_2), \ldots, (n_r, \nu_r)$ which obeys (2.7).

If one assigns the number $n = 0$ to those elementary system types which do not appear on the left side of (2.7), one can also formulate the axiom as follows: Each composite system type is uniquely characterized by positive integers $n_1, n_2, \ldots$ (with $\sum_i n_i < \infty$), where $n_i$ is the number of elementary systems of type $i$ on the left side of (2.7). For this one briefly says: The system is composed of $n_1$ elementary systems of type 1, $n_2$ elementary systems of type 2, and so forth.

This axiom made it easy to develop a non-relativistic quantum mechanics of atoms, molecules, and collisions.

In order to concretize the theory, one must introduce further axioms for the elementary system types that actually occur. It is typical for physical theories that the axioms never finally solve a problem. They only present a useable approximation in a certain field of application (fundamental domain in the sense of XIII §1 and [3] §3). Every physical theory is an approximation (see XIII §3 and [3] §§6 and 9). Here we shall not consider several approximations jointly (this would cause unnecessary complications). We only formulate an approximation very useful for atoms and molecules.

The following occur as elementary system types:

First, a system type with a small mass m (relative to the remaining masses), the spin $s=1/2$ and the electric charge $(-e)$ ($e$=elementary charge). This system type is called an "electron".

Second, there is a series of elementary system types of "larger" masses $m_i$, of different spins and charges $Ze$ ($Z$ a positive integer). One calls these elementary system types "atomic nuclei of the charge numbers $Z$".

That we describe the nuclei as elementary system types is just the approximation used. It is customary to characterize the atomic nucleus by two positive integers $Z$ and $A$ (instead of $Z$ and its mass $m$), where $A$ is called the nucleon number. It determines the mass by $m=Am_p-\Delta$, where $m_p$ is the mass of the "smallest" nucleus with $Z=1$ (the proton) and $\Delta$ (small relative to $Am_p$) is called the mass defect.

Here we have introduced the new concept of the electric charge, which has originally been defined in one of the macroscopic pretheories. In quantum mechanics the electric charge is defined indirectly, namely as a parameter in the Hamiltonian (2.6) (see $H$ in [2] VIII (5.8)). The charge of the system types is especially simply measured in "exterior fields", where the Hamiltonian has a particularly simple structure (see [2] VIII §6).

If one has completed the theory in this way by further structures (with axioms), one can uniquely characterize each composite system type by its numbers of electrons and of different nuclei.

If the composite system contains only one atomic nucleus, it is called an atom, otherwise a molecule. If the number of electrons in the system equals the sum of the nuclear charge numbers, it is called neutral, otherwise ionized. On historical grounds, it is usual to give names to atoms with various nuclear charge numbers: $Z=1$, Hydrogen; $Z=2$, Helium, etc.

In this book, not even a reference to the structure of atoms and molecules is needed. The reader interested in a logical structure analysis on the basis of prepara-

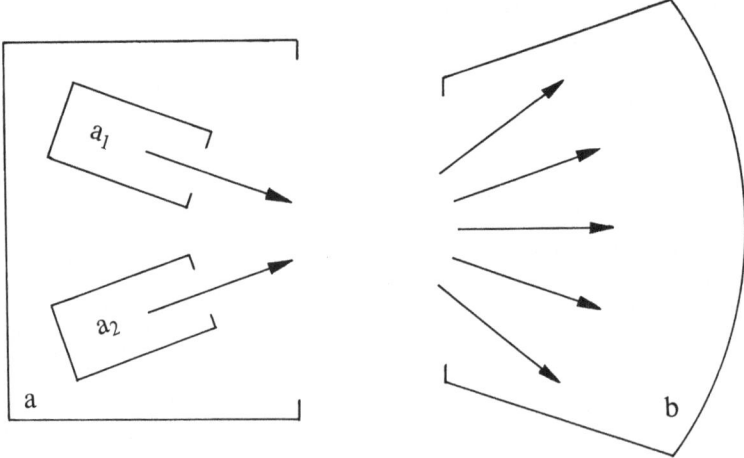

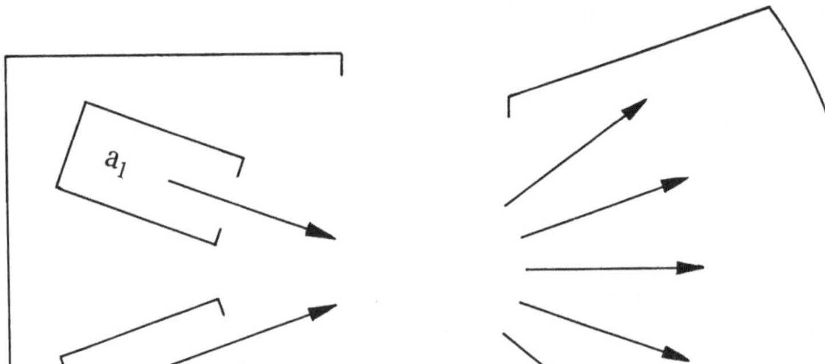


**Fig. 7**

tion and registration procedures is referred to [2] XI through XV. Let us only recall that the Hilbert space of a composite system is obtained from a product space by reducing it to the subspaces symmetric and anti-symmetric in identical subsystems (see [2] VIII).

We must go into more detail for basic concepts of scattering theory since these will be used in XI. We assume that the reader knows scattering theory; a short introduction is given in [2] XVI.

A scattering experiment is characterized by a device symbolically shown in Fig. 7. Two preparation devices produce the systems which collide. With the mapping $\gamma$ introduced above, the two devices taken together represent a preparation procedure $\gamma(a_1, a_2)$. The scattering problem consists in calculating $\varphi(\gamma(a_1, a_2))$ when $\varphi(a_1)$ and $\varphi(a_2)$ are given.

The systems produced by the composite preparation procedure $\gamma(a_1, a_2)$ are registered. The experimental registrations in may cases yield the (differential) cross section (see [2] XVI §6).

The calculation of $\varphi(\gamma(a_1, a_2))$ from $\varphi(a_1)$, $\varphi(a_2)$ is in principle based on the Hamiltonian (2.6). This happens as follows:

One first introduces $\varphi(a_1)\times\varphi(a_2)$ in the product space $\mathscr{H}_1\times\mathscr{H}_2$ and then goes over to a symmetric operator

$$w^i=\{\varphi(a_1)\times\varphi(a_2)\}_s \tag{2.8}$$

which is defined in the symmetrized resp. antisymmetrized subspace $\{\mathscr{H}_1 \times\mathscr{H}_2\}_s\subset\mathscr{H}_1\times\mathscr{H}_2$ and there has the trace 1. Here, $w^i$ intuitively represents an ensemble in which there seems to be no interaction between the systems produced according to $a_1$ and to $a_2$. Scattering theory indicates how the interaction in principle leads to a "wave operator" $\omega_-$ (on $\{\mathscr{H}_1\times\mathscr{H}_2\}_s$), so that

$$\varphi(\gamma(a_1, a_2))=\omega_-\, w^i\, \omega_-^+ \tag{2.9}$$

holds. The operator $\Omega_-$ defined by

$$\Omega_-\, w^i=\omega_-\, w^i\, \omega_-^+ \tag{2.10}$$

is a mixture-morphism which maps the set K of ensembles corresponding to $\{\mathscr{H}_1 \times\mathscr{H}_2\}_s$ into itself, so that $\Omega_- K$ is in general a proper subset of $K$! Also $\Omega_-$ is briefly called a wave operator.

The scattering experiment is completely described by $\Omega_-$ since $\mu(\Omega_- w^i, g)$ gives the probabilities for all possible effects g. But it is useful, as follows to introduce the so-called scattering operator.

The scattered systems are again without mutual interaction "after the scattering", so that one can define a second wave operator $\omega_+$. Let $w^f$ be an ensemble after scattering which behaves as if no mutual interaction of the scattered systems were present. Then $\omega_+\, w^f\, w_+^+$ is the ensemble with the interaction taken into account. Because it must coincide with the ensemble $\varphi(\gamma(a_1, a_2))$, we get

$$\varphi(\gamma(a_1, a_2))=\omega_+\, w^f\, \omega_+^+ . \tag{2.11}$$

For "complete" wave operators (see [2] XVI D4.3.2), from (2.9) and (2.11) follows

$$w^f=\omega_+^+\, \omega_-\, w^i\, \omega_-^+\, \omega_+ . \tag{2.12}$$

The operator

$$S=\omega_+^+\,\omega_- \tag{2.13}$$

is called the scattering operator. Using it, one can write (2.12) in the form

$$w^f=S\,w^i\,S^+. \tag{2.14}$$

The "motion reversal" transformation (also called time reversal) is of general interest for quantum mechanics and especially problematic for scattering theory. The form of this time reversal is treated more precisely in [2] X §4. Here let us state it briefly and then discuss some questions connected with the g.G.-closure of quantum mechanics, which are of special significance in XI and XII.

Let us define an operator $C$ in the Hilbert space for n electrons. It is not difficult to extend this definition to a system of electrons and atomic nuclei. It is well known that we can represent the Hilbert vectors $\chi$ corresponding to the n-electron system as

$$\sum_{\alpha_1,\ldots,\alpha_n} \langle \boldsymbol{r}_1,\boldsymbol{r}_2,\ldots,\boldsymbol{r}_n\rangle\,|\chi;\alpha_1,\ldots,\alpha_n\rangle\, u_{\alpha_1}\,u_{\alpha_2}\ldots u_{\alpha_n} \tag{2.15}$$

(see [2] IX §6). The $\alpha_i$ take the values $+1/2$ and $-1/2$. Here, $u_{+1/2}$ and $u_{-1/2}$ are basis vectors of the two-dimensional spin space $\mathscr{H}_{1/2}$ for an electron (see (1.6)). The $\langle \boldsymbol{r}_1,\ldots|\chi;\,\alpha_1,\ldots\rangle$ are complex-valued position functions. The representation (2.15) must be chosen so that the position operator $\boldsymbol{Q}_i$ has the form: "multiplication with $\boldsymbol{r}_i$" and the momentum operator $\boldsymbol{P}_i$ has the form: "$\frac{1}{i}$ grad$_i$".

We now define $C$ by

$$\begin{aligned} C\sum_{\alpha_1,\ldots,\alpha_n} &\langle \boldsymbol{r}_1,\ldots|\chi;\alpha_1\ldots\rangle\, u_{\alpha_1}\,u_{\alpha_2}\ldots u_{\alpha_n}\\ =\sum_{\alpha_1,\ldots,\alpha_n} &\overline{\langle \boldsymbol{r}_1,\ldots|\chi;\alpha_1,\ldots\rangle}\,(\alpha_1,\ldots,\alpha_n)\,u_{-\alpha_1}\ldots u_{-\alpha_n}. \end{aligned} \tag{2.16}$$

Here $(\alpha_1,\ldots,\alpha_n)$ equals $+1$ when the number of the $\alpha_i$ with $\alpha_i=-1/2$ is even, and equals $-1$ when this number is odd. $\overline{\langle \boldsymbol{r}_1,\ldots|\ldots\rangle}$ is the complex conjugate of $\langle \boldsymbol{r}_1,\ldots|\ldots\rangle$.

The operator $C$ is anti-unitary (for the definition of an anti-unitary operator, see [2] A IV §13).

$C$ is called the time reversal operator; it makes $C^2=(-1)^n\,1$, hence $C^{-1}=(-1)^n\,C$. We easily find

$$\begin{aligned} C\boldsymbol{Q}_i&=\boldsymbol{Q}_i\,C,\\ C\boldsymbol{P}_i&=-\boldsymbol{P}_i\,C,\\ C\boldsymbol{S}_i&=-\boldsymbol{S}_i\,C,\\ CH&=HC, \end{aligned} \tag{2.17}$$

where $H$ is the Hamiltonian without exterior fields, as given in [2] VIII (5.8).

For $g\in L$,

$$Tg=C\,g\,C^{-1} \tag{2.18}$$

defines a $\mathscr{B}$-continuous effect automorphism $T$ (see V §2 and IX §1). The dual transformation $T'$ is for $w \in K$ given by

$$T' w = C^{-1} w C = C w C^{-1}. \tag{2.19}$$

Although $T$ and $T'$ can easily be defined mathematically, the following physical questions are not easy to answer.

Let an effect procedure $(b_0, b)$ be given by a device. How can another device realize an effect procedure $(b'_0, b')$ which (approximately) makes

$$\psi(b'_0, b') = C \psi(b_0, b) C^{-1}?$$

Let a preparation procedure $a$ be given by a device. How can a device realize a preparation procedure $a'$ which obeys $\varphi(a') = C \varphi(a) C^{-1}$? We shall return to this problem in XII §3.

Here let us only explain why $C$ is also denoted as motion reversal transformation. This comes from the invariance of $H$ under $C$, as expressed by the last equation in (2.17).

From (2.17) follows

$$C e^{iH\tau} = e^{-iH\tau} C. \tag{2.20}$$

The Schrödinger picture has

$$w_\tau = e^{-iH\tau} w_0 e^{iH\tau}. \tag{2.21}$$

With

$$w'_0 = C w_\tau C^{-1}, \tag{2.22}$$

from (2.21) together with (2.20) follows

$$w'_\tau = e^{-iH\tau} w'_0 e^{iH\tau} = C e^{iH\tau} w_\tau e^{-iH\tau} C^{-1} = C w_0 C^{-1}. \tag{2.23}$$

Through a time interval $\tau$, let $w_0$ run according to (2.21). At the time $\tau$, use (2.22) to go from $w_\tau$ over to $w'_0 = C w_\tau C^{-1}$. Let this ensemble again run through the same interval $\tau$. Then by (2.23) we do not get back $w_0$, but the ensemble $C w_0 C^{-1}$ (motion reversed to $w_0$).

## §3 Measurement Scatterings and Transpreparations

In more detail we must consider the application of scattering processes to measurement and to transpreparation. Although such processes are very familiar to experimental physicists (not with this terminology), their theoretical investigation is not common among theorists. We must occupy ourselves more precisely with measurement scatterings and transpreparations, in order to generalize such processes in XI, namely to the scattering of microsystems by macrosystems.

By the structures treated in this §3 (investigated in great detail in [2] XVII), as in the description of scattering in §2 we go beyond the basic structure of coupling only one preparation with only one registration device (used as point of departure in III). We must now deal with more than two devices; but they shall not always be represented mathematically by new structure terms (as in the mapping $\gamma$ in §2).

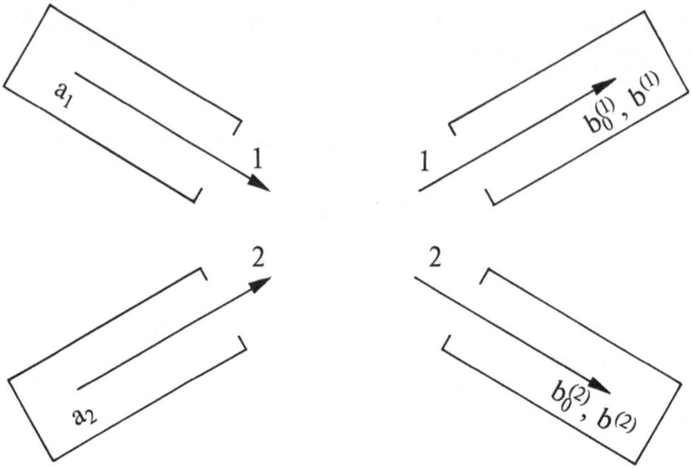


**Fig. 8**

The starting point is a device symbolically shown in Fig. 8, only that we do not compose $a_1$ and $a_2$ to a preparation procedure as in a scattering experiment.

For departure we use the formula (2.14), regarding $w^i$ as substituted form (2.8). For $\varphi(a_1)$ we briefly write $w_1$, for $\varphi(a_2)$ we write $w_2$ and for $w^f$ simply $w$. Then (2.14) becomes

$$w = S(w_1 \times w_2)\, S^+. \tag{3.1}$$

For brevity we have not symmetrized $w_1 \times w_2$ (assuming the systems from $a_1$ and $a_2$ as different).

As a registration procedure, in Fig. 8 we consider one that after the scattering registers only the systems 2, so that we can set $\psi(b_0, b) = \mathbf{1} \times g_2$ (with $g_2 \in L(\mathscr{H}_2)$). The probability for $\psi(b_0, b)$ in the ensemble (3.1) then is

$$tr(S(w_1 \times w_2)\, S^+ (\mathbf{1} \times g_2)). \tag{3.2}$$

Now let us combine the devices corresponding to $a_2$ and to $(b_0, b)$ to a device registering the systems produced by $a_1$. This is indeed possible since (3.2) is (relative to $w_1$) a linear functional $l(w_1)$ over $K(\mathscr{H}_1)$, with $0 \leq l(w_1) \leq 1$, so that there is a $g_1 \in L(\mathscr{H}_1)$ with

$$\operatorname{tr}(S(w_1 \times w_2)\, S^+ (\mathbf{1} \times g_2)) = \operatorname{tr}_1(w_1\, g_1). \tag{3.3}$$

Here, $g_1$ is the effect determined by the registration procedure composed of $a_2$ and $(b_0, b)$. The trace $\operatorname{tr}_1$ must be formed relative to the Hilbert space $\mathscr{H}_1$ of the systems prepared by $a_1$.

For a fixed $w_2$, (3.3) defines a $\mathscr{B}$-continuous effect morphism (as the mapping dual to a mixture-morphism; see V §2). This morphism $L(\mathscr{H}_2) \to L(\mathscr{H}_1)$, given by

$$\mathbf{T}(1, 2; w_2)\, g_2 = g_1, \tag{3.4}$$

is called the *measurement scattering morphism* assigned to the scattering.

We can regard the devices $a_1$, $a_2$, $b_0^{(2)}$ in Fig. 8 (with a registration of "only" the systems 2) as a whole (i.e. all three devices together) as a preparation device for system 1. In order to make this still clearer, we consider a registration device that consists of two parts (see Fig. 8), of which one part registers only the system 1 and the other only the system 2.

First of all, we again can think of $a_1$ and $a_2$ (in the sense of scattering) as an $a$. Likewise, $b_0^{(1)}, b^{(1)}, b_0^{(2)}, b^{(2)}$ form a special registration device $b_0, b$ with

$$\psi(b_0, b) = g_1 \times g_2, \tag{3.5}$$

where

$$g_1 = \psi(b_0^{(1)}, b^{(1)}) \in L(\mathscr{H}_1) \quad \text{and} \quad g_2 = \psi(b_0^{(2)}, b^{(2)}) \in L(\mathscr{H}_2).$$

Therefore, by (3.1) we easily obtain

$$\operatorname{tr}(S(w_1 \times w_2)\, S^+ (g_1 \times g_2)) \tag{3.6}$$

as the probability that $b^{(1)}$ and $b^{(2)}$ are triggered together.

We can now compose $a_1, a_2, b_0^{(2)}$ to a preparation procedure $a^{(1)}$ for the systems 1. Then $b_0^{(1)}$ is a registration method for the systems 1 prepared by the composite device. Since different indicators $b^{(2)}$ can be present on $b_0^{(2)}$, by the correspondence $(a_1, a_2, b^{(2)}) \to a^{(1)}(b^{(2)})$, the preparation procedures $a^{(1)}$ form a Boolean ring isomorphic to $\mathscr{R}(b_0^{(2)})$. What are the $\varphi(a^{(1)}(b^{(2)}))$?

When (3.6) is rewritten

$$\begin{aligned}\operatorname{tr}(S(w_1 \times w_2)\, S^+ (g_1 \times g_2)) &= \operatorname{tr}((\mathbf{1} \times g_2^{\frac{1}{2}})\, S(w_1 \times w_2)\, S^+ (\mathbf{1} \times g_2^{\frac{1}{2}})(g_1 \times \mathbf{1}))\\ &= \operatorname{tr}_1(R_1 [(\mathbf{1} \times g_2^{\frac{1}{2}})\, S(w_1 \times w_2)\, S^+ (\mathbf{1} \times g_2^{\frac{1}{2}})]\, g_1),\end{aligned} \tag{3.7}$$

$R_1$ is called the reduction operator on the system 1. This $R_1$ is a mixture-morphism $K(\mathscr{H}_1 \times \mathscr{H}_2) \to K(\mathscr{H}_1)$, defined by

$$\operatorname{tr}(w(g_1 \times \mathbf{1})) = \operatorname{tr}_1((R_1\, w)\, g_1). \tag{3.8}$$

With $\psi(b_0^{(2)}, b^{(2)}) = g_2$, (3.7) immediately implies

$$\varphi(a^{(1)}(b^{(2)})) = \frac{R_1 [(\mathbf{1} \times g_2^{\frac{1}{2}})\, S(w_1 \times w_2)\, S^+ (\mathbf{1} \times g_2^{\frac{1}{2}})]}{\operatorname{tr}(S(w_1 \times w_2)\, S^+ (\mathbf{1} \times g_2))}. \tag{3.9}$$

The equation

$$S(1; g_2)\, w = R_1 [(\mathbf{1} \times g_2^{\frac{1}{2}})\, w (\mathbf{1} \times g_2^{\frac{1}{2}})] \tag{3.10}$$

defines a mapping $\check{K}(\mathscr{H}_1 \times \mathscr{H}_2) \to \check{K}(\mathscr{H}_1)$, which is easily seen to be an operation (V, *D* 11.1). The mapping $b_2 \to S(1; \psi(b_0^{(2)}, b^{(2)}))$ defines an operation measure $\chi$ on the Boolean ring $\mathscr{R}(b_0^{(2)})$:

$$\mathscr{R}(b_0^{(2)}) \xrightarrow{\ \chi\ } \Pi \tag{3.11}$$

with $\Pi$ as in V, *D* 11.2. By completing the Boolean ring $\mathscr{R}(b_0^{(2)})$ to a complete ring $\Sigma$, from (3.11) one obtains a transpreparator

$$\Sigma \xrightarrow{\ \chi\ } \Pi \tag{3.12}$$

in the sense of V, *D* 11.3.

Therefore, the preparation device composed from the three devices corresponding to $a_1, a_2, b_0^{(2)}$ is a (special) example of the general considerations in V §11 on transpreparators. In XI §5, we shall get to know how macrosystems can also be used to define transpreparators.

The reader further interested in applying scattering processes to registration and transpreparation is referred to [2] XVII, especially to §§5, 6 and 7.

# X The Embedding Problem

After the discovery of quantum mechanics, the opinion arose that quantum mechanics is a more comprehensive theory than "classical" mechanics (for a more exact formulation of a "more comprehensive theory" see XIII §3, [3] §8, and [48]).

After the development of quantum electrodynamics, one even hoped to obtain "all" classical theories from a quantum theory. Thus the classical mechanics of mass points and continua as well as thermodynamics and electrodynamics should follow as "approximate" (less comprehensive) theories. This expectation led to great difficulties regarding the "measurement process" in quantum mechanics. It appeared impossible to interpret quantum mechanics unless one admits paradoxes or a conscious "observer" or a profound modification of logics. We cannot review all these discussions; for such a review the reader is referred to [21].

We shall neither discuss arguments against the diverse interpretations of quantum mechanics. We rather elaborate the standpoint taken in III when we established an axiomatic basic for quantum mechanics. For this standpoint, quantum mechanics is not more comprehensive than classical theories. But then what is the relation between classical theories and quantum mechanics? This and the next chapter are devoted to this question.

Before further discussions let us again expound this sharply: The author's opinion is that the notion of quantum mechanics as the "most comprehensive" theory is wrong. The "axiomatic basis" presented in this book reflects precisely *the* conception and *the* idea espoused at the beginning of quantum mechanics with an astoundingly clear intuitive view by N. Bohr (see again [21]). The axiomatic basis (set forth from III on) also does not allow us to regard the objectivating description of macroscopic systems (presented in II) as some approximation to quantum mechanics. On the contrary, the theories from II were used as *pretheories* (see XIII §§1 and 3, [3] §§3, 5 and 10.5, and [48]) for quantum mechanics.

## §1 Classical Theories as Approximations to the Quantum Mechanics of Microsystems

In order to elucidate the relations of the various theories, let us first describe several situations where classical theories in fact can approximate quantum mechanics. In this description we shall emphasize more the conceptual situations than the mathematical demonstrations.

The quantum mechanics $\mathscr{P}\mathscr{T}_q$ was introduced for the domain of those directed interactions between macrosystems which one characterize as carried over by microsystems. Therefore, let us *first* consider *only* the fundamental domain for which we have recognized quantum mechanics as a g.G.-closed theory (see XIII §4.3 and [3] §10.3).

In considering theories that are less comprehensive than $\mathscr{PT}_q$ we go back to the concepts from XIII §3, [3] §8, and [48]. In general, one thus obtains a theory $\mathscr{PT}_2$ less comprehensive than $\mathscr{PT}_q$ by the scheme

$$\mathscr{PT}_q \to \mathscr{PT}_1 \rightsquigarrow \mathscr{PT}_2, \tag{1.1}$$

where $\to$ means a restriction and $\rightsquigarrow$ an embedding.

We first consider restrictions: Most often in quantum mechanics, one goes over to subsets $\mathscr{Q}_s$ of $\mathscr{Q}$ and $\mathscr{R}_{0s}$ of $\mathscr{R}_0$ and $\mathscr{R}_s$ of $\mathscr{R}$. To this there corresponds a transition to subsets $K_s$ of $K$ and $L_s$ of $L$. Therefore one restricts the fundamental domain to part of the experiments (with respect to preparation as well as to registration).

To construct an accelerator for electrons, for instance, we are not interested in "all possible" experiments with electrons. Rather we examine only a certain part, relevant for that accelerator. Therefore, it would be much too involved to invoke the quantum mechanics of electrons when we *only* (!) want to describe how they behave in the accelerator.

For this purpose, we restrict $K_s$ to ensembles with no electrons of low energies (intuitively, no "wave lengths" exceeding the scale on which the exterior fields change noticeably). The coexistent effects included in $L_s$ should only represent (imprecise) measurements of positions and velocities (see [43]). The theory obtained this way is an example of a restriction $\mathscr{PT}_1$.

The theories $\mathscr{PT}_2$ in (1.1) are chosen mostly for the purpose of "mathematically simpler" representations of $\mathscr{PT}_1$, i.e. as "standard embeddings" in the sense of XIII §3 and [3] §8. In the example just mentioned, one makes $\mathscr{PT}_2$ a "classical point mechanics" of individual charged mass points in exterior fields. The embedding of $\mathscr{PT}_1$ in $\mathscr{PT}_2$ in this case shows in which fundamental domain of experiments such a "classical point mechanics" is useable for electrons. One finds sketched in [1] XI §1.2, how the mathematics of restriction and embedding enters the just described transition from $\mathscr{PT}_q$ to $\mathscr{PT}_2$ (the classical mechanics of charged mass points). For such transitions there also are mathematically more exact methods (as presented in [23]).

Therefore, in a situation like (1.1), $\mathscr{PT}_2$ in fact is a less comprehensive theory than $\mathscr{PT}_q$. Of course, for quantum mechanics there are not only approximations in the form of the described example (where $\mathscr{PT}_2$ is a classical point mechanics). Many books on quantum mechanics mainly describe approximations for various purposes. But precisely this problem will *not* be the content of this chapter. We have only mentioned these approximations for quantum mechanics in order to prevent misunderstandings of the problem we shall investigate.

One such error is the notion that also the Newtonian mechanics of our planetary system is only an approximation of a "quantum theory" of the planetary system, just as the classical mechanics of charged mass points is *really* an approximation for the quantum mechanics of electrons.

Also the description of tennis balls via a classical mechanics is not an approximation of a "quantum mechanics" for tennis balls. A dreamt-up quantum mechanics for tennis balls at least is no g.G.-closed theory, since it is physically impossible to make experiments with tennis balls similar to the interference experiments with electrons.

But what is then really the relation between quantum mechanics and macroscopic physics? First that theories of macroscopic systems are used as pretheories for quantum mechanics (as described in III). On the other hand, quantum mechanics is used in "statistical mechanics" in order to say something about macroscopic systems. What does "to use" mean in this connection?

## §2 Macroscopic Systems and an Extrapolated Quantum Mechanics

First we must recognize that no comprehensive theory yet exists for macroscopic systems. Such a theory ought to describe all (at least the not too energetic) processes on macrosystems. But we are still far from such a comprehensive theory. Instead we possess many different theories (most not even g.G.-closed), such as classical point mechanics, hydrodynamics, thermodynamics, electrodynamics. By these we describe one or another aspect of macro-systems, mostly with ad hoc axioms for material behavior such as equations of state. As a general feature of all these macroscopic theories $\mathscr{PT}_m$, in II we emphasized that there are different theories for different areas of application.

In order not to lose track of the various theories, it is useful to imagine that a theory for all macro-systems would exist. Since we do not know it, let us briefly call it $\mathscr{PT}_?$. Then each of the known macroscopic theories $\mathscr{PT}_{m1}, \mathscr{PT}_{m2}, \ldots$ would be less comprehensive than $\mathscr{PT}_?$. In the sense of XIII §3, [3] §8 or [48], we can indicate this by

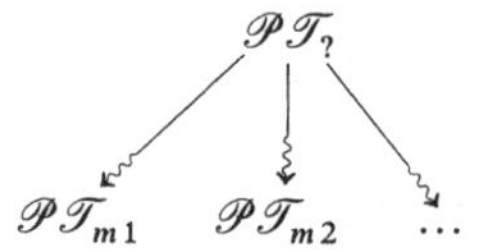


(2.1)

In II we described structures as they occur in each of the theories $\mathscr{PT}_{m1}$, $\mathscr{PT}_{m2}, \ldots$. As in II, let us consider a particular $\mathscr{PT}_{m\nu}$, denoting it by $\mathscr{PT}_m$. Concerning the presentation in II, we thus ascribe a physical interpretation to the state space $Z$. Similarly, suppose that a fundamental domain is known on which this theory can be applied.

In II §3.3, we described how the dynamics are determined by the set $K_m(\hat{Y})$. But in II we did not consider any special form of dynamical laws. In the various theories $\mathscr{PT}_m$, the dynamics is more or less known. For example, in Newton's theory $\mathscr{PT}_m$ (of mass points moving under gravity), the dynamical laws are completely known. In a mechanics of continua, the dynamics is determined up to equations of state (and material coefficents like those of friction or heat conduction), which are theoretically not established.

As in the example (of the mechanics of continua), often the dynamical laws are just poorly known in practice (to the dynamics there also belongs the behavior of systems in electric and magnetic fields!). The "statistical mechanics" now asserts that one can determine the dynamics in a $\mathscr{PT}_m$ by an extrapolated quantum mechanics. But for our initial purposes it is conceptually simpler to act as if the dynamics were known, and on this basis to investigate the relation of $\mathscr{P}\mathscr{T}_m$ to an extrapolated quantum mechanics. Conversely, such investigations can be used to find (or at least to narrow down) the dynamical laws. These methods of tracking down the dynamics will not be detailed in this book. A clear description of this procedure (without mathematical rigor) may be found in [1] XV. How rigor could be supplied, will follow from the presentations in this section.

## §2.1 An Extrapolated Quantum Mechanics

We assume the reader familiar with the structures (sketched in IX §2) of the quantum mechanical description of "composite" systems. We can think of all the systems considered as composed of atomic nuclei (as approximately elementary systems) and electrons.

It is characteristic of the mathematical structure of composites in quantum mechanics that a non-elementary system can be composed of arbitrarily many elementary systems. Formally, one can compose systems from a larger and larger number of subsystems, letting this number tend to infinity. This fact alone indicates (by [3] §9) that such a composition must be an idealization (the number of elementary systems in the entire universe is bounded). Therefore, it is impossible that the formal theoretical construction of systems from a larger and larger number of elementary systems yields a realistic theory of these systems.

What does one do in such a situation? Extrapolating the theory into a domain where it probably is no longer realistic, one compares it with experience, in order to approach more and more the limits of validity of the theory. This procedure is entirely legitimate and represents one of the fruitful developmental methods of physics. After the discovery of electrons it was entirely legitimate, first to extrapolate classical point mechanics into the new region, in order to recognize the limits of the "particle picture" (see [1] XI §1.1). We now apply the same method to our problem.

Therefore we formally extend quantum mechanics (as established for microsystems and their composites such as atoms and simple molecules), to systems of "very many" particles. We briefly denote this extrapolated quantum mechanics by $\mathscr{P}\mathscr{T}_{q\,\mathrm{exp}}$.

In quantum mechanics one is used to consider only the sets $K(\mathscr{H})$ and $L(\mathscr{H})$ and to "forget" the set $\mathscr{Q}$ of preparation procedures and the set $\mathscr{F}$ of effect procedures. One obtains $\mathscr{H}$ for many particle systems by symmetrizations from a product space $\mathscr{H}_1 \times \mathscr{H}_2 \times \ldots \times \mathscr{H}_N$ with $N$ a very large number ($N > 10^{20}$). If $\mathscr{P}\mathscr{T}_{q\,\mathrm{exp}}$ were a realistic theory then the image $\varphi(\mathscr{Q}')$ ought to be dense in $K(\mathscr{H})$ (in physical approximation). Likewise, $\psi(\mathscr{F})$ ought to be dense in $L(\mathscr{H})$. Now let us not only think of the abstract sets $K(\mathscr{H})$ and $L(\mathscr{H})$, but also of the corresponding sets $\mathscr{Q}$ and $\mathscr{F}$ (of preparations and effects). Then it becomes clear that this "physical" denseness is in *no* way guaranteed. Hence a realistic theory need not be obtainable by formal extrapolation to many particles.

On the contrary, experience as well as theoretical considerations quickly convey that it is impossible to realize the fictive preparation and registration possibilities from $\mathscr{Q}$ and $\mathscr{R}$. Instead of going into a far-reaching discussion, let us only give some indications (also see [1] XV).

By the drastic example of the planetary system, already in §1 we alerted the reader to an absurdity. For this planetary system (including the sun), by $\mathscr{P}\mathscr{T}_{q\,\mathrm{exp}}$ described as a "many particle" system, it appears absurd to regard as realistic all the preparation and registration "possibilities" occurring in $\mathscr{Q}'$ and $\mathscr{R}$. From which parts of the universe should one construct all the "registration devices" from $\mathscr{R}_0$? How to measure, for example, the precise positions of all nucleons and electrons of this many-particle system up to "only" $10^{-8}$ cm?

But already much smaller systems in our every day surroundings indicate that not all the elements from $\mathscr{Q}$ and $\mathscr{R}$ of a $\mathscr{P}\mathscr{T}_{q\,\mathrm{exp}}$ can be realistic. For example, simply think of the preparation possibilities of a waterdrop. We shall quickly have at hand many realistic preparation possibilities. But the construction of a preparation device $a \in \mathscr{Q}$ such that $\varphi(a)$ is the projector onto an eigenvector of the energy operator presents an unsolvable problem. Even if we imagine to have invented a construction possibility for such a preparation device, the most insignificant heat radiated from the surroundings would destroy its application. Of course, also the preparation device must not radiate any heat.

This example demonstrates one of the reasons for which not all preparation procedures from $\mathscr{Q}$ can be realized: For macrosystems it is *not* possible to remove all influences of the environment not coming from the intended preparation device. Only the truly realistic possibilities of preparing a macrosystem in its actual surroundings are at our disposal. Asking for the realistic preparation possibilities, we thus come up against the question of the realistic interaction possibilities of a macrosystem with its environment. In particular, for the preparation procedures we encounter the question: "Which variety of structures of macrosystems can be prepared by realistic interactions with the surroundings?"

The depth of this problem becomes still clearer if we adjoin organisms as parts of the macrosystem: Which possibilities are there really in order to prepare an aquarium with fish as a "many particle system"? Obviously the whole historic evolution of organisms on Earth has been necessary for the aquarium to exist today. But in this case, how can the many other elements of $Q$ be "realized"?

But also large parts of $\mathscr{R}$ (from $\mathscr{P}\mathscr{T}_{q\,\mathrm{exp}}$) can only represent fictive registration possibilities. For example, how could one to within $10^{-10}$ cm register the positions of all atomic nuclei and electrons of the above-mentioned waterdrop? One would not only need a monstrously energetic X-ray blitz (more than $10^{20}$ quanta of wave lengths below $10^{-10}$ cm), but also a super-microscope to analyze this blitz. On the other hand, to measure the discrete energy levels of the waterdrop one ought to analyze an enormous number of transitions whose wave lengths exceed billions of light years.

If we would even take $\mathscr{P}\mathscr{T}_{q\,\mathrm{exp}}$ so serious that all the operations formally appearing in $\mathscr{P}\mathscr{T}_{q\,\mathrm{exp}}$ (see IX §3 and XI §5) were realizable, then one could also conclude that a dead cat can be transformed into a live cat (as Piron drastically made clear during one of the discussions in Reisenburg Castle in summer 1979).

Therefore, it is not just in cosmic dimensions, but already in "normal" macroscopic regions that $\mathscr{P}\mathscr{T}_{q\,\mathrm{exp}}$ cannot represent a g.G.-closed theory.

Thus, asking for registration possibilities of macrosystems we again encounter the quest for the realistic interaction possibilities of a macrosystem with its surroundings. In particular, the registration procedures raise the question: "By which actions on the environment can the macrosystem be recognized?"

After these brief indications to the problem of preparation and registration possibilities, let us summarize a result of importance for the connection of $\mathscr{P}\mathscr{T}_{q\,\exp}$ with $\mathscr{P}\mathscr{T}_m$.

The theory $\mathscr{P}\mathscr{T}_{q\,\exp}$ contains many purely fictive elements. The set $\mathscr{Q}$ of preparation procedures introduced formally in $\mathscr{P}\mathscr{T}_{q\,\exp}$ contains many purely fictive preparation procedures; they are unrealizable and hence have nothing to do with the real structure of the world. The formally introduced set $\mathscr{R}_0$ correspondingly contains purely fictive registation methods and $\mathscr{R}$ contains purely fictive registration procedures. The fictive partition of $Q$ into ensembles $\varphi(a) \in K(\mathscr{H})$ and of $\mathscr{F}$ into effects $\psi(b_0, b) \in L(\mathscr{H})$ is not realistic. Not all possibilities for distinguishing the $a \in \mathscr{Q}'$ by means of the $f \in \mathscr{F}$ are really at one's disposal. Therefore, the elements of $K(\mathscr{H})$ and $L(\mathscr{H})$ can only be viewed as fictive ensembles and fictive effects. In a realistic theory, something else must take their place.

Do there remain any realistic features of $\mathscr{P}\mathscr{T}_{q\,\exp}$? And if so, which are they? To just this question let us turn now.

## §2.2 The Embedding of $\mathscr{P}\mathscr{T}_m$ in $\mathscr{P}\mathscr{T}_{q\,\exp}$

Even if we completely disregard that some elements of $\mathscr{P}\mathscr{T}_{q\,\exp}$ are fictive, this $\mathscr{P}\mathscr{T}_{q\,\exp}$ has the same disadvantage as the quantum mechanics of microsystems. In fact, no theory is yet known for deducing the mappings $\varphi$ and $\psi$ from concretely prescribed constructions of the preparation and registration devices. In this context see the detailed presentations in [2]. In practice one proceeds in the application of $\mathscr{P}\mathscr{T}_{q\,\exp}$ similarly to the application of $\mathscr{P}\mathscr{T}_q$ and "divines" certain observables (i.e. mappings $\Sigma \xrightarrow{F} L$) corresponding to registrable quantities. This cannot be denied; but it shows the still unsatisfactory status of a theory of macrosystems. It thus appears not meaningful (in searching for a systematic way to a theory of macrosystems) to proceed solely from the theory $\mathscr{P}\mathscr{T}_{q\,\exp}$. But we are not in the fatal situation of knowing only $\mathscr{P}\mathscr{T}_{q\,\exp}$. We rather know very realistic theories $\mathscr{P}\mathscr{T}_m$ of macroscopic systems. Having described some structures of these $\mathscr{P}\mathscr{T}_m$ in II, we have just before §2.1 recapitulated them with the simplified assumption that also the dynamics are known in $\mathscr{P}\mathscr{T}_m$.

Our basic idea (as described at the beginning of §2) says that neither is $\mathscr{P}\mathscr{T}_m$ more comprehensive than $\mathscr{P}\mathscr{T}_{q\,\exp}$ nor $\mathscr{P}\mathscr{T}_{q\,\exp}$ more comprehensive than $\mathscr{P}\mathscr{T}_m$. We now must approach the relation between these two theories and later see whether this approach reflects the correspondence between these theories and experience. For this purpose, we first return to (2.1) from which we select the relation between $\mathscr{P}\mathscr{T}_?$ and one of the $\mathscr{P}\mathscr{T}_{m\nu}$:

$$\mathscr{P}\mathscr{T}_? \rightsquigarrow \mathscr{P}\mathscr{T}_m. \tag{2.2.1}$$

This we now supplement by the notion that $\mathscr{P}\mathscr{T}_?$ is more comprehensive than $\mathscr{P}\mathscr{T}_{q\,\exp}$:

$$\mathscr{P}\mathscr{T}_? \rightsquigarrow \mathscr{P}\mathscr{T}_{q\,\exp}. \tag{2.2.2}$$

Initially, the two relations (2.2.1) and (2.2.2) are not correlated. Nevertheless they would cause a relation between $\mathscr{P}\mathscr{T}_m$ and $\mathscr{P}\mathscr{T}_{q\,\exp}$ if the restrictions and embeddings in (2.2.1) and (2.2.2) were known. But since we even do not know $\mathscr{PT}_?$, these relations can only help us to guess a direct relation between $\mathscr{PT}_m$ and $\mathscr{PT}_{q\,\mathrm{exp}}$. We base this guess for implementing (2.2.1) and (2.2.2) "imagined" diagram

$$\begin{array}{ccc} \mathscr{PT}_? & \rightsquigarrow & \mathscr{PT}_{q\,\mathrm{exp}} \\ \downarrow & & \downarrow \\ \mathscr{PT} & \rightsquigarrow & \mathscr{PT}'_{q\,\mathrm{exp}}, \end{array} \tag{2.2.3}$$

with a $\mathscr{PT}'_{q\,\mathrm{exp}}$ to be discussed soon. Then $\mathscr{PT}_?$ is more comprehensive than $\mathscr{PT}_{q\,\mathrm{exp}}$ because from an embedding (in the sense of the theorem from XIII (3.2.4) or [3] §8 or [48]) one obtains that part of $\mathscr{PT}_{q\,\mathrm{exp}}$ which reflects realistic structures. But we cannot construct this embedding because of our ignorance of $\mathscr{PT}_?$. Hence let us by restrictions and embeddings go over to $\mathscr{PT}_m$ and $\mathscr{PT}'_{q\,\mathrm{exp}}$, of course assuming analogous restrictions on the left and the right of (2.2.3).

We can only indicate such restrictions by examples: If the restriction of $\mathscr{PT}_?$ to $\mathscr{PT}_m$ is that to the hydrodynamics of water, then on the right side one must restrict $\mathscr{PT}_{q\,\mathrm{exp}}$ to hydrogen and oxygen nuclei and electrons in appropriate ratios, and to such energies that one need not deal with vapor or ice. With corresponding restrictions of $\mathscr{PT}_?$ and $\mathscr{PT}_{q\,\mathrm{exp}}$, the embedding of $\mathscr{PT}_?$ into $\mathscr{PT}_{q\,\mathrm{exp}}$ should as in (2.2.3) go over into an embedding of $\mathscr{PT}_m$ into $\mathscr{PT}'_{q\,\mathrm{exp}}$. The only distinction from a "normal" embedding (see below in XIII §3.2) may lie in those structures of $\mathscr{PT}'_{q\,\mathrm{exp}}$ which reflect the real situations and the non-fictive possibilities for experiments.

These structures are possibly more comprehensive than the embedding $\mathscr{PT}_m \rightsquigarrow \mathscr{PT}'_{q\,\mathrm{exp}}$ says. Whether $\mathscr{PT}'_{q\,\mathrm{exp}}$ in fact has more real structures than the image of $\mathscr{PT}_m$, depends on how comprehensive $\mathscr{PT}_m$ is. For example, if one has the diagram

$$\begin{array}{ccc} \mathscr{PT}_m & & \\ & \searrow\!\!\!\rightsquigarrow & \\ \downarrow & & \mathscr{PT}'_{q\,\mathrm{exp}} \\ & \nearrow\!\!\!\rightsquigarrow & \\ \mathscr{PT}'_m & & \end{array}$$

then it can happen that the image of $\mathscr{PT}_m$ in $\mathscr{PT}'_{q\,\mathrm{exp}}$ yields precisely the realistic part of $\mathscr{PT}'_{q\,\mathrm{exp}}$. However, the image of the theory $\mathscr{PT}'_m$ (which is less comprehensive than $\mathscr{PT}_m$) may in this embedding yield less than the realistic part of $\mathscr{PT}'_{q\,\mathrm{exp}}$. Such an example occurs when $\mathscr{PT}_m$ is the theory of the Boltzmann collision equation and $\mathscr{PT}'_m$ is aerodynamics. Then the transition from $\mathscr{PT}_m$ to $\mathscr{PT}'_m$ is sketched briefly in [3] pages 99–100, and in more detail in [1] XV §10.6.

Having thus critically discussed the meaning of the bottom line in (2.2.3), let us implement the embedding of $\mathscr{PT}_m$ into $\mathscr{PT}'_{q\,\mathrm{exp}}$ by a special approach, naturally suggested by physical considerations. Since obviously $\mathscr{Q}$, $\mathscr{R}_0$ and $\mathscr{R}$ contain a set of fictive elements, let us view the sets $\mathscr{Q}_m$, $\mathscr{R}_{0m}$, $\mathscr{R}_m$ of $\mathscr{PT}_m$ as subsets of $\mathscr{Q}$, $\mathscr{R}_0$, $\mathscr{R}$, namely as the subsets of the "realizable" preparation and registration procedures. Corresponding to the general embedding theorem in XIII (3.2.4), [3] (8.14) or [48] (3.4), we can express this by an embedding

$$M_m \xrightarrow{\;i\;} M. \tag{2.2.4}$$

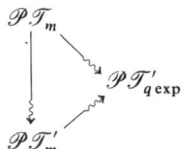

Since in the sequel we shall consider only this embedding (2.2.4) of $\mathscr{P}\mathscr{T}_m$, let us drop the prime on $\mathscr{P}\mathscr{T}'_{q\,\mathrm{exp}}$. Then $M$ denotes the set of systems in $\mathscr{P}\mathscr{T}_{q\,\mathrm{exp}}$ while $M_m$ is the set of systems in $\mathscr{P}\mathscr{T}_m$. Further let us assume that the mapping $i$ in (2.2.4) is bijective. One could simply say that we identify $M_m$ with $M$.

We can extend $i$ (in a canonical way) to a mapping of the subsets:

$$\mathscr{P}(M_m) \xrightarrow{\;i\;} \mathscr{P}(M).$$

In particular let us require the following relations (2.2.5) and (2.2.6) for a further concretization of XIII (3.24) or [3] (8.14):

$$\mathscr{Q}_m \xrightarrow{\;i\;} \mathscr{Q}, \qquad \mathscr{R}_m \xrightarrow{\;i\;} \mathscr{R}, \qquad \mathscr{R}_{0m} \xrightarrow{\;i\;} \mathscr{R}_0. \tag{2.2.5}$$

This of course means that one obtains the same probabilities in $\mathscr{P}\mathscr{T}_m$ as in $\mathscr{P}\mathscr{T}_{q\,\mathrm{exp}}$:

$$\lambda_{\mathscr{S}_m}(a \cap b_0, a \cap b) = \lambda((ia) \cap (ib_0), (ia) \cap (ib)). \tag{2.2.6}$$

One must reckon with the possibility that (2.2.6) is mathematically *not exact* (see §3) but gives such a good approximation that the difference between the left and right sides remains experimentally undetectable (for instance smaller than $10^{-100}$). But we let (2.2.6) stand as an equation in order to avoid too great mathematical difficulties.

Therefore, (2.2.5) together with (2.2.6) expresses our notion that $\mathscr{Q}_m$, $\mathscr{R}_m$ can be regarded as subsets of $\mathscr{Q}$, $\mathscr{R}$. For this embedding $i$ we furthermore demand that structures of the "same physical meaning" in $\mathscr{P}\mathscr{T}_m$ and in $\mathscr{P}\mathscr{T}_{q\,\mathrm{exp}}$ are also mathematically related to each other. Thus (2.2.6) already expresses the coincidence of probabilities interpreted in the same way in $\mathscr{P}\mathscr{T}_m$ and $\mathscr{P}\mathscr{T}_{q\,\mathrm{exp}}$.

There surely can be many further relations that are "interpreted in the same way" in $\mathscr{P}\mathscr{T}_m$ and $\mathscr{P}\mathscr{T}_{q\,\mathrm{exp}}$. The introduction of such relations restricts the mapping i more and more, as we shall presently demonstrate by the example of the time displacement. Perhaps the investigation of equally interpreted relations could be a first step on the way toward solving the problem of "macroscopic observables" (see the discussion after (2.3.13) and in §3).

In II (4.2.11) we defined a mapping $R_\tau$ of $\mathscr{R}_m$ into itself. Its physical interpretation was that the registration procedures are translated by a time $\tau$. The time displacement operator $\delta_\tau$ in quantum mechanics (considered in IX §1) has the same physical interpretation. For this reason, for the mapping $i$ from $\mathscr{R}_m$ into $\mathscr{R}$ it would be natural to require

$$i\,R_\tau = \delta_\tau\, i. \tag{2.2.7a}$$

While $R_\tau$ on the left side is a mapping of $\mathscr{R}_m$ into itself, the $\delta_\tau$ on the right side maps $\mathscr{R}$ into itself. According to IX §§1 and 2, in quantum mechanics the mapping $\delta_\tau$ is given by

$$\psi\, \delta_\tau(b_0, b) = \psi(\delta_\tau b_0, \delta_\tau b) = U_\tau\, \psi(b_0, b)\, U_\tau^+.$$

By IX (2.5), the operator $U_\tau$ (in IX written $U(\tau)$) is determined by the Hamiltonian IX (2.6).

We must recognize that the obvious requirement (2.2.7a) is still not entirely realistic (this makes it understandable, that the equation (2.2.8) set forth below need hold only in physical approximation). That (2.2.7a) is not realistic, follows from the fact that the $\delta_\tau$ from $\mathscr{PT}_{q\,\mathrm{exp}}$ are not at all realizable in the domain of macrosystems! Why not?

By preparation and registration procedures, the quantum mechanics $\mathscr{PT}_q$ (and hence the extrapolated theory $\mathscr{PT}_{q\,\mathrm{exp}}$) describes the "interaction possibilities" of the microsystems with the surroundings. Correspondingly, $\mathscr{PT}_{q\,\mathrm{exp}}$ depicts the partially fictive interaction possibilities of the macrosystems with the surroundings (see §2.1). The mutual interaction of the microsystems is described by $\mathscr{PT}_q$ in the form presented briefly in IX §2. This description only holds in the time interval "between" preparation and registration. About the physical meaning of $\delta_\tau$ in $\mathscr{PT}_q$, this description says that the time interval between a preparation procedure $a$ and the registration method $\delta_\tau b_0$ equals $\tau$ plus the time interval between $a$ and $b_0$. Therefore, the $\delta_\tau$ in $\mathscr{PT}_q$ means that the time interval of no interaction with the environment is increased by $\tau$.

In fact one can in a very good approximation isolate microsystems between $a \in \mathscr{Q}$ and $b_0 \in \mathscr{R}_0$ from the surroundings (frequently with great experimental effort). But prepared macrosystems in practice continually interact with the environment (as described in §2.1). Therefore, $R_\tau$ in $\mathscr{PT}_m$ does not mean that from the time $t=0$ (the system was prepared before $t=0$; see II §1) till $t=\tau$ no significant interaction with the surroundings occurs (as one would require according to $\mathscr{PT}_{q\,\mathrm{exp}}$!). Rather, $R_\tau$ only means that $R_\tau b_0$ measures the trajectories just from $\tau$.

Therefore, $i R_\tau b_0$ in contrast to (2.2.7a) is not identical with $\delta_\tau i b_0$. In $\mathscr{PT}_m$ we only require that the unavoidable interaction of the macrosystem with the environment is till $t=\tau$ so slight that the trajectories are not noticeably changed. This means nothing but a certain stability of the *macroscopic* dynamics under the unavoidable slight disturbances from the surroundings. In contrast, the considerations in §2.1 showed a great *microscopic* change in $\mathscr{PT}_{q\,\mathrm{exp}}$, already by the most insignificant heat radiation.

For the macroscopic behavior of the really preparable macrosystems, the macroscopic stability also means that one can do as if (in the sense of $\mathscr{PT}_{q\,\mathrm{exp}}$) no interaction is present until $\tau$. For the macrosystems prepared in reality, $i R_\tau b_0$ and $\delta_\tau i b_0$ therefore are not distinguishable *macroscopically*. Mathematically, this means that the probabilities

$$\lambda((ia)\cap(iR_\tau b_0),\ (ia)\cap(iR_\tau b)) \quad \text{and} \quad \lambda((ia)\cap(\delta_\tau i b_0),\ (ia)\cap(\delta_\tau ib))$$

practically coincide for all $a \in \mathscr{Q}_m$. We thus come to weaken the requirement (2.2.7a) to

$$\lambda((ia)\cap(iR_\tau b_0),\ (ia)\cap(iR_\tau b)) = \lambda((ia)\cap(\delta_\tau i b_0),\ (ia)\cap(\delta_\tau ib)) \tag{2.2.7b}$$

for all $a \in \mathscr{Q}_m$. As explained above, (2.2.7b) is really just an approximation (though a very good one). In order to avoid mathematical complications from the beginning, we nevertheless have formulated (2.2.7b) as an equation.

The weakening to (2.2.7b) is essential. There surely can exist elements $a \in \mathscr{Q}$ with $a \notin i\mathscr{Q}_m$ for which $\lambda(a\cap(iR_\tau b_0),\ a\cap(iR_\tau b))$ essentially differs from $\lambda(a\cap(\delta_\tau i b_0),\ a\cap(\delta_\tau ib))$! This must even be expected, though we shall not explain it here. In fact, a measurement method of the form $\delta_\tau i b_0$ is anyhow unrealistic, just because an interaction of the macrosystems with the environment (from $t=0$ until $t=\tau$) is unavoidable.

With $R_\tau b_0 \in R_{0m}$ and $R_\tau b \in R_m$, from (2.2.6) with $R_\tau b_0$ instead of $b_0$ (and $R_\tau b$ instead of $b$) follows

$$\lambda_{\mathscr{S}_m}(a \cap R_\tau b_0,\ a \cap R_\tau b) = \lambda((ia) \cap (iR_\tau b_0),\ (ia) \cap (iR_\tau b)).$$

With (2.2.7b) this implies

$$\lambda_{\mathscr{S}_m}(a \cap R_\tau b_0,\ a \cap R_\tau b) = \lambda((ia \cap (\delta_\tau i b_0),\ (ia) \cap (\delta_\tau i b)),$$

which again holds only approximately.

Finally, the function $\mu$ from $\mathscr{PT}_{q\,\text{exp}}$ and the special form of $\delta_\tau$ in $\mathscr{PT}_{q\,\text{exp}}$ (see above) yield

$$\lambda_{\mathscr{S}_m}(a \cap R_\tau b_0,\ a \cap R_\tau b) = \mu(\varphi(ia),\ U_\tau \psi(ib_0, ib)\, U_\tau^+). \tag{2.2.8}$$

Since $R_\tau$ is in $\mathscr{PT}_m$ defined only as a semigroup for $\tau \geq 0$, also (2.2.8) is only a requirement for $\tau \geq 0$; for $\tau = 0$ it goes over into (2.2.6). For the embedding (2.2.4), (2.2.5) we therefore only require the approximate validity of (2.2.8) for $\tau \geq 0$.

But (2.2.8) also need not hold for "very large" $\tau$. This follows already from the general considerations in XIII §2.5 and [3] §9. Also the "recurrence theorem" (see (2.3.26) and [1] XV §§10, 11) shows that (2.2.8) cannot be meaningful for too large $\tau$. Here let us only mention that on the right side of (2.2.8) the system is assumed as "isolated"; only then is $\delta_\tau$ in the given way representable by $U_\tau$. But the macrosystems are really not isolated; hence (2.2.8) can indeed hold for finite macroscopic times $\tau$, but only for the $a \in \mathscr{Q}_m$ and $b \in \mathscr{R}_m$. For too large $\tau$ (some or millions of years), (2.2.8) could well be false.

With the canonical extension of $i$, from (2.2.5) follows the relation

$$\mathscr{F}_m \xrightarrow{\ i\ } \mathscr{F}. \tag{2.2.9}$$

With $\mu_m$ as in II (2.4.4), then (2.2.8) can be written

$$\mu_m(a, R_\tau f) = \mu(\varphi(ia),\ U_\tau \psi(if)\, U_\tau^+) \tag{2.2.10}$$

for $\tau \geq 0$ but not for $\tau$ of physically meaningless magnitudes.

Again, let us emphasize that in (2.2.10) the "equality sign" need not be mathematically exact; only *physically* the two sides must be indistinguishable.

If an embedding theorem holds in the form just sketched, let us briefly say that $\mathscr{PT}_m$ is compatible with $\mathscr{PT}_{q\,\text{exp}}$. In this context, we have in mind that a sufficiently comprehensive theory $\mathscr{PT}_m$ is more comprehensive than $\mathscr{PT}_{q\,\text{exp}}$. Since $\mathscr{PT}_m$ is often known only in broad outline, one conversely uses the embedding theorem in order to obtain a better (more comprehensive) theory $\mathscr{PT}_m$. In this sense, one could say that (2.2.10) is the basis of "statistical" mechanics. Of course, we cannot develop an entire statistical mechanics. In [1] XV one finds a presentation (not with complete mathematical rigor) of statistical mechanics, which rests on the notions just sketched (of embedding $\mathscr{PT}_m$ in $\mathscr{PT}_{q\,\text{exp}}$). In order at least to point out which problems are raised by an embedding theorem of the form (2.2.10), in the following §§2.3 through 2.6 let us draw consequences from it.

## §2.3 General Consequences of the Embedding Theorem

In §2.2 we did not prove the embedding theorem, but only formulated it. Here let us proceed as if it were established and deduce consequences. Then one may try to prove the embedding theorem by first proving its "essential" consequences.

We first consider the connection of the embedding $i$ with the mappings $\varphi_m$ from II §3.2 and $\varphi$ from III D5.1.1. In a diagram (already using the considerations from II §3.3) we can summarize this by

$$\begin{array}{ccc} \mathcal{Q}'_m & \xrightarrow{\varphi_m} & K(\hat{S}_m) \\ {\scriptstyle i}\Big\downarrow & & \\ \mathcal{Q}' & \xrightarrow{\varphi} & K. \end{array} \tag{2.3.1}$$

Analogously, the connection of $i$ with the mappings $\psi_{ms}$ from II (3.3.3) and $\psi$ from III D5.1.2 yields

$$\begin{array}{ccc} \mathcal{F}_m & \xrightarrow{\psi_{ms}} & L(\hat{S}_m) \\ {\scriptstyle i}\Big\downarrow & & \\ \mathcal{F} & \xrightarrow{\psi} & L. \end{array} \tag{2.3.2}$$

By means of the mappings $\varphi_m$ and $\psi_{ms}$, we can rewrite $\mu_m$ as

$$\mu_m(a,f) = \langle \varphi_m(a), \psi_{ms}(f) \rangle$$

so that (2.2.10) becomes

$$\langle \varphi_m(a), \psi_{ms}(R_\tau f) \rangle = \mu(\varphi(ia), U_\tau \psi(if)\, U_\tau^+). \tag{2.3.3}$$

For $U_\tau g U_\tau^+$ we briefly write $\mathcal{U}_\tau g$ with $\mathcal{U}_\tau$ a mapping of $\mathcal{B}'$ into itself. In analogy to classical mechanics one may call $\mathcal{U}_\tau$ the Liouville operator. Finally, with $V_\tau^{(s)}$ from the diagram II (4.2.14), from (2.3.3) follows

$$\langle \varphi_m(a), V_\tau^{(s)} \psi_{ms}(f) \rangle = \mu(\varphi(ia), \mathcal{U}_\tau \psi(if)). \tag{2.3.4a}$$

The set $\psi_{ms}(\mathcal{F}_m)$ separates (i.e. $la\, \psi_{ms}(\mathcal{F}_m)$ is norm-dense in $L(\hat{S}_m)$) since we assumed (see II §3.1) $la\, \psi_m(\phi)$ dense in $L(\hat{Y})$. Therefore, from (2.3.4a) follows that to each $\varphi(ia)$ there uniquely corresponds a $\varphi_m(a)$; hence there is a mapping $j$ with $j\varphi(ia) = \varphi_m(a)$. In the special case of $\tau = 0$, the relation (2.3.4a) implies that $j$ is continuous relative to the norm-topology in $\varphi\, i(\mathcal{Q}'_m)$ and to the $\sigma(\varphi_m(\mathcal{Q}'_m), C(\hat{S}_m))$-topology since $\sigma(\varphi_m(\mathcal{Q}'_m), C(\hat{S}_m))$ and $\sigma(\varphi_m(\mathcal{Q}'_m), \psi_{ms}(\mathcal{F}_m))$ coincide on the precompact set $\varphi_m(\mathcal{Q}'_m)$. Therefore one can extend $j$ as a mapping of the norm-closure $K_m$ of $\varphi i(\mathcal{Q}'_m)$ into the set $K(\hat{S}_m)$. The existence of direct mixtures makes $K_m$ convex and $K_m \xrightarrow{j} K(\hat{S}_m)$ affine.

Thus (2.3.4a) takes the form

$$\langle jw, V_\tau^{(s)} \psi_{ms}(f) \rangle = \mu(w, \mathcal{U}_\tau \psi(if)) \tag{2.3.4b}$$

for all $w \in K_m$.

To deduce consequences from (2.3.4b) is very difficult without introducing ideal-izations. The first idealization is only a mathematical one. It has no physical signifi-cance, especially because we allow that (2.3.4b) holds only in "very good" approxima-tion. This idealization is to say that $j$ is continuous also relative to the norm-topolo-gies in $\varphi i(\mathcal{Q}'_m)$ and $\varphi_m(\mathcal{Q}'_m)$. If $\psi_{ms}(\mathscr{F}_m)$ (and not only $la\,\psi_{ms}(\mathscr{F}_m)$) were norm-dense in $L(S_m)$, this idealization would be a theorem. When $j$ is norm-continuous, then $jK_m$ is a subset of the norm-closure $K_m(\hat{S}_m)$ of $\varphi_m(\mathcal{Q}'_m)$, and $jK_m$ is norm-dense in $K_m(\hat{S}_m)$. Therefore, the diagram (2.3.1) can be completed to

$$K_m(\hat{S}_m) \subset K(\hat{S}_m)$$

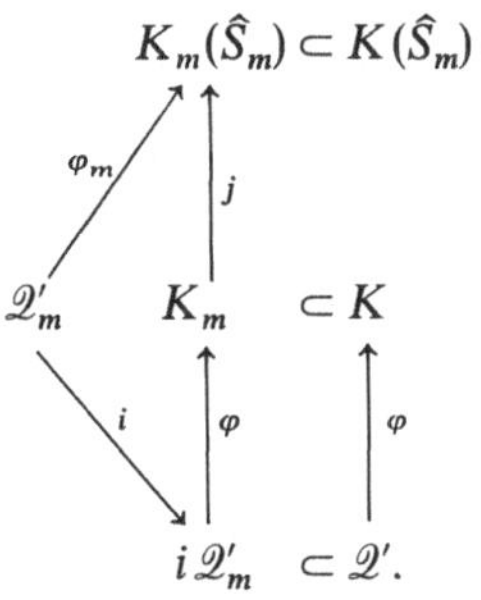

$$(2.3.5)$$

The completion of (2.3.2) is not so easy since $K_m(\hat{S}_m)$ need not separate the elements of $L(\hat{S}_m)$!

By means of the considerations from II §3.3 and V §3, let us go over from $C(\hat{S}_m)$, $C'(\hat{S}_m)$ to another pair of dual Banach spaces.

With $\Sigma_m = \mathscr{B}(\hat{S}_m)/\mathscr{I}(\hat{S}_m)$ from II §3.3, we can form (according to V §3.1) the space $\mathscr{B}(\Sigma_m)$ and the space $\mathscr{B}'(\Sigma_m)$ dual to it. By II §3.3 one can identify $K_m(\hat{S}_m)$ and $\tilde{K}_m(\hat{S}_m)$ with subsets of $K(\Sigma_m)$; for this reason let us write $K_m(\hat{S}_m) = K_m(\Sigma_m)$ and $\tilde{K}_m(\hat{S}_m) = \tilde{K}_m(\Sigma_m)$. With $\mathscr{B}_m(\Sigma_m)$ as the Banach subspace of $\mathscr{B}(\Sigma_m)$ spanned by $K_m(\Sigma_m)$, there follows $\tilde{K}_m(\Sigma_m) = K(\Sigma_m) \cap \mathscr{B}_m(\Sigma_m)$. The elements of $C(\hat{S}_m)$ uniquely define norm-continuous linear forms over $K(\Sigma_m)$; hence one can identify $C(\hat{S}_m)$ with a subset of $\mathscr{B}'(\Sigma_m)$. In this sense $L(\hat{S}_m)$ becomes a $\sigma(\mathscr{B}'(\Sigma_m), \mathscr{B}(\Sigma_m))$-dense subset of $L(\Sigma_m)$. The embeddings just described allow us to extend certain mappings. As a precaution, let us mention that $K(\hat{S}_m)$ must *not* be interpreted as a subset of $K(\Sigma_m)$, since the set $\mathscr{I}(\hat{S}_m)$ (of the sets of $\bar{u}$-measure zero) depends on the set $K_m(\hat{S}_m)$!

Therefore, in (2.3.5) we can put $K_m(\hat{S}_m) = K_m(\Sigma_m)$ and replace $K_m(\hat{S}_m) \subset K(\hat{S}_m)$ by $K_m(\Sigma_m) \subset K(\Sigma_m)$. Then, instead of (2.3.2) we can write

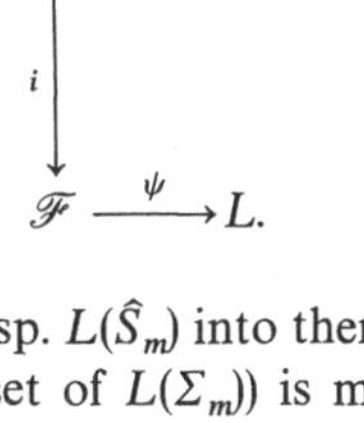

$$\begin{array}{ccc} \mathscr{F}_m & \xrightarrow{\psi_{ms}} & L(\Sigma_m) \\ \downarrow{\scriptstyle i} & & \\ \mathscr{F} & \xrightarrow{\psi} & L. \end{array} \qquad (2.3.6)$$

The mappings $V_\tau^{(s)}$ of $C(\hat{S}_m)$ resp. $L(\hat{S}_m)$ into themselves can be extended to $\mathscr{B}'(\Sigma_m)$ resp. $L(\Sigma_m)$, where $\Sigma_m$ (as a subset of $L(\Sigma_m)$) is mapped into itself. One recognizes this easily from the facts that $V_\tau$ carries an element of $\mathscr{B}(\hat{Y})$ into an element of $\mathscr{B}(\hat{Y})$, and that a set of measure zero is mapped into a set of measure zero (as at the end of II §4.2 in the proof of $sf_1 = sf_2 \Rightarrow sV_\tau f_1 = sV_\tau f_2$). Using (2.3.5), we can thus write (2.3.4b) in the form

$$\langle jw, V_\tau^{(s)} \psi_{ms}(f)\rangle = \mu(w, \mathscr{U}_\tau \psi(if)) \quad \text{for all } w \in K_m, \tag{2.3.7}$$

where $\langle \ldots, \ldots \rangle$ now is the canonical bilinear form of $\mathscr{B}(\Sigma_m)$ and $\mathscr{B}'(\Sigma_m)$.

In (2.3.6) let us introduce $L_m \subset L$ as the $\sigma(\mathscr{B}', \mathscr{B})$-closure of the set $\psi i(\mathscr{F}_m)$, and $L_m(\Sigma_m) \subset L(\Sigma_m)$ as the $\sigma(\mathscr{B}'(\Sigma_m), \mathscr{B}(\Sigma_m))$-closure of the set $\psi_{ms}(\mathscr{F}_m)$. Though $la\psi_{ms}(\mathscr{F}_m)$ is $\sigma(\mathscr{B}'(\Sigma_m), \mathscr{B}(\Sigma_m))$-dense in $L(\Sigma_m)$ since $la\psi_{ms}(\mathscr{F}_m)$ is norm-dense in $L(\hat{S}_m)$, in general neither a unique mapping of $L_m$ into $L(\Sigma_m)$ nor of $L(\Sigma_m)$ into $L_m$ follows from (2.3.6). This makes an evaluation of (2.3.6) difficult. For this reason we go over to partial mappings with *fixed* $b_0 \in \mathscr{R}_{0m}$. Then (2.3.6) implies the diagram

$$\begin{array}{ccc} \mathscr{R}_m(b_0) & \xrightarrow{\psi_{msb_0}} & L(\Sigma_m) \\ {\scriptstyle i}\big\downarrow & & \\ \mathscr{R}(ib_0) & \xrightarrow{\psi_{b_0}} & L_m \subset L, \end{array} \tag{2.3.8}$$

with

$$\psi_{msb_0}(b) = \varphi_{ms}(b_0, b) \quad \text{and} \quad \psi_{b_0}(b) = \psi(b_0, b).$$

Since $i\mathscr{R}_m(b_0) \subset \mathscr{R}(ib_0)$, the Boolean rings $\mathscr{R}_m(b_0)$ and $i\mathscr{R}_m(b_0)$ are isomorphic. From (2.3.8) follows

$$\begin{array}{ccc} & & L(\Sigma_m) \\ & \nearrow^{\psi_{msb_0}} & \\ \mathscr{R}_m(b_0) & & \\ & \searrow_{\psi_{ib_0} i} & \\ & & L_m \subset L, \end{array} \tag{2.3.9}$$

where $\psi_{msb_0}$ and $\psi_{ib_0} i$ are additive measures over $\mathscr{R}_m(b_0)$. By (2.3.7) the relation

$$\langle jw, \psi_{msb_0}(b)\rangle = \mu(w, \psi_{ib_0} i(b)) \tag{2.3.10}$$

holds for all $w \in K_m$.

Since $\varphi_m(\mathscr{Q}'_m)$ is norm-dense in $K_m(\hat{S}_m) = K_m(\Sigma_m)$, a $w \in K_m$ exists (easily shown by mixing) such that $jw$ is an effective measure from $K(\Sigma_m)$. Now think of such a $w = w_0$ as chosen. Then $\langle jw_0, \psi_{msb_0}(b)\rangle = 0$ implies $b = \emptyset$. With this $w_0$ we can on $\mathscr{R}_m(b_0)$ define a metric

$$d(b_1, b_2) = \langle jw_0, \varphi_{msb_0}(b_1 \backslash b_1 \cap b_2) + \psi_{msb_0}(b_2 \backslash b_1 \cap b_2)\rangle.$$

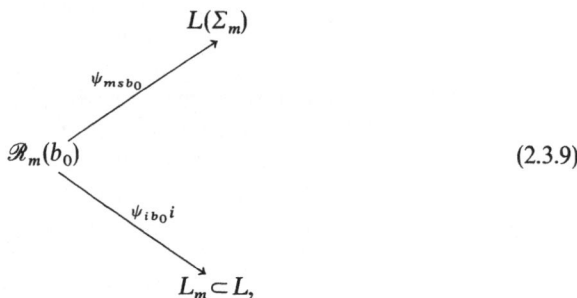

Because of (2.3.10), both measures $\psi_{msb_0}$ and $\psi_{ib_0}i$ extend to the completion $\Sigma_{b_0}$ *of* $\mathscr{R}_m(b_0)$ (see V T 1.3.2); hence instead of (2.3.9) one obtains

$$\begin{array}{ccc} & & L(\Sigma_m) \\ & \overset{\psi_{msb_0}}{\nearrow} & \\ \Sigma_{b_0} & & \\ & \underset{\psi_{ib_0}i}{\searrow} & \\ & & L_m \subset L. \end{array} \tag{2.3.11}$$

Then $\Sigma_{b_0} \xrightarrow{\psi_{msb_0}} L(\Sigma_m)$ and $\Sigma_{b_0} \xrightarrow{\psi_{ib_0}i} L$ are two observables in the sense of V D 1.3.1.

Since all elements of $L(\Sigma_m)$ coexist, it is natural to assume (in the sense of the "classical" way of registering trajectories) that there is a registration method $\tilde{b}_0 \in \mathscr{R}_{0m}$ so that one can register (approximately) *all* trajectories. This just means

$$\overline{\text{co}}^\sigma\, \psi_{ms\tilde{b}_0}\, \Sigma_{\tilde{b}_0} = L(\Sigma_m), \tag{2.3.12}$$

since the left side is all one can attain approximately (in the $\sigma(\mathscr{B}'(\Sigma_m), \mathscr{B}(\Sigma_m))$-topology) with registrations of the method $\tilde{b}_0$.

Therefore, as a second idealization let us *require* the existence of a $\tilde{b}_0 \in \mathscr{R}_{0m}$ such that (2.3.12) holds. This does not mean that there can be *only one* such $\tilde{b}_0$.

This second idealization is not only a mathematical one: The existence of a $\tilde{b}_0$ with (2.3.12) corresponds to the classical assumption that one can measure the "total" trajectories, i.e. measure without perturbing the trajectories essentially. The existence of such a $\tilde{b}_0$ (the physical *possibility* to construct a device to measure the total trajectories, see [3] §10 or XIII §4.6) does not mean that one could not construct other devices that partially disturb the statistics of the trajectories (see II §3.1). The presumption of a $\tilde{b}_0$ to exist with (2.3.12) is therefore stronger than the presumption "$la\, \psi_m(\phi)$ dense in $L(\hat{Y})$" introduced in II §3.1.

The presumption (2.3.12) can also be said to state the possibility to "continually measure" the trajectories. Therefore, (2.3.12) expresses a certain stability of the dynamics against unavoidable perturbations by the measuring process (see again II §3.2).

We shall maintain the presumption (2.3.12) until §2.6 (for most of the deductions). In §2.7 we shortly shall discuss so "sensitive" macro-systems that (2.3.12) cannot be fulfilled (hence only "$la\, \psi_m(\phi)$ dense in $L(\hat{Y})$" is possible).

We can apply V T 3.3.1 (ii) to the observable $\Sigma_{\tilde{b}_0} \xrightarrow{\psi_{ms\tilde{b}_0}} L(\Sigma_m)$ where we can identify $\partial_e L(\Sigma_m)$ with $\Sigma_m$ (see V §3.1). Hence for each $\eta \in \Sigma_m$ there is exactly one $\sigma \in \Sigma_{\tilde{b}_0}$ with $\eta = \psi_{ms\tilde{b}_0}(\sigma)$. Therefore, there is a subset $\Sigma'_{\tilde{b}_0}$ of $\Sigma_{\tilde{b}_0}$ for which

$$\Sigma'_{\tilde{b}_0} \xrightarrow{\psi_{ms\tilde{b}_0}} \Sigma_m = \partial_e L(\Sigma_m)$$

is a bijective mapping. From the $\sigma$-additivity of the measure $\psi_{ms\tilde{b}_0}$ over $\Sigma_{\tilde{b}_0}$, it follows that $\Sigma'_{\tilde{b}_0}$ is a Boolean ring isomorphic to $\Sigma_m$ under the mapping $\psi_{ms\tilde{b}_0}$. From (2.3.11)

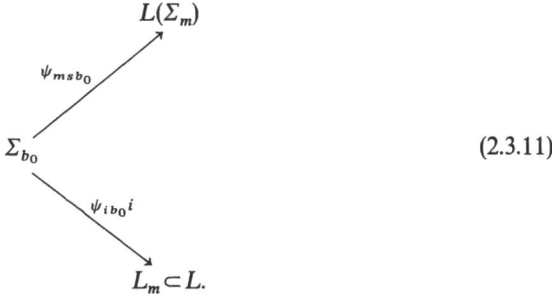

thus follows

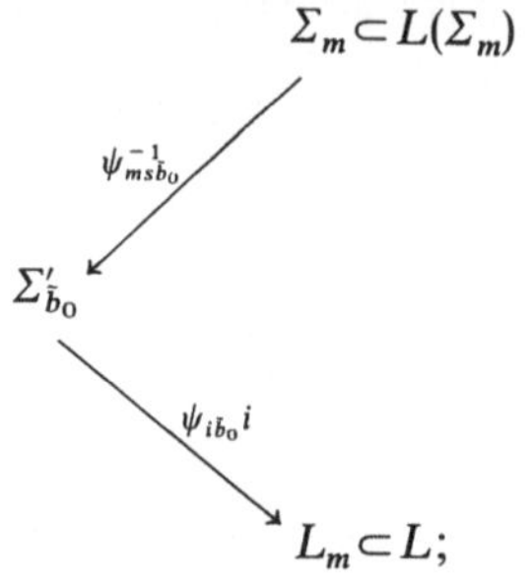


hence (2.3.11) implies the existence of an observable

$$\Sigma_m \xrightarrow{F_{\tilde{b}_0}} L_m \subset L \tag{2.3.13}$$

with

$$F_{\tilde{b}_0} = \psi_{i\tilde{b}_0} i \, \psi_{ms\tilde{b}_0}^{-1}.$$

One often calls (2.3.13) *the* "macroscopic observable" of the macrosystems described by $\mathcal{PT}_m$ although, according to the above remark, in general there can be various observables (2.3.13). As observables in $\mathcal{PT}_{q\,\mathrm{exp}}$, these $F_{\tilde{b}_0}$ in general need *not* coexist.

According to V §3.2, $F_{\tilde{b}_0}$ is the restriction of $S'_{\tilde{b}_0}$. $S_{\tilde{b}_0}$ is the mixture-morphism corresponding to the observable (2.3.13):

$$\mathcal{B} \xrightarrow{S_{\tilde{b}_0}} \mathcal{B}(\Sigma_m). \tag{2.3.14}$$

Therefore, (2.3.10) implies

$$\langle j w, y \rangle = \mu(w, S'_{\tilde{b}_0} y) = \langle S_{\tilde{b}_0} w, y \rangle \tag{2.3.15}$$

for all $y \in \mathcal{B}'(\Sigma_m)$ and for all $w \in K_m$, and hence $S_{\tilde{b}_0}|_{K_m} = j$. Thus, the diagram (2.3.5) can also be drawn as

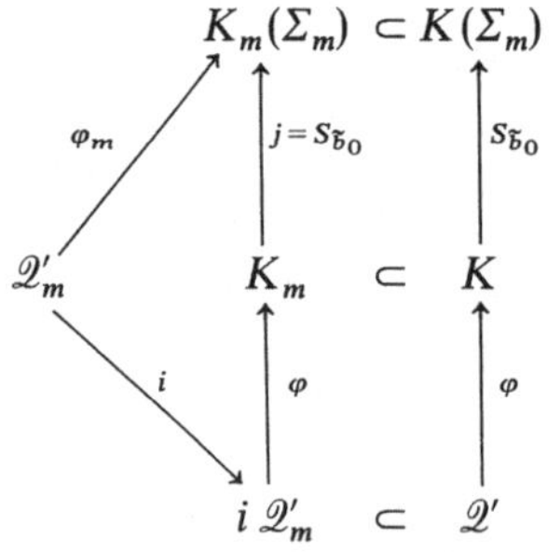


(2.3.16)

Although there can be different $\tilde{b}_0$ obeying (2.3.12), the ensuing mappings $S_{\tilde{b}_0}$ must of course agree with $j$ on $K_m$ (but not necessarily on all of $K$!).

The existence of a $\tilde{b}_0$ with (2.3.12) is therefore equivalent to the statement that the mapping $j$ can be extended to all of $K$ as a mixture-morphism. Such extensions are not unique.

That $j$ cannot be extended as a mixture-morphism means that every linear and norm-continuous extension of $j$ does not map $K$ into $K(\Sigma_m)$. In other words, for every extension there are such elements $w \in K$, $w \notin K_m$ which by the extension are mapped on elements outside $K(\Sigma_m)$! We cannot exclude this possibility (see §2.7).

From (2.3.7) follows

$$\langle V_\tau^{(s)\prime} j w, \psi_{ms}(f)\rangle = \mu(\mathscr{U}'_\tau w, \psi(i f)) \tag{2.3.17}$$

for all $w \in K_m$. Since $\psi_{ms}(\mathscr{F}_m)$ separates, it follows as above that to each $\mathscr{U}'_\tau w$ there corresponds a unique $V_\tau^{(s)\prime} j w$. Thus $j$ can be extended to the set $\bigcup_{\tau \geq 0} \mathscr{U}'_\tau K_m$ with

$$V_\tau^{(s)\prime} j w = j \mathscr{U}'_\tau w \tag{2.3.18}$$

for all $w \in K_m$.

If a $\tilde{b}_0$ with (2.3.12) exists, we may replace $j$ in (2.3.17) by $S_{\tilde{b}_0}$ and choose $f = (\tilde{b}_0, b)$, with $b \in \mathscr{R}_m(\tilde{b}_0)$. Then $L(\Sigma_m) = \overline{\mathrm{co}}^\sigma\, \psi_{ms\tilde{b}_0}(\Sigma'_{\tilde{b}_0})$ implies

$$\langle V_\tau^{(s)\prime} S_{\tilde{b}_0} w, y\rangle = \mu(w, \mathscr{U}_\tau S'_{\tilde{b}_0} y)$$

for all $w \in K_m$ and all $y \in L(\Sigma_m)$, i.e.

$$V_\tau^{(s)\prime} S_{\tilde{b}_0} w = S'_{\tilde{b}_0} \mathscr{U}'_\tau w \tag{2.3.19}$$

for all $w \in K_m$. Therefore $S_{\tilde{b}_0}$ equals $j$ also on $\bigcup_{\tau \geq 0} \mathscr{U}'_\tau K_m$.

With $w_1 = \mathscr{U}_{\tau_1} w$ and $w \in K_m$, from (2.3.18) follows

$$V_{\tau_2}^{(s)\prime} j w_1 = V_{\tau_2}^{(s)\prime} j \mathscr{U}'_{\tau_1} w = V_{\tau_2}^{(s)\prime} V_{\tau_1}^{(s)\prime} j w = V_{\tau_1+\tau_2}^{(s)\prime} j w = j \mathscr{U}_{\tau_1+\tau_2} w = j \mathscr{U}_{\tau_2} w_1 .$$

Thus we have proved that (2.3.18) even holds for all $w \in \bigcup_{\tau \geq 0} \mathscr{U}'_\tau K_m$.

With $K_m(\hat{Y}) = K_m(\hat{S}_m) = K_m(\Sigma_m)$, from II (4.2.9) follows

$$V_\tau^{(s)\prime} K_m(\Sigma_m) \subset K_m(\Sigma_m).$$

All this (together with the existence of direct mixtures) implies that $j$ can be extended (as an affine mapping) to

$$\tilde{K}_m = \overline{\mathrm{co}} \bigcup_{\tau \geq 0} \mathscr{U}'_\tau K_m$$

and that

$$j \tilde{K}_m \subset K_m(\Sigma_m)$$

resp.

$$S_{\tilde{b}_0} \tilde{K}_m \subset K_m(\Sigma_m).$$

Also, (2.3.18) and (2.3.19) hold for all $w \in \tilde{K}_m$.

Therefore (2.3.18) and (2.3.19) mean that

$$V_\tau^{(s)\prime} j = j \mathscr{U}'_\tau, \qquad \text{resp. } V_\tau^{(s)\prime} S_{\tilde{b}_0} = S_{\tilde{b}_0} \mathscr{U}'_\tau \tag{2.3.20}$$

hold as mappings of $\tilde{K}_m$ into $K_m(\Sigma_m)$, where $j \tilde{K}_m$ resp. $S_{\tilde{b}_0} \tilde{K}_m$ are dense in $K_m(\Sigma_m)$.

We can regard (2.3.20) as *the* basic relation for the embedding of $\mathscr{P}\mathscr{T}_m$ into $\mathscr{P}\mathscr{T}_{q\,\mathrm{exp}}$. Since $S_{\tilde{b}_0}$ is uniquely determined by $F_{\tilde{b}_0}$, through $S_{\tilde{b}_0}$ the macro-observable (2.3.13) and the set $\tilde{K}_m$ of ensembles essentially enter (2.3.20). Also the dynamics of the macrosystems are determined by $\tilde{K}_m$ and $S_{\tilde{b}_0}$, since $S_{\tilde{b}_0} K_m$ is dense in $K_m(\Sigma_m)$ $= K_m(\hat{S}_m) = K_m(\hat{Y})$ (see II §3.3).

The considerations immediately following (2.3.19) show that (with fixed $F_{\tilde{b}_0}$) there is a largest subset $K(\tilde{b}_0)$ of $K$ on which (2.3.20) is satisfied. Moreover, one sees that $K(\tilde{b}_0)$ is a closed convex subset of $K$ and that $\mathscr{U}'_\tau K(\tilde{b}_0) \subset K(\tilde{b}_0)$ for $\tau \geq 0$. Therefore, in particular, $\tilde{K}_m \subset K(\tilde{b}_0)$.

We often simply put $\tilde{K}_m = K(\tilde{b}_0)$. Since $K(\tilde{b}_0)$ is uniquely determined by $\hat{S}_m$ and $F_{\tilde{b}_0}$, then $K_m(\Sigma_m)$ as the norm-closure of $S_{\tilde{b}_0}\tilde{K}_m = S_{\tilde{b}_0}K(\tilde{b}_0)$ is also uniquely determined by $\hat{S}_m$ and $F_{\tilde{b}_0}$. In this sense, $S_{\tilde{b}_0}K(\tilde{b}_0)$ determines the "finest" dynamics compatible with the macro-observable $F_{\tilde{b}_0}$. If the norm-closure of $S_{\tilde{b}_0}\tilde{K}_m$ differs from the norm-closure of $S_{\tilde{b}_0}K(\tilde{b}_0)$, one still can "imagine" that the macrosystems satisfy the "finest" dynamics determined by $S_{\tilde{b}_0}K(\tilde{b}_0)$. The actual preparation possibilities, however, do not allow us experimentally to work out the details of this finest dynamics, since only the trajectory ensembles from $S_{\tilde{b}_0}\tilde{K}_m$ are realizable. We shall return to this dynamical problem in §2.5.

The mathematical structure of the fundamental relation (2.3.20) is well known. It means that the representation of the time translation semigroup in $K_m(\Sigma_m)$ by the $V_\tau^{(s)\prime}$ is homomorphic to the representation in $\tilde{K}_m$ by the $\mathscr{U}'_\tau$, and that $j$ resp. $S_{\tilde{b}_0}$ are homomorphic mappings of these representations.

First of all the question arises whether (2.3.20) can at all be satisfied for suitable $S_{\tilde{b}_0}$ and $\tilde{K}_m$. In this book we cannot go into this question in mathematical generality. Only some indications for such an analysis shall be given here, and an example follows in §3.

In the applications, the macrosystems occupy finite space regions (in II we did not point this out explicitly). This finiteness is attained in the macroscopic description by prescribing "boundaries" and boundary conditions. In $\mathscr{P}\mathscr{T}_{q\,\mathrm{exp}}$ the given boundaries are special structures of the Hamiltonian $H$ (e.g. an external "box potential"). For the systems restricted to finite regions, one thus obtains a discrete spectrum of the operator $H$. This has the consequence that the operator $\mathscr{U}_\tau$ also has a discrete frequency spectrum.

Therefore, writing $H = \sum_\nu \varepsilon_n E_\nu$ with the eigenvalues $\varepsilon_\nu$ and the projection operators $E_\nu$, we obtain

$$\mathscr{U}'_\tau w = U_\tau^+ w U_\tau = e^{-iH\tau} w e^{iH\tau} = \sum_{\nu,\mu} e^{i(\varepsilon_\mu - \varepsilon_\nu)\tau} E_\nu w E_\mu . \tag{2.3.21}$$

Therefore, $\mathscr{U}'_\tau w$ is an almost periodic function, which leads to the well-known "recurrence theorem" (for suitable $\tau$, $\mathscr{U}'_\tau w$ repeatedly comes arbitrarily close to the value $w$ in the sense of the $\sigma(\mathscr{B}, \mathscr{B}')$-topology).

When $\mathscr{U}'_\tau$ has a discrete frequency spectrum as in (2.3.21), we easily see that (2.3.20) cannot hold exactly and not at all for very large $\tau$. In fact, the frequency spectrum of $V_\tau^{(s)}$ is (except at the frequency zero) most often continuous, corresponding to the fact that the trajectories in $\hat{S}_m$ exhibit an irreversible behavior. Thus, most often the left side of (2.3.20) shows an aperiodic behavior, the right side an almost periodic behavior. Therefore, (2.3.20) need not hold at all for $\tau$ of the magnitude of the recurrence time.

In the almost periodic expression

$$\mathscr{U}'_\tau w = \sum_{\nu,\mu} e^{i(\varepsilon_\mu - \varepsilon_\nu)\tau} E_\nu w E_\mu = \sum_\nu E_\nu w E_\nu + \sum_{\nu \neq \mu} e^{i(\varepsilon_\mu - \varepsilon_\nu)\tau} E_\nu w E_\mu ,$$

the frequencies $\varepsilon_\mu - \varepsilon_\nu$ can be so dense that with suitable $w \in K$ we may replace the second summand on the right side by

$$\sum_{\substack{\nu,\mu \\ \nu \neq \mu}} e^{i(\varepsilon_\mu - \varepsilon_\nu)\tau} E_\nu w E_\mu \sim \int_{-\infty}^{+\infty} e^{i\omega\tau} w(\omega)\, d\omega.$$

This approximation is useable in the sense of the $\sigma(\mathscr{B}, \mathscr{B}')$-topology, so that suitable $g \in L$ make

$$\mu(\mathscr{U}'_\tau w, g) = \operatorname{tr}([\mathscr{U}'_\tau w]\, g) \sim \sum_\nu \operatorname{tr}(E_\nu w E_\nu) + \int_{-\infty}^{+\infty} e^{i\omega\tau} \operatorname{tr}(w(\omega)\, g)\, d\omega. \qquad (2.3.22)$$

The approximation (2.3.22) is very plausible and can easily be verified in artificially constructed examples. As such examples, one can choose the special almost periodic functions

$$f(t) = \sum_\nu a_\nu e^{i\omega_\nu t}, \qquad (2.3.23)$$

where the $\omega_\nu$ are distributed very densely and the $a_\nu$ only "slowly" change with $\omega_\nu$. Then for (2.3.23) one can find an approximation

$$f(t) = \int_{-\infty}^{+\infty} a(\omega)\, e^{i\omega t}\, d\omega,$$

which is very good for $t \ll (\Delta\omega)^{-1}$ (with $\Delta\omega$ the maximal distance between adjacent $\omega_\nu$). For real macro-systems, the distances of the frequencies in (2.3.21) are in fact so small that $(\Delta\omega)^{-1}$ exceeds $10^6$ years. Hence the main question is: "For which $w \in K$ and $g \in L$ is the approximation (2.3.22) useable?"

One possible way of handling this problem consists in trying to formulate a mathematically correct transition to "infinitely large" systems. For macroscopic theories, such transitions are very well known. They are used when one is less interested in the boundary effects than in the propagation processes. For example, for the heat conduction equation one can seek solutions for an "infinitely" extended body with the boundary condition that the temperature at infinity is constant. Similarly one can try to find limits in $\mathscr{PT}_{q\,\mathrm{exp}}$ for infinitely large systems when one only considers local effects (local observables) and special ensembles which "at infinity" correspond to the well-known thermodynamic equilibrium.

In this way, we certainly can obtain limits for which the dynamics is characterized by a continuous frequency spectrum. But it is an error to think that thereby the embedding problem is solved or that at least $L_m$ and $\tilde{K}_m$ are determined.

Another possible transition to a "continuous" frequency spectrum consists in replacing (2.3.22) by

$$\mu(w[\mathscr{U}_\tau g]) \sim \sum_\nu \operatorname{tr}(w E_\nu g E_\nu) + \sum_{\substack{\nu,\mu \\ \nu \neq \mu}} \operatorname{tr}(w E_\mu g E_\nu)\, e^{i(\varepsilon_\mu - \varepsilon_\nu)\tau - \tau/T}$$

for $\tau$ not too large, where $T$ is large relative to all macroscopic "observation times" of the system but still small relative to $(\Delta\omega)^{-1}$. But then one must investigate whether for appropriate $w$ and $g$ the right side in practice does not depend on $T$ as long as this $T$ obeys the conditions just given.

But in any case there remains the decisive question: "For what $S_{\tilde{b}_0}$ can the fundamental relation (2.3.20) be satisfied on a set $K(\tilde{b}_0) \neq \emptyset$?" Let us again emphasize that this question refers only to the semi-group of time translations with $\tau \geq 0$.

This question is not trivial, since the $\mathscr{U}'_\tau$ represent isomorphic mappings of the convex set $K$ into itself, whereas one knows from examples of macrosystems that the $V_\tau^{(s)\prime}$ are "contractive" (represent mixture-morphisms of $K(\Sigma_m)$ into itself, where $V_\tau^{(s)} K(\Sigma_m)$ is a proper subset of $K(\Sigma_m)$ for $\tau > 0$). But B.Sz. Nagy [31] has given examples where the $\mathscr{U}_\tau$ are homomorphic to a contractive semigroup for $\tau \geq 0$.

The theorems given by B.Sz. Nagy have the following structure: Let $Q$ be a mixture morphism of $K$ into itself with $Q^2 = Q$. Let $QK$ be properly smaller than $K$. If $\mathscr{U}_\tau$ has a continuous frequency spectrum from $-\infty$ to $+\infty$, there exists a $Q$ such that the $V'_\tau = Q\,\mathscr{U}'_\tau\,Q$ form a contractive semigroup ($V'_{\tau_1+\tau_2} = V'_{\tau_1} V'_{\tau_2}$ for $\tau_1$, $\tau_2 \geq 0$ and $V'_\tau\, QK$ is properly smaller than $QK$ for $\tau > 0$).

If such a "Nagy case" occurs, one can perform the following construction. We set

$$\tilde{K}_m = \overline{\text{co}} \bigcup_{\tau \geq 0} \mathscr{U}'_\tau\, QK$$

$$K_Q = QK. \tag{2.3.24}$$

Let the present $\tilde{K}_m$ correspond to the set called $\tilde{K}_m$ on page 27, while $K_Q$ corresponds to the set denoted by $K_m(\Sigma_m)$. We assert that

$$V'_\tau\, Q = Q\,\mathscr{U}'_\tau \tag{2.3.25}$$

then holds as a mapping from $\tilde{K}_m$ into $K_Q$. Therefore, if one thinks of $Q$ as the analogue of $S_{\tilde{b}_0}$ and of $V'_\tau$ as the analogue of $V_\tau^{(s)\prime}$, one has in (2.3.25) a mathematical model for the fact that a contractive semigroup $V'_\tau$ can be homomorphic to the semigroup $\mathscr{U}'_\tau$.

(2.3.25) can easily be shown: For a $w = \mathscr{U}'_{\tau_1} Q \tilde{w}$, with arbitrary $\tilde{w} \in K$, we have

$$\begin{aligned} Q\,\mathscr{U}'_\tau\, w &= Q\,\mathscr{U}'_{\tau+\tau_1} Q\,\tilde{w} = Q\,\mathscr{U}'_\tau\, Q\,\mathscr{U}'_{\tau_1} Q\,\tilde{w} \\ &= Q\,\mathscr{U}'_\tau\, Q\,Q\,\mathscr{U}'_{\tau_1} Q\,\tilde{w} = V'_\tau\, Q\,\mathscr{U}'_{\tau_1} Q\,\tilde{\omega} \\ &= V'_\tau\, Q\,\mathscr{U}'_{\tau_1} Q\,\tilde{w} = V'_\tau\, Q\,w. \end{aligned}$$

These discussions make it mathematically thinkable that (2.3.20) can be satisfied "approximately for $\tau$ that are not too large". "Approximately" because one must (for not too large $\tau$) approximate the discrete frequency spectrum by a continuous one.

In §3 we shall give a more specialized example for an approximate embedding. This will prove in a mathematically rigorous way that (2.3.20) can be approximately fulfilled.

The given considerations in no way show how the trajectory space $\hat{S}_m$ and the mapping $S_{\tilde{b}_0}$ can be chosen for real systems. It appears hopeless to solve this problem "in generality" if one recalls that systems as the mentioned aquarium with fish ought to be included. On the other hand, it appears conceivable that one could indeed find a general mathematical formulation for determining $\hat{S}_m$ and $S_{\tilde{b}_0}$, although this problem is practically unsolvable in complicated cases.

Certainly it is initially more meaningful to ask for $\hat{S}_m$ and $S_{\tilde{b}_0}$ in special simple cases, as one does in "statistical mechanics" with more or (most often) less exact methods. In this connection, one generally is not aware that the question of $\hat{S}_m$ and $S_{\bar{b}_0}$ is concerned. In §2.5 we shall for illustration give some initial indications.

For now we can conclude our considerations by saying that $\mathscr{PT}_m$ and $\mathscr{PT}_{q\,\exp}$ most likely are compatible as formulated by (2.3.20).

## §2.4 Partitioning of the Macroscopic Effects into Classes

Above, we already mentioned that $K_m(\Sigma_m)$ need not separate the elements of $L(\Sigma_m)$. Let us investigate more closely the partitioning into classes by $K_m(\Sigma_m)$ of the effects from $L(\Sigma_m)$. We would have done this already in II; but in the description of macro-systems by $\mathscr{PT}_m$ it has never been customary to consider such a partitioning into classes. For the embedding problem this partitioning presents an interesting structure. We can describe the partitioning of $L(\Sigma_m)$ in the structure of the dual pair $\mathscr{B}(\Sigma_m)$, $\mathscr{B}'(\Sigma_m)$.

The set $K_m(\Sigma_m)^\perp$ of all elements in $\mathscr{B}'(\Sigma_m)$ that are orthogonal to $K_m(\Sigma_m)$ is a $\sigma(\mathscr{B}'(\Sigma_m), \mathscr{B}(\Sigma_m))$-closed subspace of $\mathscr{B}'(\Sigma_m)$. The set $K_m(\Sigma_m)^{\perp\perp}$ is the norm-closed (hence also $\sigma(\mathscr{B}(\Sigma_m), \mathscr{B}'(\Sigma_m))$-closed) subspace of $\mathscr{B}(\Sigma_m)$ spanned by $K_m(\Sigma_m)$. Thus $K_m(\Sigma_m)^{\perp\perp}$ is the subspace $\mathscr{B}_m(\Sigma_m)$ already introduced in §2.3. Then $\mathscr{B}_m(\Sigma_m)^\perp = K_m(\Sigma_m)^\perp$, while $\mathscr{B}'(\Sigma_m)/\mathscr{B}_m(\Sigma_m)^\perp$ can be identified with the Banach space $\mathscr{B}'_m(\Sigma_m)$ dual to $\mathscr{B}_m(\Sigma_m)$. The canonical mapping $\mathscr{B}'(\Sigma_m)\to\mathscr{B}'(\Sigma_m)/\mathscr{B}_m(\Sigma_m)^\perp=\mathscr{B}'_m(\Sigma_m)$ is the mapping $s'$ dual to the injection $\mathscr{B}_m(\Sigma_m)\xrightarrow{s}\mathscr{B}(\Sigma_m)$.

In the sequel, the set $\tilde{K}_m(\Sigma_m)=K(\Sigma_m)\cap\mathscr{B}_m(\Sigma_m)$ introduced in §2.3 will occur more and more frequently. We have $K_m(\Sigma_m)\subset\tilde{K}_m(\Sigma_m)$, while one can identify $\tilde{K}_m(\Sigma_m)$ with $\tilde{K}_m(\hat{S}_m)$ (see §2.3) and $\tilde{K}_m(\hat{S}_m)$ with $\tilde{K}_m(\hat{Y})$. Hence $\tilde{K}_m(\Sigma_m)$ describes an "idealized" dynamics (as in II §3.3).

Therefore, we can write the partition of $L(\Sigma_m)$ as the mapping $L(\Sigma_m)\xrightarrow{s'}\mathscr{B}'_m(\Sigma_m)$. Let the image set $s'\,L(\Sigma_m)$ be denoted by $\tilde{L}(\Sigma_m)$ and $s'\,L_m(\Sigma_m)$ by $\tilde{L}_m(\Sigma_m)$. This $\tilde{L}(\Sigma_m)$ is a $\sigma(\mathscr{B}'_m(\Sigma_m), \mathscr{B}_m(\Sigma_m))$-compact, convex subset of $\mathscr{B}'_m(\Sigma_m)$. Therefore $\tilde{L}(\Sigma_m)$ is generated by the set $\partial_e\tilde{L}(\Sigma_m)$ of its extreme points.

For $g\in\partial_e\tilde{L}(\Sigma_m)$, $s'^{-1}g$ is a closed face of $L(\Sigma_m)$. Hence there is an extreme point of $L(\Sigma_m)$, i.e. a $\sigma\in\Sigma_m$ with $s'\sigma=g$. If there were two extreme points $\sigma_1$, $\sigma_2$ with $s'\sigma_1=s'\sigma_2=g$, then

$$g=s'\sigma_1=s'(\sigma_1\wedge\sigma_2^*)+s'(\sigma_1\wedge\sigma_2)$$
$$g=s'\sigma_2=s'(\sigma_1^*\wedge\sigma_2)+s'(\sigma_1\wedge\sigma_2)$$

would imply

$$g=\tfrac{1}{2}s'(\sigma_1\vee\sigma_2)+\tfrac{1}{2}s'(\sigma_1\wedge\sigma_2).$$

Since $g$ is an extreme point, we find $s'(\sigma_1\wedge\sigma_2)=g$ and hence $s'(\sigma_1\wedge\sigma_2^*)=s'(\sigma_1^*\wedge\sigma_2)=0$, i.e. $\sigma_1\wedge\sigma_2^*=\sigma_1^*\wedge\sigma_2=0$ and thus $\sigma_1=\sigma_2$ ($s'$ is effective on $\Sigma_m$ according to the definition of $\Sigma_m$!).

Therefore, for the subset $\Sigma_{mk}=s'^{-1}\,\partial_e\tilde{L}(\Sigma_m)$ of $\Sigma_m$ the mapping $\Sigma_{mk}\xrightarrow{s'}\partial_e\tilde{L}(\Sigma_m)$ is bijective. In general, $\Sigma_{mk}$ is not a Boolean subring of $\Sigma_m$! If $\Sigma_{mk}$ is not a Boolean ring, then the two dual Banach spaces $\mathscr{B}_m(\Sigma_m)$, $\mathscr{B}'_m(\Sigma_m)$ together with their sets of ensembles $\tilde{K}_m(\Sigma_m)$ and of effects $\tilde{L}(\Sigma_m)$ do not describe "classical systems" in the sense of VII §5.3. The objectivating description of macrosystems does not imply automatically that these systems are classical!

From $\Sigma_{mk} \xrightarrow{s'} \partial_e \tilde{L}(\Sigma_m)$ bijective, with $L(\Sigma_{mk}) \overset{\text{def}}{=} \overline{\text{co}}^\sigma \Sigma_{mk}$ follows that the mapping

$$L(\Sigma_{mk}) \xrightarrow{s'} \tilde{L}(\Sigma_m) \tag{2.4.1}$$

is surjective.

Let us introduce a sort of inverse to the mapping (2.4.1) by choosing, for $g \in L(\Sigma_m)$, an element $h$ from the set $(s'^{-1} g) \cap L(\Sigma_{mk})$ and setting $r^* g = h$. This defines a mapping

$$L(\Sigma_m) \supset L(\Sigma_{mk}) \xleftarrow{r^*} \tilde{L}(\Sigma_m), \tag{2.4.2}$$

where

$$\Sigma_{mk} \xleftarrow{r^*} \partial_e \tilde{L}(\Sigma_m)$$

is the inverse to the bijective mapping

$$\Sigma_{mk} \xrightarrow{s'} \partial_e L(\Sigma_m).$$

One obtains $s' r^* = 1$ as a mapping from $\tilde{L}(\Sigma_m)$ into itself and $p^2 = p$ for $p = r^* s'$ as a mapping from $L(\Sigma_m)$ into itself. This implies $pL(\Sigma_m) \subset L(\Sigma_{mk})$ and that $p$ is the identity mapping on $\Sigma_{mk}$. For $\sigma \in \Sigma_m$ and for all $u \in K_m(\Sigma_m)$, we therefore have $\langle u, \sigma \rangle = \langle u, p\sigma \rangle$.

The mapping $r^*$ is uniquely defined if and only if (2.4.1) is also injective. If one can define $r^*$ so that it is affine, then (2.4.1) is injective and $r^*$ is the mapping inverse to (2.4.1). Then $L(\Sigma_{mk})$ and $L(\Sigma_m)$ are two isomorphic covex sets. Below we shall investigate a special case in which $r^*$ is affine.

Because of II (4.2.9), i.e. $V'_\tau K_m(\Sigma_m) \subset K_m(\Sigma_m)$, from $s' f_1 = s' f_2$ follows

$$\langle u, V_\tau^{(s)} f_1 \rangle = \langle V_\tau^{(s)\prime} u, f_1 \rangle = \langle V_\tau^{(s)\prime} u, f_2 \rangle = \langle u, V_\tau^{(s)} f_2 \rangle$$

for all $u \in K_m(\Sigma_m)$ and hence $s' V_\tau^{(s)} f_1 = s' V_\tau^{(s)} f_2$. Therefore, a mapping $\tilde{V}_\tau$ is defined by the diagram

$$\begin{array}{ccc} L(\Sigma_m) & \xrightarrow{s'} & \tilde{L}(\Sigma_m) \\ \Big\downarrow V_\tau^{(s)} & & \Big\downarrow \tilde{V}_\tau \\ L(\Sigma_m) & \xrightarrow{s'} & \tilde{L}(\Sigma_m). \end{array} \tag{2.4.3}$$

From $V_{\tau_1+\tau_2}^{(s)} = V_{\tau_1}^{(s)} V_{\tau_2}^{(s)}$ (see II §4.2), with $s' V_\tau^{(s)} = \tilde{V}_\tau s'$ follows

$$\tilde{V}_{\tau_1+\tau_2} s' = s' V_{\tau_1+\tau_2}^{(s)} = s' V_{\tau_1}^{(s)} V_{\tau_2}^{(s)} = \tilde{V}_{\tau_1} s' V_{\tau_2}^{(s)} = \tilde{V}_{\tau_1} \tilde{V}_{\tau_2} s'$$

and hence

$$\tilde{V}_{\tau_1+\tau_2} = \tilde{V}_{\tau_1} \tilde{V}_{\tau_2}, \tag{2.4.4}$$

i.e. the $\tilde{V}_\tau$ form a semigroup.

Whereas $V_\tau^{(s)}$ has only a kinematic meaning in $\Sigma_m$ and $L(\Sigma_m)$, by the mapping on $\tilde{L}(\Sigma_m)$ one obtains time displacement operators $\tilde{V}_\tau$ in $\tilde{L}(\Sigma_m)$ which reflect something of the dynamics. For instance, for $g \in s' \Sigma_{mk}$ the element $s' V_\tau^{(s)} r^* g$ specifies (by its convex combination from the elements of $s' \Sigma_{mk}$) the "partition" over the various $\sigma \in \Sigma_{mk}$ after a time $\tau$. For that reason we call

$$s' V_\tau^{(s)} r^* = \tilde{V}_\tau s' r^* = \tilde{V}_\tau \tag{2.4.5}$$

the dynamical operators of time displacement.

Let us also consider a partition into classes of the effects in $\mathscr{P}\mathscr{T}_{q\,\mathrm{exp}}$. Let $K_m^*$ be a norm-closed, convex subset of $\tilde{K}_m$ with $jK_m^*$ resp. $S_{\tilde{b}_0} K_m^*$ dense in $K_m(\Sigma_m)$ (e.g. $K_m^* = K_m$). Let the Banach subspace $K_m^{*\perp\perp}$ of $\mathscr{B}$ generated by $K_m^*$ be denoted by $\mathscr{B}_m$. As a norm continuous affine mapping, $j$ can be extended to all of $\mathscr{B}_m$. The canonical mapping $\mathscr{B}' \to \mathscr{B}'/\mathscr{B}_m^\perp = \mathscr{B}'_m$ is the dual $l'$ to the injection $\mathscr{B}_m \xrightarrow{l} \mathscr{B}$. Let the $\sigma(\mathscr{B}'_m, \mathscr{B}_m)$-compact set $l' L$ be denoted by $\tilde{L}$ and the compact set $l' L_m$ be called $\tilde{L}_m$.

Because of $jK_m^* \subset K_m(\Sigma_m)$ we have $j\mathscr{B}_m \subset \mathscr{B}_m(\Sigma_m)$, so that we have the diagram

$$\begin{array}{ccc} B(\Sigma_m) & \xleftarrow{s} & \mathscr{B}_m(\Sigma_m) \\ \uparrow S_{\tilde{b}_0} & & \uparrow j \\ \mathscr{B} & \xleftarrow{l} & \mathscr{B}_m, \end{array} \tag{2.4.6}$$

since $S_{\tilde{b}_0}$ coincides with $j$ on $\mathscr{B}_m$. Since $jK_m^*$ is dense in $K_m(\Sigma_m)$, the set $j\mathscr{B}_m$ is dense in $\mathscr{B}_m(\Sigma_m)$ so that the mapping $j'$ dual to $j$ is injective. The diagram dual to (2.4.6) is

$$\begin{array}{ccc} \mathscr{B}'(\Sigma_m) & \xrightarrow{s'} & \mathscr{B}'_m(\Sigma_m) \\ \uparrow S'_{\tilde{b}_0} & & \uparrow j' \\ \mathscr{B} & \xrightarrow{l'} & \mathscr{B}'_m. \end{array} \tag{2.4.7}$$

Also if $S_{\tilde{b}_0}$ does not exist but $j$ is norm-continuous, the mappings $\mathscr{B}_m \xrightarrow{j} \mathscr{B}_m(\Sigma_m)$ and $\mathscr{B}'_m(\Sigma_m) \xrightarrow{j'} \mathscr{B}'_m$ are well defined ($j'$ is injective). From (2.3.17) (valid for all $w \in \tilde{K}_m$), i.e. from

$$\langle jw, \psi_{ms}(f) \rangle = \mu(w, \psi(if))$$

for all $w \in K_m^*$ and all $f \in F_m$, then follows

$$L_m(\Sigma_m) \xrightarrow{s'} \tilde{L}_m(\Sigma_m) \xrightarrow{j'} \tilde{L}_m \subset \tilde{L}.$$

It need not be $j' \tilde{L}(\Sigma_m) \subset \tilde{L}$! Defining

$$L_m^i(\Sigma_m) = s'^{-1}\{j'^{-1}[\tilde{L} \cap j' \tilde{L}(\Sigma_m)]\},$$

we have

$$L_m(\Sigma_m) \subset L_m^i(\Sigma_m) \subset L(\Sigma_m).$$

The elements of $L_m(\Sigma_m)$ are the registrable trajectory effects. But we can imagine that also the elements of $L^i_m(\Sigma_m)$ can be registered in an "idealized" form. If $L^i_m(\Sigma_m) \neq L(\Sigma_m)$, the set $L^i_m(\Sigma_m)$ describes a *finite* imprecision of the measurements of the trajectories in the following sense: As a compact convex set, $L^i_m(\Sigma_m)$ is generated by $\partial_e L^i_m(\Sigma_m)$. The elements of $\partial_e L^i_m(\Sigma_m)$ give the finest possibilities to distinguish trajectories. The elements of $\partial_e L^i_m(\Sigma_m)$ replace the elements of $\partial_e L(\Sigma_m)$ which are the characteristic functions of the elements of $\Sigma_m$. The elements of $\partial_e L^i_m(\Sigma_m)$ describe how much these characteristic functions must be "smeared out", i.e. how imprecisely the boundaries of the elements of $\Sigma_m$ are determined by measurements. If $L^i_m(\Sigma_m) \neq L(\Sigma_m)$, the "usual" classical idealization (that one can measure more and more precisely without finite limits) is then not applicable.

Let us not discuss such a finite imprecision any more: Only in §2.6 and §3 we shall remark on this possibility. Now let us use the idealization that $S_{\tilde{b}_0}$ exists. Then $j'\tilde{L}(\Sigma_m) = \tilde{L}_m \subset \tilde{L}$ implies $L^i_m(\Sigma_m) = L(\Sigma_m)$. Proof: Because of $S'_{\tilde{b}_0}\, L(\Sigma_m) \subset L_m$, it is first guaranteed that $j'\tilde{L}(\Sigma_m) \subset \tilde{L}_m$. The mapping $\tilde{L}(\Sigma_m) \xrightarrow{j'} \tilde{L}_m$ is surjective if for each $g \in L_m$ there is a $g' \in L(\Sigma_m)$ with $l'g = j's'g' = l'S'_{\tilde{b}_0}g'$, i.e. with $\mu(w, g) = \mu(w, S'_{\tilde{b}_0}g') = \langle jw, g'\rangle$ for all $w \in K^*_m$. But each $g \in L_m$ is representable approximately (in the $\sigma(\mathscr{B}', \mathscr{B})$-topology) by a $\psi(ib_0, ib)$ (since $\overline{\mathrm{co}}^\sigma\, \psi_{ms\tilde{b}_0}\, \Sigma_{\tilde{b}_0} = L(\Sigma_m)$). For $\psi(ib_0, ib)$ and all $w \in \tilde{K}_m$, as above the validity of (2.3.17) (at $\tau = 0$) moreover implies

$$\langle S_{\tilde{b}_0} w, \psi_{ms}(b_0 \cdot b)\rangle = \mu(w, \psi(ib_0, ib)) \qquad \text{for all } w \in K^*_m. \tag{2.4.8}$$

Therefore, since $L(\Sigma_m)$ and $L_m$ are compact and $S'_{\tilde{b}_0}$ is continuous, for each $g \in L_m$ there is a $g' \in L(\Sigma_m)$ with $l'g = l'\tilde{S}'_{\tilde{b}_0}g'$.

Since the mapping $\tilde{L}(\Sigma_m) \xrightarrow{j'} \tilde{L}_m$ is bijective, its inverse $\tilde{T} = j'^{-1}$ exists:

$$\tilde{L}_m \xrightarrow{\tilde{T}} \tilde{L}(\Sigma_m).$$

From the diagram (2.4.7) one obtains

$$\begin{array}{ccc} L(\Sigma_m) & \xrightarrow{s'} & \tilde{L}(\Sigma_m) \\ \Big\downarrow S'_{\tilde{b}_0} & & j' \Big\downarrow \Big\uparrow \tilde{T} \\ L_m & \xrightarrow{l'} & \tilde{L}_m . \end{array} \tag{2.4.9}$$

As the bijective image of the convex set $\tilde{L}(\Sigma_m)$, also $\tilde{L}_m$ is convex. Thus $\hat{L}_m = l'^{-1}\tilde{L}_m \cap L$ is a convex and $\sigma(\mathscr{B}', \mathscr{B})$-compact subset of $L$. In (2.4.9) one also can replace $L_m$ by $\hat{L}_m$.

(2.3.19) implies

$$l'S'_{\tilde{b}_0}V^{(s)}_\tau = l'\mathscr{U}_\tau S'_{\tilde{b}_0} \tag{2.4.10}$$

as a mapping of $\mathscr{B}'(\Sigma_m)$ into $\mathscr{B}'_m$ or as a mapping of $L(\Sigma_m)$ into $\tilde{L}_m$.

By multiplication of (2.4.10) with $\tilde{T}$, because of $s' = \tilde{T}l'S'_{\tilde{b}_0}$ (see (2.4.9)) we find

$$s'V^{(s)}_\tau = \tilde{V}_\tau s' = \tilde{T}l'\mathscr{U}_\tau S'_{\tilde{b}_0} \tag{2.4.11}$$

as a mapping of $L(\Sigma_m)$ into $\tilde{L}(\Sigma_m)$.

By multiplying (2.4.11) from the right by $r^*$, for the dynamical operators in $\tilde{L}(\Sigma_m)$ we get

$$\tilde{V}_\tau = \tilde{T} l' \mathcal{U}_\tau S'_{\tilde{b}_0} r^*. \tag{2.4.12}$$

We abbreviate:

$$\hat{T} = \tilde{T} l' \tag{2.4.13}$$

and

$$P = S'_{\tilde{b}_0} r^* \hat{T}, \tag{2.4.14}$$

thus defining $\hat{T}$ as a mapping of $\hat{L}_m$ onto $\tilde{L}(\Sigma_m)$ and $P$ as a mapping of $\hat{L}_m$ into itself. Because of $s' = \tilde{T} l' S'_{\tilde{b}_0}$, we obtain $p = r^* \tilde{T} l' S'_{\tilde{b}_0}$ and $\tilde{T} l' S'_{\tilde{b}_0} r^* = 1$. With (2.4.14) follows

$$\hat{T} P = \hat{T}$$

and

$$S'_{\tilde{b}_0} r^* = P S'_{\tilde{b}_0} r^*$$

as well as

$$\begin{aligned} P^2 &= S'_{\tilde{b}_0} r^* \hat{T} S'_{\tilde{b}_0} r^* \hat{T} \\ &= S'_{\tilde{b}_0} r^* \tilde{T} l' \tilde{S}_{\tilde{b}_0} r^* \hat{T} = S'_{\tilde{b}_0} r^* \hat{T} = P. \end{aligned}$$

Therefore $P$ is a projector of $L_m$ into $PL_m \subset S'_{\tilde{b}_0} L(\Sigma_{mk})$. Thus (2.4.12) can also be written

$$\begin{aligned} \tilde{V}_\tau &= \hat{T} \mathcal{U}_\tau P S'_{\tilde{b}_0} r^* \\ &= \hat{T} P \mathcal{U}_\tau P S'_{\tilde{b}_0} r^*. \end{aligned}$$

With (2.4.12), from $\tilde{V}_{\tau_1+\tau_2} = V_{\tau_1} V_{\tau_2}$ follows

$$\begin{aligned} \tilde{V}_{\tau_1+\tau_2} &= \tilde{T} l' \mathcal{U}_{\tau_1} S'_{\tilde{b}_0} r^* \tilde{T} l' \mathcal{U}_{\tau_2} S'_{\tilde{b}_0} r^* \\ &= \tilde{T} l' \mathcal{U}_{\tau_1} P \mathcal{U}_{\tau_2} \tilde{S}'_{\tilde{b}_0} r^* \\ &= \tilde{T} l' \mathcal{U}_{\tau_1+\tau_2} S'_{\tilde{b}_0} r^* \end{aligned}$$

and from this

$$P \mathcal{U}_{\tau_1} P \mathcal{U}_{\tau_2} P = P \mathcal{U}_{\tau_1+\tau_2} P. \tag{2.4.15}$$

We abbreviate:

$$\tilde{\mathcal{U}}_\tau = P \mathcal{U}_\tau P. \tag{2.4.16}$$

The $\tilde{\mathcal{U}}_\tau$ form a semigroup, which by

$$\tilde{V}_\tau = \hat{T} \tilde{\mathcal{U}}_\tau S'_{\tilde{b}_0} r^* \tag{2.4.17}$$

yield the semigroup of the $\tilde{V}_\tau$.

The $\tilde{\mathcal{U}}_\tau$ can be called operators of the "reduced" dynamics since (by means of the projection $P$) they are formed from the $\mathcal{U}_\tau$ according to (2.4.16).

These considerations do not depend on the choice of $K^*_m \subset \tilde{K}_m$, provided only that $S_{\tilde{b}_0} K^*_m$ is dense in $K_m(\Sigma_m)$. Hence the partition of $L_m$ into classes (i.e. into the elements of $\tilde{L}_m$) does not depend on $K^*_m$ since $j'$ maps the set $\tilde{L}(\Sigma_m)$ (which does not depend on $K^*_m$) bijectively onto $\tilde{L}_m$. Because of $s' = \tilde{T} l' S'_{\tilde{b}_0}$ and $s' L(\Sigma_{mk})$

$= s' L(\Sigma_m)$, we have $l' S'_{\tilde{b}_0} L(\Sigma_{mk}) = \tilde{L}_m$. Thus from the diagram (2.4.9) we can obtain

$$\begin{array}{ccc} L(\Sigma_m) & \overset{r^*}{\underset{s'}{\leftrightarrows}} & \tilde{L}(\Sigma_m) \\ \Big\downarrow S'_{\tilde{b}_0} & & j' \Big\uparrow\Big\downarrow \tilde{T} \\ S'_{\tilde{b}_0} L(\Sigma_m) & \xrightarrow{l'} & \tilde{L}_m, \end{array} \tag{2.4.18}$$

where $l'$ is surjective. This in particular shows that

$$\Sigma_{mk} \xrightarrow{j' s'} \partial_e \tilde{L}_m$$

is a bijection. We have $l' \partial_e S'_{\tilde{b}_0} L(\Sigma_m) \supset \partial_e \tilde{L}_m$ and

$$S'_{\tilde{b}_0} \Sigma_{mk} \supset \partial_e S'_{\tilde{b}_0} L(\Sigma_{mk}),$$

and $S'_{\tilde{b}_0} l'$ is a bijection of $\Sigma_{mk}$ onto $\partial_e \tilde{L}_m$ (because $j' s' = S'_{\tilde{b}_0} l'$). This implies that

$$\Sigma_{mk} \xrightarrow{S'_{\tilde{b}_0}} \partial_e S'_{\tilde{b}_0} L(\Sigma_{mk}) \xrightarrow{l'} \partial_e \tilde{L}_m \tag{2.4.19}$$

are bijections. Therefore, $P$ is the identity mapping on $\partial_e S'_{\tilde{b}_0}(\Sigma_{mk})$. Since $\Sigma_{mk}$ is uniquely determined as the set of those elements from $\Sigma_m$ which are mapped by $l' S'_{\tilde{b}_0}$ onto $\partial_e L_m$, with $N$ as the kernel of the observable $\Sigma_m \xrightarrow{F_{\tilde{b}_0}} L_m \subset L$ (see V D3.3.4), we conclude $N \supset \Sigma_{mk}$.

Whereas in the preceding derivations no additional assumptions about the structure of $K_m(\Sigma_m)$ entered, let us now consider a very frequent *special case*. In this context, we first investigate the macrotheory for its own sake (without reference to embedding).

Let us take the special case where $\Sigma_{mk}$ is a Boolean subring of $\Sigma_m$. Then $\Sigma_{mk}$ is also complete since $s'^{-1} \partial_e \tilde{L}(\Sigma_m) \cap \Sigma_m$ is closed in the $\sigma(\mathscr{B}(\Sigma_m), \mathscr{B}'(\Sigma_m))$-topology.

If $\Sigma'$ is a complete Boolean ring and $\Sigma' \xrightarrow{f} \Sigma_m$ an isomorphic mapping of $\Sigma'$ onto the Boolean subring $f\Sigma'$ of $\Sigma_m$, then (for generality see V §3.2) a mixture-morphism $K(\Sigma_m) \xrightarrow{r} K(\Sigma')$ is defined by $\langle u, f\sigma \rangle = \langle v, \sigma \rangle$ and $u \to v$. On $\Sigma'$, the dual mapping $L(\Sigma') \xrightarrow{r'} L(\Sigma_m)$ is identical with the isomorphism $\Sigma' \xrightarrow{f} f\Sigma' \subset \Sigma_m$.

If $f\Sigma' = \Sigma_{mk}$, the mapping $L(\Sigma') \xrightarrow{s'r'} \tilde{L}(\Sigma_m)$ is bijective. From $r'\Sigma' = \Sigma_{mk}$ follows immediately $s'r'\Sigma' = s'\Sigma_{mk} = \partial_e \tilde{L}(\Sigma_m)$ and from this that

$$L(\Sigma') \xrightarrow{s'r'} \tilde{L}(\Sigma_m)$$

is surjective. Let us show that $\mathscr{B}'(\Sigma') \xrightarrow{s'r'} \mathscr{B}'_m(\Sigma_m)$ is injective. If there were a $y \in \mathscr{B}'(\Sigma')$ with $s'r'y = 0$, then one could choose $\|y\| = 1$ and decompose $y$ into $y = y_+ - y_-$ (see [2] IV §2.1) so that the supports $\sigma_\pm$ of the positive and negative parts $y_\pm$ obey $\sigma_+ \wedge \sigma_- = 0$. We have $y_+, y_- \in L(\Sigma')$ and $y_+ \leq \sigma_+$. Therefore, $\tilde{y} = \sigma_+ - y = (\sigma_+ - y_+) + y_-$ is an element of $L(\Sigma')$. Because of $s'r'y = 0$ we get $s'r'\tilde{y} = s'r'\sigma_+ \in \partial_e \tilde{L}(\Sigma_m)$. If we had $\tilde{y} \in \Sigma'$, we would have $\tilde{y} - \sigma_+ \in \Sigma'$ and hence

$s'r'y \in \partial_e \tilde{L}(\Sigma_m)$, contradicting $y \neq 0$ and $s'r'y=0$. For $\tilde{y} \in L(\Sigma')$ the spectral representation of $\tilde{y}$ (see [2] IV §2.1) shows that there exist a $\sigma \in \Sigma'$ ($\sigma \neq 0$) and an $\varepsilon > 0$ such that $\tilde{y} \pm \varepsilon\sigma \in L(\Sigma')$ holds. From $\tilde{y} = 1/2(\tilde{y}+\varepsilon\sigma) + 1/2(\tilde{y}-\varepsilon\sigma)$ follows $s'r'\tilde{y} = s'r'\sigma_+ = \frac{1}{2}s'r'(\tilde{y}+\varepsilon\sigma) + \frac{1}{2}s'r'(\tilde{y}-\varepsilon\sigma)$. Since $s'r'\sigma_+ \in \partial_e \tilde{L}$, we must have $s'r'(\tilde{y}+\varepsilon\sigma) = s'r'\sigma_+ = s'r'\tilde{y}$ and hence $s'r'\sigma = 0$, in contradiction to $\sigma \neq 0$.

$L(\Sigma') \xrightarrow{s'r'} \tilde{L}(\Sigma_m)$ bijective implies $\mathscr{B}_m(\Sigma_m) \xrightarrow{rs} \mathscr{B}(\Sigma')$ injective and $rs\mathscr{B}_m(\Sigma_m)$ dense in $\mathscr{B}'(\Sigma')$. From $s'r'L(\Sigma') = \tilde{L}(\Sigma_m)$ also follows that the mapping $\mathscr{B}_m(\Sigma_m) \xrightarrow{rs} \mathscr{B}(\Sigma')$ preserves the norm; therefore, $\mathscr{B}_m(\Sigma_m) \xrightarrow{rs} \mathscr{B}(\Sigma')$ is an isomorphism of Banach spaces. Hence we can identify $\mathscr{B}_m(\Sigma_m)$ with $\mathscr{B}(\Sigma')$ in the following way. $\mathscr{B}(\Sigma') \xrightarrow{(rs)^{-1}} \mathscr{B}_m(\Sigma_m) \xrightarrow{s} \mathscr{B}(\Sigma_m)$ defines a mixture-morphism $\mathscr{B}(\Sigma') \xrightarrow{\tilde{s}} \mathscr{B}(\Sigma_m)$ whose adjoint mapping $\mathscr{B}'(\Sigma_m) \xrightarrow{\tilde{s}'} \mathscr{B}'(\Sigma')$ is given by

$$\mathscr{B}'(\Sigma_m) \xrightarrow{s'} \mathscr{B}'_m(\Sigma_m) \xrightarrow{(r's')^{-1}} \mathscr{B}'(\Sigma').$$

Then we get $r\tilde{s}=1$, $\tilde{s}'r'=1$ and $\tilde{s}K(\Sigma') = \tilde{K}_m(\Sigma_m)$.

The mapping $\tilde{K}_m(\Sigma_m) = K(\Sigma_m) \cap \mathscr{B}_m(\Sigma_m) \xrightarrow{r} K(\Sigma')$ is a bijection. But $rK_m(\Sigma_m)$ can be a proper subset of $K(\Sigma')$.

Above, we proceeded from the assumption $f(\Sigma') = \Sigma_{mk}$. This is equivalent to saying that $\tilde{K}_m(\Sigma_m) \xrightarrow{rs} K(\Sigma')$ is a bijective mapping. To finish the proof we now need only show that $\tilde{K}_m(\Sigma_m) \xrightarrow{rs} K(\Sigma')$ bijective implies $f(\Sigma') = \Sigma_{mk}$.

A mixture-morphism $\mathscr{B}(\Sigma') \xrightarrow{t} \mathscr{B}_m(\Sigma_m)$ is defined by $K(\Sigma') \xrightarrow{(rs)^{-1}} K(\Sigma_m)$. It can easily be shown that $\mathscr{B}(\Sigma') \xrightarrow{t} \mathscr{B}_m(\Sigma_m)$ and $\mathscr{B}_m(\Sigma_m) \xrightarrow{rs} \mathscr{B}(\Sigma')$ are bijections with $trs = rst = 1$. Then, $\mathscr{B}(\Sigma') \xrightarrow{\tilde{s}=st} \mathscr{B}(\Sigma_m)$ can be defined as above, with $K(\Sigma') \xrightarrow{\tilde{s}} \tilde{K}_m(\Sigma_m)$ as a bijection. From this follows $r\tilde{s}=1$. For the adjoint mappings

$$\mathscr{B}'(\Sigma') \underset{r'}{\overset{\tilde{s}'}{\leftrightarrows}} \mathscr{B}'(\Sigma_m) \tag{2.4.20}$$

we find $\tilde{s}'r'=1$. From $\Sigma' \xrightarrow{r'} \Sigma_m$ thus follows $\tilde{s}'(r'\Sigma') = \Sigma'$. Therefore, $r'\Sigma'$ equals $\tilde{s}'^{-1}\Sigma'$. From

$$\mathscr{B}(\Sigma') \xrightarrow{t} \mathscr{B}_m(\Sigma_m) \xrightarrow{s} \mathscr{B}(\Sigma_m)$$

we obtain

$$L(\Sigma') \xleftarrow{t'} \tilde{L}(\Sigma_m) \xleftarrow{s'} L(\Sigma_m),$$

where $t'$ must be a bijection and hence $t'^{-1}\Sigma' = \partial_e \tilde{L}(\Sigma_m)$. Then, with $\Sigma_{mk} = s'^{-1}\partial_e \tilde{L}(\Sigma_m)$ we conclude $\Sigma_{mk} = s'^{-1}\Sigma$ and finally $\Sigma_{mk} = r'\Sigma'$.

Under the assumption made, we can use all the preceding results when we replace $\mathscr{B}_m(\Sigma_m)$ by $\mathscr{B}(\Sigma')$, $\mathscr{B}'_m(\Sigma_m)$ by $\mathscr{B}'(\Sigma')$, $\tilde{L}(\Sigma_m)$ by $L(\Sigma')$, $s$ by $\tilde{s}$, and finally $r^*$ by the mapping $r'$ dual to $r$. Putting $\tilde{s}r=q$, we get $q^2=q$ and $p=r^*\tilde{s}'=r'\tilde{s}'=q'$.

Therefore, the systems are "classical" in the sense of VII §5.3 if $\Sigma_{mk}$ is a Boolean ring and if we take $\tilde{K}_m(\Sigma_m)$ as the set of ensembles. If $K_m(\Sigma_m)$ is a proper subset of $\tilde{K}_m(\Sigma_m)$, the systems are classical in a restricted form, namely, we may "think" of these systems as classical but *in reality* not all ensembles in $\tilde{K}(\Sigma_m)$ can be prepared.

The mapping

$$A_\tau = rV_\tau^{(s)'}\tilde{s} \tag{2.4.21}$$

defines mixture-morphisms of $K(\Sigma')$ into itself. From (2.4.5) follows

$$\tilde{V}_\tau = A'_\tau = \tilde{s}' V_\tau^{(s)} r' \tag{2.4.22}$$

while $\tilde{s}' V_\tau^{(s)} = \tilde{V}_\tau \tilde{s}'$ (see the diagram (2.4.3)) yields

$$\tilde{s} A_\tau = V_\tau^{(s)} \tilde{s}. \tag{2.4.23}$$

Since the $\tilde{V}_\tau$ form a semigroup, the $A_\tau$ do also. The $A_\tau$ will be called the dynamical operators in $K(\Sigma')$. This designation rests on the following considerations:

The set $K_m(\Sigma_m)$ determines the mapping $s$ and hence the set $\Sigma_{mk}$. If $\Sigma_{mk}$ is a Boolean subring, the mappings $\tilde{s}, r$ are determined by $\Sigma' = \Sigma_{mk}$. Then $r, \tilde{s}$ and the kinematic mappings $V_\tau^{(s)}$ determine $A_\tau$. But under certain assumptions a sort of converse also holds:

If the mapping $\Sigma' \xrightarrow{f} \Sigma_m$ is given as an isomorphism of $\Sigma'$ onto $f(\Sigma') = \Sigma_{mk}$ (and hence $r$), then $s$ is restricted by the fact that we must have $r\tilde{s} = 1$. Via $\tilde{K}_m(\Sigma_m) = \tilde{s}K(\Sigma')$, the set $\tilde{K}_m(\Sigma_m)$ and hence the dynamics are determined by $\tilde{s}$, while the subset $K_m(\Sigma_m)$ of $\tilde{K}_m(\Sigma_m)$, is not determined by $\tilde{s}$. But $K_m(\Sigma_m)$ must be so large that $K_m(\Sigma_m)^\perp = \tilde{K}(\Sigma_m)^\perp$.

If we prescribe $A_\tau$, then $\tilde{s}'$ is further restricted since (2.4.21) must be satisfied between the kinematic mappings $V_\tau^{(s)}$ and the dynamic mappings $A_\tau$. How does this restriction look?

Let the complete subring of $\Sigma_m$ generated by $\bigcup_{\tau \geq 0} V_\tau^{(s)} \Sigma_{mk}$ be briefly denoted by $\bar{\Sigma}_{mk}$. Therefore, the kinematic transformations map $\bar{\Sigma}_{mk}$ into itself. We will show that $\tilde{s}'$ is uniquely defined on $\bar{\Sigma}_{mk}$ by the $A_\tau$.

The elements of the Boolean subring of $\Sigma_m$ generated by $\bigcup_{\tau \geq 0} V_\tau^{(s)} \Sigma_{mk}$ are finite disjoint unions of elements of the form

$$(V_{\tau_1}^{(s)} \sigma_1) \wedge (V_{\tau_2}^{(s)} \sigma_2) \wedge \ldots \wedge (V_{\tau_n}^{(s)} \sigma_n), \quad \text{with} \quad \sigma_1, \sigma_2, \ldots, \sigma_n \in \Sigma_{mk}.$$

Therefore, by the continuity of $\tilde{s}'$, it suffices to determine

$$\tilde{s}' [(V_{\tau_1}^{(s)} \sigma_1) \wedge (V_{\tau_2}^{(s)} \sigma_2) \wedge \ldots].$$

We proceed step by step:

With $\sigma \in \Sigma_{mk}$ (i.e. $\sigma = r\rho$ for $\rho \in \Sigma'$), from (2.4.22) follows

$$\tilde{s}' V_\tau^{(s)} \sigma = \tilde{s}' V_\tau^{(s)} r\rho = A'_\tau \rho.$$

In order to calculate $\tilde{s}'[\sigma_1 \wedge V_\tau^{(s)} \sigma_2]$ with $\sigma_1 = r\rho_1, \sigma_2 = r\rho_2$, we proceed from

$$V_\tau^{(s)} \sigma_2 = (\sigma_1 \wedge V_\tau^{(s)} \sigma_2) \vee (V_\tau^{(s)} \sigma_2 \backslash \sigma_1 \wedge V_\tau^{(s)} \sigma_2) \tag{2.4.24}$$

and

$$\sigma_1 \vee V_\tau^{(s)} \sigma_2 = \sigma_1 \vee (V_\tau^{(s)} \sigma_2 \backslash \sigma_1 \wedge V_\tau^{(s)} \sigma_2). \tag{2.4.25}$$

For $\tilde{s}\sigma_1 = \rho_1$, from (2.4.25) follows

$$\rho_1 + \tilde{s}(V_\tau^{(s)} \sigma_2 \backslash \sigma_1 \wedge V_\tau^{(s)} \sigma_2) \leq 1$$

(1 is the unit element in $L(\Sigma')$), hence

$$\tilde{s}(V_\tau^{(s)} \sigma_2 \backslash \sigma_1 \wedge V_\tau^{(s)} \sigma_2) \leq \rho_1^* = 1 - \rho_1.$$

From (2.4.24) follows

$$\tilde{s}'\, V_\tau^{(s)}\, \sigma_2 = \tilde{s}'(\sigma_1 \wedge V_\tau^{(s)}\, \sigma_2) + \tilde{s}'(V_\tau^{(s)}\, \sigma_2 \backslash \sigma_1 \wedge V_\tau^{(s)}\, \sigma_2).$$

This is the unique decomposition of the left side into a summand $\leq \rho_1$ and a summand $\leq \rho_1^*$ (see VII §5). Therefore we get

$$\begin{aligned} \tilde{s}'(\sigma_1 \wedge V_\tau^{(s)}\, \sigma_2) &= P_{\rho_1}\, \tilde{s}'\, V_\tau^{(s)}\, \sigma_2 \\ &= P_{\rho_1}\, A'_\tau\, \rho_2, \end{aligned} \tag{2.4.26}$$

where $P_\rho$ is the operator that projects each $y \in \mathscr{B}'(\Sigma')$ onto the summand $\leq \rho$. For $\tau_2 \geq \tau_1$, from

$$V_{\tau_1}^{(s)}\, \sigma_1 \wedge V_{\tau_2}^{(s)}\, \sigma_2 = V_{\tau_1}^{(s)}\, [\sigma_1 \wedge V_{\tau_2-\tau_1}^{(s)}\, \sigma_2]$$

with $\tilde{s}'\, V_{\tau_1}^{(s)} = A'_\tau\, \tilde{s}'$ we obtain

$$\tilde{s}'(V_{\tau_1}^{(s)}\, \sigma_1 \wedge V_{\tau_2}^{(s)}\, \sigma_2) = A'_{\tau_1}\, [P_{\rho_1}\, A'_{\tau_2-\tau_1}\, \rho_2]. \tag{2.4.27}$$

After two similar steps, for $(\tau_3 \geq \tau_2 \geq \tau_1)$ we get

$$\tilde{s}'(V_{\tau_1}^{(s)}\, \sigma_1 \wedge V_{\tau_2}^{(s)}\, \sigma_2 \wedge V_{\tau_3}^{(s)}\, \sigma_3) = A'_{\tau_1}\, [P_{\rho_1}\, A'_{\tau_2-\tau_1}\, [P_{\rho_2}\, A'_{\tau_3-\tau_2}\, \rho_3]]. \tag{2.4.28}$$

We thus perceive that $\bar{\Sigma}_{mk} \xrightarrow{\tilde{s}'} L(\Sigma')$ is determined by $A_\tau$. If $\bar{\Sigma}_{mk} = \Sigma_m$, then $\tilde{s}'$ is completely determined by $A_\tau$; hence the dynamics is determined by $\tilde{K}_m(\Sigma_m) = \tilde{s}K(\Sigma')$ (see the beginning of §2.4 and II §3.3).

What is the significance of this structure analysis of the macro-theory $\mathscr{PT}_m$ for its embedding in $\mathscr{PT}_{q\,\exp}$?

In §3 we shall see that only an imprecise embedding is possible if $\Sigma_{mk}$ is a Boolean ring, since $\mathscr{B}_m$ in general has a high but finite dimension. Nevertheless let us (as so far) discuss the idealization of a precise embedding. It is simpler to take $\Sigma_{mk}$ as Boolean ring than to say that $\Sigma_{mk}$ is only approximately a Boolean ring. This idealization can be advantageous also for the problems discussed in §3.

We first rewrite the diagram (2.4.9) in the form

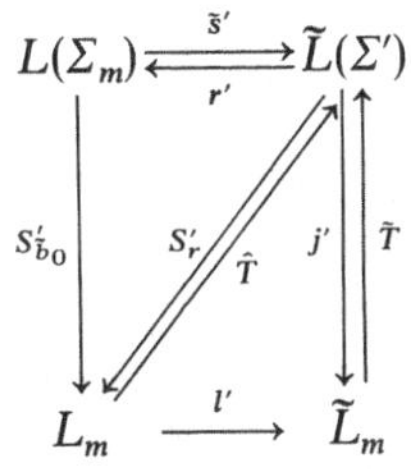


(2.4.29)

where $\hat{T}$ is defined in (2.4.13) and $S_r$ by

$$S_r = r S_{\tilde{b}_0}. \tag{2.4.30}$$

Since $\tilde{T}$ is the mapping inverse to $j'$, from (2.4.29) follows

$$\hat{T} S'_r = 1, \qquad \hat{T} S'_{\tilde{b}_0} = \tilde{s}', \qquad P = S'_r\, \hat{T} \tag{2.4.31}$$

with $P$ as in (2.4.14).

From (2.4.12) and (2.4.17) follows

$$A'_\tau = \tilde{V}_\tau = \hat{T}\, \mathscr{U}_\tau\, S'_r = \hat{T}\, \tilde{\mathscr{U}}_\tau\, S'_r \tag{2.4.32}$$

with $\tilde{\mathscr{U}}_\tau = P\, \mathscr{U}_\tau\, P$.

In (2.4.29) one can forget $\tilde{L}_m$ and with $L(\tilde{b}_0) \overset{\text{def}}{=} \overline{\text{co}} \bigcup_{\tau \geq 0} \mathscr{U}_\tau S'_{\tilde{b}_0} L(\Sigma_m)$ consider the diagram

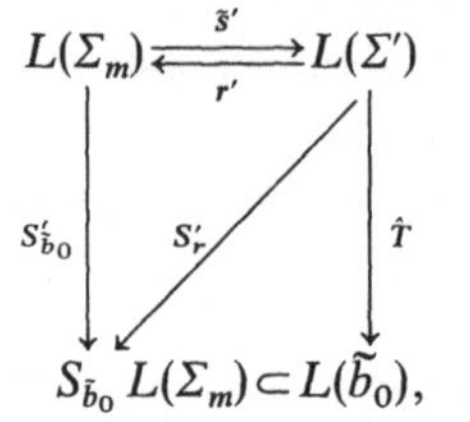

$$\begin{array}{ccc} L(\Sigma_m) & \overset{\tilde{s}'}{\underset{r'}{\rightleftarrows}} & L(\Sigma') \\ \Big\downarrow S'_{\tilde{b}_0} & \swarrow S'_r & \Big\downarrow \hat{T} \\ S_{\tilde{b}_0} L(\Sigma_m) & \subset & L(\tilde{b}_0), \end{array} \tag{2.4.33}$$

where $PL(b_0) = S'_r L(\Sigma')$. The map $S'_r$ must be injective since $\hat{T}S'_r = 1$. This is equivalent to the fact that $S_r \mathscr{B}$ is dense in $\mathscr{B}(\Sigma')$, i.e. that the Banach subspace of $\mathscr{B}(\Sigma')$ generated by $S_r K$ is all of $\mathscr{B}(\Sigma')$. But $S_r K$ must not be dense in $K(\Sigma')$. The affine mapping $L(\tilde{b}_0) \xrightarrow{\hat{T}} L(\Sigma_{mk})$ is not determined solely by $S'_r$, but rather depends in an essential way on $K^*_m$.

When $K^*_m$ has been chosen, $K_m(\Sigma_m)$ follows as the closure of $S_{\tilde{b}_0} K^*_m$. Let the subset $rK_m(\Sigma_m)$ of $K(\Sigma')$ be called $K_m(\Sigma')$. Then $rS_{\tilde{b}_0} K^*_m = S_r K^*_m$ is dense in $K_m(\Sigma')$. Because $r\tilde{s} = 1$ and $\tilde{s}K(\Sigma') = \tilde{K}_m(\Sigma_m)$, we have $\tilde{s}K_m(\Sigma') = K_m(\Sigma_m)$. Moreover, it follows that $q = \tilde{s}r$ is the identity mapping on $\tilde{K}_m(\Sigma_m)$. Hence we get $S_{\tilde{b}_0} K^*_m = qS_{\tilde{b}_0} K^*_m = \tilde{s}S_r K^*_m$.

$\hat{T}$ is determined by $K^*_m$: Because $\hat{T}S'_r = 1$, for $\tilde{y} = Py = S'_r \hat{T}y$ follows $\hat{T}\tilde{y} = Ty$ and moreover $l'\tilde{y} = l'S'_r \tilde{T}l'y = j'\tilde{T}l'y = l'y$. Conversely, if for $y \in \mathscr{B}'$ there is a $\tilde{y} \in S'_r \mathscr{B}'(\Sigma')$ such that $l'y = l'\tilde{y}$, then $\tilde{T}l' = \hat{T}$ also yields $\hat{T}y = \hat{T}\tilde{y}$. Here, $x$ is uniquely determined by $\tilde{y} = S'_r x$ and hence $\hat{T}y = \hat{T}S'_r x = x$. Thus one obtains $\hat{T}$, with the largest domain of definition, as follows: With $l'^{-1}(l'S'_r \mathscr{B}'(\Sigma'))$ as the domain of definition of $\hat{T}$, with $l'y = l'S'_r x$ one puts $\hat{T}y = x$.

Since $l'$ is determined by $K^*_m$, thus $\hat{T}$ is also determined. Then $A_\tau$ is also determined by $\hat{T}$ according to (2.4.32). By $K^*_m$ and $S_r$, the set $K_m(\Sigma')$ is given as the closure of $S_r K^*_m$. If $\tilde{s}$ is known, it yields $K_m(\Sigma_m) = \tilde{s}K_m(\Sigma')$.

The observable $\Sigma_m \xrightarrow{F_{\tilde{b}_0}} L$ is also "essentially" determined, i.e. insofar as the form of $F_{\tilde{b}_0}$ manifests itself macroscopically. To this end, let us show that instead of $F_{\tilde{b}_0}$ one can just as well use $F_{\tilde{b}_0} q' = F_{\tilde{b}_0} r'\tilde{s}'$, i.e. instead of $S_{\tilde{b}_0}$ just as well $qS_{\tilde{b}_0} = \tilde{s}S_r$. As we have seen above, $S_{\tilde{b}_0} K^*_m = qS_{\tilde{b}_0} K^*_m$.

If the embedding condition (2.4.10) is satisfied, (2.4.32) follows. From (2.4.32) with (2.4.23) follows

$$\tilde{s}' V_\tau^{(s)} = A'_\tau \tilde{s}' = \hat{T}\mathscr{U}_\tau S'_r s' = \hat{T}\mathscr{U}_\tau (qS_{\tilde{b}_0})'.$$

Because of $\tilde{s}' = \hat{T}S_{\tilde{b}_0}$ and $\tilde{s}'q' = \tilde{s}'r'\tilde{s}' = \tilde{s}'$, also $\tilde{s}' = \hat{T}(qS_{\tilde{b}_0})$ holds; this implies

$$\tilde{T}l'(qS_{\tilde{b}_0})' V_\tau^{(s)} = \tilde{T}l'\mathscr{U}_\tau (qS_{\tilde{b}_0})'.$$

Multiplication by $j'$ yields

$$l'(qS_{\tilde{b}_0})' V_\tau^{(s)} = l'\mathscr{U}_\tau (qS_{\tilde{b}_0})',$$

i.e. the embedding condition (2.4.10) for $qS_{\tilde{b}_0}$ instead of $S_{\tilde{b}_0}$. Therefore, we can always choose

$$\Sigma_m \xrightarrow{S'_r \tilde{s}'} L$$

as the macroscopic observable $\Sigma_m \to L$.

Therefore, if $\bar{\Sigma}_{mk}=\Sigma_m$ holds for the macro-theory, then the dynamics is determined by $S_r$, $K_m^*$, $\mathscr{U}_\tau$. This is the exact formulation of the intuitive notion that "the dynamics of the macroscopic system is given by the microscopic Hamiltonian" (to be explained more clearly in the examples of §2.5).

Hence, since the macro-observable $\Sigma_m \xrightarrow{\tilde{F}_{b0}} L$ can be replaced by the equivalent $S'_r \tilde{s}'$, for $\bar{\Sigma}_{mk}=\Sigma_m$ this observable is determined by $A_\tau$.

For example, by means of (2.4.28) follows

$$(S'_r \tilde{s}')[V_{\tau_1}^{(s)}\sigma_1 \wedge V_{\tau_2}^{(s)}\sigma_2 \wedge V_{\tau_3}^{(s)}\sigma_3]=S'_r A'_{\tau_1}[P_{\rho_1} A'_{\tau_2-\tau_1}[P_{\rho_2} A'_{\tau_3-\tau_2}\rho_3]],$$

with the $A'_\tau$ given by (2.4.32).

Till now we have not assumed that $K_m(\Sigma_m)\xrightarrow{r} K(\Sigma')$ is surjective. We carefully distinguished between $K(\Sigma')$ and $K_m(\Sigma')=rK_m(\Sigma_m)$. But in the macroscopic theory $\mathscr{P}\mathscr{T}_m$, one often makes the *idealizing* (see XIII §2.5 and [3]) assumption $\tilde{K}_m(\Sigma_m)=K_m(\Sigma_m)$. For the embedding problem, this can be an unnecessarily "stringent" idealization. Very likely, just by means of the embedding one obtains an estimate of how "close" the set $K_m(\Sigma_m)$ can come to $\tilde{K}_m(\Sigma_m)$. We shall call the difference between $K_m(\Sigma_m)$ and $\tilde{K}_m(\Sigma_m)$ the *macroscopic imprecision*. In fact, $K_m(\Sigma_m)\subsetneqq\tilde{K}_m(\Sigma_m)$ with $r'\tilde{K}_m(\Sigma_m)=K(\Sigma')$ means that not each $\sigma\in\Sigma'$ permits us to prepare ensembles $u$ with the probability $u(\sigma)=1$.

For the case $K_m(\Sigma_m)=\tilde{K}_m(\Sigma_m)$, i.e. $rK_m(\Sigma_m)=K_m(\Sigma')=K(\Sigma')$, we call the macrosystems described in $\mathscr{P}\mathscr{T}_m$ *classical*, while $\Sigma'$ is named the set of their objective properties (completely analogous to II §5 and VII §5.3).

For such classical systems, a further embedding condition suggests itself:

Since $\Sigma'$ is a complete Boolean ring, we can introduce an additive measure by

$$\Sigma' \xrightarrow{W_k} \breve{K}(\Sigma') \tag{2.4.34}$$

with $W_k(\rho)=m_\rho$ and $m_\rho(\sigma)=m_0(\sigma\wedge\rho)$. Here we have $\rho$, $\sigma\in\Sigma'$ and $m_0$ is an effective measure from $K(\Sigma')$. Then, with $\varepsilon$ as the unit element of $\Sigma'$, we get $W_k(\varepsilon)=m_0$.

$W_m=\tilde{s}W_k$ is a preparator (V D6.4):

$$\Sigma' \xrightarrow{W_m} \breve{K}_m(\Sigma_m)\subset\breve{K}(\Sigma_m), \tag{2.4.35}$$

where we have used $\tilde{s}K(\Sigma')=K_m(\Sigma_m)$. Since this preparator yields ensembles from $K_m(\Sigma_m)$ only, it should be realizable (in the sense of a requirement analogous to *APr* in V §8), i.e. there should be an $\tilde{a}\in\mathscr{Q}'_m$ such that one can identify $\Sigma'$ with $\mathscr{Q}_m(\tilde{a})$ and $W_m$ with $\lambda_{\mathscr{Q}_m}(\tilde{a},a)\,\varphi_m(a)$. Then a second preparator is defined by $W(a)=\lambda_{\mathscr{Q}}(i\tilde{a},ia)\,\varphi(ia)$:

$$\Sigma' \xrightarrow{W} \breve{K}. \tag{2.4.36}$$

Because of $\lambda_{\mathscr{Q}_m}(\tilde{a},a)=\lambda_{\mathscr{Q}}(i\tilde{a},ia)$, then $\|W(\rho)\|=m_0(\rho)$ holds with $m_0=W_k(\varepsilon)$.

Therefore, according to V T6.3, there is a mixture-morphism

$$K(\Sigma') \xrightarrow{T\tilde{a}} K \tag{2.4.37}$$

with

$$\varphi(ia)=T_{\tilde{a}}\,r\,\varphi_m(a) \quad \text{for } a\in\mathscr{Q}_m(\tilde{a}).$$

According to (2.3.16), we have

$$\varphi_m(a) = S_{\tilde{b}_0}\, \varphi(i\,a)$$

which implies

$$\varphi_m(a) = S_{\tilde{b}_0}\, T_{\tilde{a}}\, r\, \varphi_m(a) \tag{2.4.38}$$

Since (due to [2] IV Theorem 2.1.11) the $\varphi_m(a)$ convexly generate all of $K_m(\Sigma_m)$, (2.4.38) holds on all of $K_m(\Sigma_m) = \tilde{s}' K(\Sigma')$. Thus we get $S_{\tilde{b}_0}\, T_{\tilde{a}}\, r\, \tilde{s} = \tilde{s}$, hence

$$S_{\tilde{b}_0}\, T_{\tilde{a}} = s, \tag{2.4.39a}$$

$$r S_{\tilde{b}_0}\, T_{\tilde{a}} = r\tilde{s} = 1. \tag{2.4.39b}$$

The operator

$$P_{\tilde{a}} \overset{\text{def}}{=} T_{\tilde{a}}\, r\, S_{\tilde{b}_0} \tag{2.4.40}$$

gives a mixture-morphism of $K$ into itself. With (2.4.39 b), it immediately yields

$$P_{\tilde{a}}^2 = P_{\tilde{a}}. \tag{2.4.41}$$

With $K_m(\tilde{a}) = T_{\tilde{a}}\, r K(\Sigma_m)$, from $S_{\tilde{b}_0} K \subset K(\Sigma_m)$ follows

$$P_{\tilde{a}} K = T_{\tilde{a}}\, r S_{\tilde{b}_0} K \subset K_m(\tilde{a}).$$

Moreover,

$$S_{\tilde{b}_0} K_m(\tilde{a}) = S_{\tilde{b}_0}\, T_{\tilde{a}}\, r K(\Sigma_m) = \tilde{s} r K(\Sigma_m) = K_m(\Sigma_m), \tag{2.4.42}$$

which implies

$$P_{\tilde{a}} K_m(\tilde{a}) = T_{\tilde{a}}\, r S_{\tilde{b}_0} K_m(\tilde{a}) = T_{\tilde{a}}\, r \tilde{s} r K(\Sigma_m)$$
$$= T_{\tilde{a}}\, r K(\Sigma_m) = K_m(\tilde{a});$$

therefore $P_{\tilde{a}} K = K_m(\tilde{a})$.

Because of $\varphi(i\,a) \in K_m$, by (2.4.37) we have

$$K_m(\tilde{a}) \subset K_m \subset \tilde{K}_m.$$

With (2.3.42), from $S_{\tilde{b}_0} \tilde{K}_m \subset K_m(\Sigma_m)$ follows

$$S_{\tilde{b}_0} K_m(\tilde{a}) = S_{\tilde{b}_0} \tilde{K}_m = K_m(\Sigma_m). \tag{2.4.43}$$

Therefore, according to the discussion at the beginning of this subsection we can choose the set $K_m^* = K_m(\tilde{a})$.

Because $K_m(\tilde{a}) = T_{\tilde{a}} K(\Sigma')$ holds with the choice $K_m^* = K_m(\tilde{a})$, a mapping $T$ is defined by the diagram

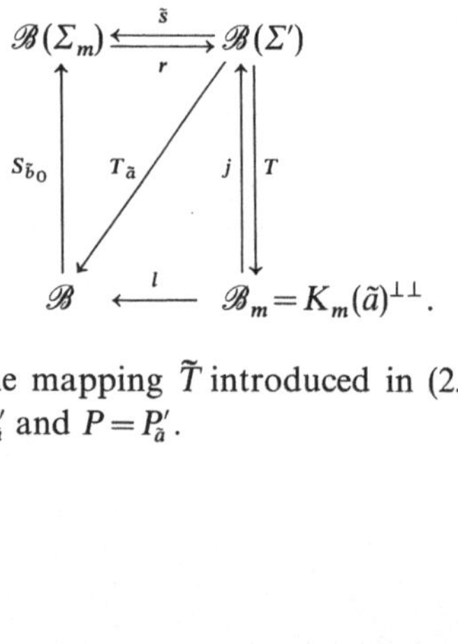


(2.4.44)

This implies $jT = 1$. For the mapping $\tilde{T}$ introduced in (2.4.9), thus we have $\tilde{T} = T'$. Moreover, we obtain $\hat{T} = T_{\tilde{a}}'$ and $P = P_{\tilde{a}}'$.

Therefore, one can use all the preceding formulas if $\hat{T}$ is replaced everywhere by $T'_{\tilde{a}}$. In particular, from (2.4.32) follows

$$A_\tau = S_r \mathcal{U}'_\tau T_{\tilde{a}} = S_r \tilde{\mathcal{U}}'_\tau T_{\tilde{a}} \tag{2.4.45}$$

with

$$\tilde{\mathcal{U}}'_\tau = P_{\tilde{a}} \mathcal{U}'_\tau P_a.$$

Then the embedding problem can be formulated as follows:

There exist mixture-morphisms $T_{\tilde{a}}$, $S_{\tilde{b}_0}$ such that (2.4.39a) holds and the condition

$$V_\tau^{(s)'} \tilde{s} = S_{\tilde{b}_0} \mathcal{U}'_\tau T_{\tilde{a}} \tag{2.4.46}$$

following from (2.4.11) is satisfied.

One can weaken this condition somewhat: There exist mixture-morphisms $T_{\tilde{a}}$ and $S_r$ (where $S_r$ is determined by the observable $\Sigma' \xrightarrow{S'_r} L$) such that (2.4.31) yields

$$S_r T_{\tilde{a}} = 1 \tag{2.4.47}$$

while (2.4.45) gives

$$A_\tau = S_r \mathcal{U}'_\tau T_{\tilde{a}}. \tag{2.4.48}$$

The embedding problem in this weakened form was investigated in [44]. In §3 we shall report results from [44].

If a mixture-morphism

$$K(\Sigma') \xrightarrow{T_{\tilde{a}}} K$$

is given, then (according to V T6.4) each effective measure $m \in K(\Sigma')$ yields a preparator

$$\Sigma' \xrightarrow{W} \check{K}, \tag{2.4.49}$$

with $W(\sigma) = T_{\tilde{a}} m_\sigma$. This implies

$$\begin{aligned} \mu(W(\sigma), 1 - S'_r \sigma) &= \langle m_\sigma, 1 - T'_{\tilde{a}} S'_r \sigma \rangle \\ &= \langle m_\sigma, \sigma^* \rangle = 0 \end{aligned}$$

and correspondingly

$$\mu(W(\sigma^*), S'_r \sigma) = 0.$$

Therefore (by [2] IV D5.6) the preparator (2.4.49) is dispersion-free with respect to the observable $\Sigma \xrightarrow{S'_r} L$.

Then (due to [2] IV §5) we find $\Sigma'$ atomic. With $\bar{e}$ as the support of the ensemble $w = T_{\tilde{a}} m$ (for an effective $m$), we get

$$S'_r \sigma = e(\sigma) + f(\sigma), \tag{2.4.50}$$

with $e(\sigma) \in G$ and $e(\sigma) \leq \bar{e} = e(\varepsilon)$. Moreover, $f(\sigma) \leq \bar{e}^\perp$. The support of all of $T_{\tilde{a}} K(\Sigma')$ is also $\bar{e}$, while $\sigma \to e(\sigma)$ is a decision observable $\Sigma \to G$. We have $\mu(T_{\tilde{a}} m, S'_r \sigma) = \mu(T_{\tilde{a}} m, e(\sigma)) = \langle m, T'_{\tilde{a}} S'_r \sigma \rangle = m(\sigma)$ for each $m \in K(\Sigma_m)$ and $\sigma \in \Sigma'$, where $T_{\tilde{a}} m$ commutes with all $e(\sigma)$.

It suffices to consider $e_\nu = e(\sigma_\nu)$ and $f_\nu = f(\sigma_\nu)$ for the atoms $\sigma_\nu$ of $\Sigma'$. There follows $\sum_\nu e_\nu = \bar{e}$. Each $m \in K(\Sigma')$ is a convex combination $m = \sum_\nu \lambda_\nu m_\nu$, with $m_\nu(\sigma_\mu) = \delta_{\nu\mu}$. It follows $T_{\tilde{a}} m_\nu = w_\nu$ with $e_\nu w_\nu e_\nu = w_\nu$ and $T_{\tilde{a}} m = \sum_\nu \lambda_\nu w_\nu$.

For $S_r$, from (2.4.50) follows

$$S_r w = \sum_\nu \mu(w, e_\nu + f_\nu)\, m_\nu. \qquad (2.4.51)$$

Hence, for $P_{\tilde{a}}$ we finally get

$$P_{\tilde{a}} w = \sum_\nu \mu(w, e_\nu + f_\nu)\, w_\nu. \qquad (2.4.52)$$

For $A_\tau$, from (2.4.48) follows

$$A_\tau m_\nu = \sum_\mu \mu(U_\tau^+ w_\nu U_\tau, e_\mu + f_\mu)\, m_\mu. \qquad (2.4.53)$$

The $A_\tau$ form a semigroup provided the $\mathscr{U}'_\tau$ form one, where

$$\tilde{\mathscr{U}}'_\tau w = \sum_{\nu,\mu} \mu(w, e_\nu + f_\nu)\, \mu(U_\tau^+ w_\nu U_\tau, e_\mu + f_\mu)\, w_\mu. \qquad (2.4.54)$$

We shall become acquainted with an application of these considerations in §2.5.

There remains the problem of systematically investigating the structure of the dynamics for a macroscopic system. Above, we became familiar with part of such a structure analysis: $\Sigma_{mk}$ is a Boolean ring; $\tilde{K}_m(\Sigma_m) = K_m(\Sigma_m)$. In the next subsection, we shall become acquainted with another such structure, namely, $\Sigma_Z$ (defined in §2.5) and $\Sigma_{mk}$ are isomorphic. But problems such as the following remain open: For which known macrosystems is $\Sigma_{mk}$ not a Boolean ring? Till now, physicists systematically investigated only systems for which $\Sigma_Z$ and $\Sigma_{mk}$ are isomorphic.

## §2.5 Dynamics

We have mentioned already (end of II §3.4 and end of the preceding §2.4) that in this book we shall not develop a general dynamics of macro-systems. In this §2.5 let us at least illustrate the relations between embeddings and dynamical laws. By examples one more easily recognizes the significance of the derivations from §2.4 for the macroscopic dynamics.

Let us proceed from the relations in §2.3, in order to see how the set $\hat{S}_m$ is determined as a subset of $\hat{Y}$ and how $K_m(\hat{S}_m)$ is determined. From many examples of statistical mechanics, one knows that in fact the Hamiltonian $H$ from $\mathscr{P}\mathscr{T}_{q\,\exp}$ decisively determines the dynamics of the system. But what has $H$ to do with $\hat{S}_m$ and $K_m(\hat{S}_m)$?

First we extend the macroscopic observable $\Sigma_m \xrightarrow{F_{\tilde{b}_0}} L$ as a mapping on the Borel ring $\mathscr{B}(\hat{Y})$ of $\hat{Y}$. For $\sigma \in \mathscr{B}(\hat{Y})$ we have $\sigma \cap \hat{S}_m \in \mathscr{B}(\hat{S}_m)$. For $\sigma \in \mathscr{B}(\hat{Y})$ we put

$$\tilde{F}_{\tilde{b}_0}(\sigma) = F_{\tilde{b}_0}(\sigma \cap \hat{S}_m), \qquad (2.5.1)$$

where we already think of $F_{\tilde{b}_0}$ as carried over naturally as a mapping from $\Sigma$ onto $\mathscr{B}(\hat{S}_m)$.

Then $\tilde{F}_{\tilde{b}_0}$ is a $\sigma$-additive measure on $\mathscr{B}(\hat{Y})$, and the support of $\tilde{F}_{\tilde{b}_0}$ is just $\hat{S}_m$ (since the support of $F_{\tilde{b}_0}$ was all of $\hat{S}_m$ !).

Since (2.3.21) holds for all $w \in \tilde{K}_m$, its application to a $y \in \sigma \cap \hat{S}_m$ because of $(V_\tau \sigma) \cap \hat{S}_m = V_\tau^{(s)}(\sigma \cap \hat{S}_m)$ yields

$$\mu(w, \tilde{F}_{\tilde{b}_0}(V_\tau \sigma)) = \mu(w, \mathscr{U}_\tau \tilde{F}_{\tilde{b}_0}(\sigma)) \tag{2.5.2}$$

for all $w \in \tilde{K}_m$.

For $\tau = 0$, (2.5.2) defines a mapping $\tilde{S}_{\tilde{b}_0}$ of $K$ into the set $K(\hat{Y})$ with

$$\tilde{S}_{\tilde{b}_0} w = u, \qquad \text{where } u(\sigma) = \mu(w, \tilde{F}_{\tilde{b}_0}(\sigma)). \tag{2.5.3}$$

The support of $\tilde{S}_{\tilde{b}_0} K$ is just $\hat{S}_m$, and the norm-closure of $\tilde{S}_{\tilde{b}_0} \tilde{K}_m$ is just the $K_m(\hat{Y})$ from II §3.2.

Thus (2.5.2) implies

$$V_\tau' \tilde{S}_{\tilde{b}_0} w = S_{\tilde{b}_0} \mathscr{U}_\tau' w \tag{2.5.4}$$

for all $w \in \tilde{K}_m$, which is equivalent to (2.3.20).

Therefore, the $\sigma$-additive measure $\tilde{F}_{\tilde{b}_0}$ over $\mathscr{B}(\hat{Y})$ determines $\tilde{S}_{\tilde{b}_0}$, and makes $\hat{S}_m$ the support of $\tilde{F}_{\tilde{b}_0}$ (and of $\tilde{S}_{\tilde{b}_0} K$). But from where does one obtain $\tilde{F}_{\tilde{b}_0}$? Till now, we only know the embedding condition (2.5.4) as a restriction for $\tilde{F}_{\tilde{b}_0}$, into which the Hamiltonian from $\mathscr{P}\mathscr{T}_{q\,\text{exp}}$ enters through $\mathscr{U}_\tau'$.

For simplicity, in the following let us also call $\mathscr{B}(\hat{Y}) \xrightarrow{F_{\tilde{b}_0}} L$ a macro-observable and often imprecisely switch between $\mathscr{B}(\hat{Y})$ and $\Sigma_m$. It will be easy for the reader, always to recognize which mappings are concerned.

The "optimistic" opinion, saying that the embedding condition (2.5.4) determines the state space $Z$ as well as the dynamics in $Z$, is scarcely credible. A so-called thermodynamic limit can neatly eliminate problems with the boundary conditions for finite systems, but it cannot solve the problems of the state space $Z$ and of the macro-observables. Still another structure besides that of the thermodynamic limit must be joined to $\mathscr{P}\mathscr{T}_{q\,\text{exp}}$ in order to find $Z$, the macro-observables, and the dynamics in $Z$.

It appears that one must know $Z$ from other sources; one must know exactly what one can really measure on the macrosystems. Of course, $Z$ itself is not at all uniquely determined, a point we shall take up briefly in §2.6.

In applications of statistical mechanics to concrete systems (e.g. gases, conductors, semiconductors) one proceeds from fixed state spaces. Therefore let us regard $Z$ and hence $Y$ as prescribed. Then $\tilde{F}_{\tilde{b}_0}$ and $\tilde{K}_m$ are no longer arbitrary, but rather must satisfy (2.5.4). Whether we can (with given $Z$) always find appropriate $\tilde{F}_{\tilde{b}_0}$ and $\tilde{K}_m$ which satisfy (2.5.4), is unknown.

How does one arrive at $\tilde{F}_{\tilde{b}_0}$ and $\tilde{K}_m$? Statistical mechanics is a theory tested in many applications. Hence till now one must have applied (2.5.4) in a more intuitive than conscious way; but how? No previous theory has existed which tells us how, for given $Z$, one can really find the macro-observables $\tilde{F}_{\tilde{b}_0}$. Hence only the condition (2.5.4) is available to us.

In all applications one prescribed $Z$ from experience. Then one has tried (by the correspondence principle, well known from the rudiments of quantum mechanics) to guess a part of $\tilde{F}_{\tilde{b}_0}$ which is not determined by (2.5.4). Besides, one also uses to prescribe a $K_m^*$ as a subset of $\tilde{K}_m$. Such a $K_m^*$ (see its use in §2.4) is guessed by the procedures known as "coarse graining" and "microscopic equipartitioning".

One often employs the procedure of "random phases" (equivalent mathematically(!) to microscopic equi-partitioning). But it is physically meaningless to invoke the "random phases" as physical(!) argument for the "foundation" of this procedure. Such a foundation rests on a physical misunderstanding of quantum mechanics. In Hilbert space, the sets $K$ and $L$ have a physical meaning but the phases of vectors do not (see XIII §2.3 and 4.8)!

After giving a part of $\tilde{F}_{\tilde{b}_0}$ and the set $K_m^*$ in such a way, till now one sought to find the macroscopic dynamics by means of $\mathscr{U}_\tau$. Thus one intuitively applied the "usual interpretation" of quantum mechanics. Precisely by (2.5.4) we have given this usual interpretation an exact form. We shall show just this by the following examples. For these, let us assume that the systems are dynamically continuous (see II D3.4.1). For this dynamic continuity it is necessary to assume $\tilde{F}_{\tilde{b}_0}(\hat{Y}\setminus Y)=0$ (more exactly, that there is a $\sigma\in\mathscr{B}(\hat{Y})$ with $\tilde{F}_{\tilde{b}_0}(\sigma)=0$ and $\hat{Y}\setminus Y\subset\sigma$). Then

$$\mathscr{B}\xrightarrow{F_{\tilde{b}_0}} L$$

is equivalent to the observable

$$\mathscr{B}(\hat{Y})\xrightarrow{\tilde{F}_{\tilde{b}_0}} L.$$

For dynamically continuous systems, $\sigma(\rho;\tau)$ with $\rho\in\mathscr{B}(Z)$ is (due to II §3.4) an element of $\mathscr{B}(Y)$. Here $\sigma(\rho;\tau)$ is the set of those trajectories which at time $\tau$ have a state $z$ from $\rho$. Then we have

$$\tilde{F}_{\tilde{b}_0}(\sigma(\rho;\tau))=F_{\tilde{b}_0}(\sigma_m(\rho;\tau))$$

with $\sigma_m(\rho;\tau)$ as in II §3.4. Abbreviating

$$\chi_\tau(\rho)=\tilde{F}_{\tilde{b}_0}(\sigma(\rho;\tau)), \tag{2.5.5}$$

we find $\chi_\tau(\rho)$ defined for all $\tau>0$. One could consider a special $\tau_0$ (physically not distinguishable from 0) as the "beginning" of the dynamics. For simplicity, let us assume that $\chi_\tau(\rho)$ is also defined for $\tau=0$. For $\chi_0(\rho)$ we simply write $\chi(\rho)$.

From $V_{\tau_1}\,\sigma(\rho;\tau_2)=\sigma(\rho;\tau_1+\tau_2)$ follows

$$\tilde{F}_{\tilde{b}_0}\,V_{\tau_1}\,\sigma(\rho;\tau_1)=\chi_{\tau_1+\tau_2}(\rho). \tag{2.5.6}$$

For $\sigma=\sigma(\rho,\tau_2)$ the condition (2.5.2), which is equivalent to (2.5.4), becomes

$$\mu(w,\chi_{\tau_1+\tau_2}(\rho))=\mu(w,\mathscr{U}_{\tau_1}\,\chi_{\tau_2}(\rho)) \tag{2.5.7}$$

for all $w\in\tilde{K}_m$. In particular, from (2.5.7) follows

$$\mu(w,\chi_\tau(\rho))=\mu(w,\mathscr{U}_\tau\,\chi(\rho)) \tag{2.5.8}$$

for all $w\in\tilde{K}_m$; this already shows that $\chi_\tau(\rho)$ is not independent of $\mathscr{U}_\tau$. But it would be false to put $\chi_\tau(\rho)=\mathscr{U}_\tau\,\chi(\rho)$ since $\chi_\tau(\rho)$ is an effect that coexists with all $\tilde{F}_{\tilde{b}_0}(\sigma)$. Hence in particular it coexists with all $\chi(\rho)$, which $\mathscr{U}_\tau\,\chi(\rho)$ in general will not do! But (2.5.8) suggests that the $\chi(\rho)$ themselves do not depend on the dynamics of the system; hence they belong to that part of $\tilde{F}_{\tilde{b}_0}$ which one must guess. After all, $\chi(\rho)$ is intuitively the effect of measuring a state $z\in\rho$ at the time 0 (i.e. when the macrosystem starts moving). But despite this, $\chi$ is not entirely arbitrary since it is conceivable that one can guess a $\chi$ for which (2.5.4) cannot be satisfied!

Therefore the observable

$$\mathscr{B}(Z) \xrightarrow{\chi} L, \tag{2.5.9}$$

which describes the measurement of the state at time 0, must be guessed.

Let us note already that the guessing rules of the correspondence principle do not suffice to determine (2.5.9). Most often, the correspondence principle yields several "microscopic" decision observables with scales corresponding to the "quantities" of interest. Thus it yields self-adjoint operators which do not even commute and hence cannot yield the observable (2.5.9) "exactly". Rather one must somehow modify these scale observables, so that the "imprecision of the macroscopic measurement" is included (also see below the discussion of macroscopic cells).

In the literature on the foundations of statistical mechanics, one does not find the "macro-observable" $\Sigma_m \xrightarrow{\tilde{F}\tilde{b}_0} L$ but *only* the observable (2.5.9). Even this occurs in special forms (e.g. characterized by *several* commuting self-adjoint operators; see below). Correspondingly, (2.5.9) is also called the macro-observable.

Let us further specialize our example so that only systems "without memory" are considered. By this one means that their distribution over the trajectories is already determined by the *initial* distribution, i.e. of the systems over the states at time $t=0$. For a mathematical formulation of this assumption, we introduce the Boolean ring $\Sigma_Z = \mathscr{B}(Z)/\mathscr{J}(Z)$, where $\mathscr{J}(Z)$ is the set of all sets $\rho \in \mathscr{B}(Z)$ with $u(\sigma(\rho;0))=0$ for all $u \in K(\Sigma_m)$. Toward the end of II §3, we assumed that the support of the set of all measures $u_\tau$ equals $Z$. Now let us particularly assume that the support of the set of all measures $u_0(\rho)=u(\sigma(\rho;0))$ with $u \in K_m(\hat{Y})$ equals $Z$. This just means that one can produce all possible states from $Z$ at time $t=0$.

A mapping $u \to u_0$ for each $u \in K(\Sigma_m)$, i.e. a mixture-morphism

$$K(\Sigma_m) \xrightarrow{r} K(\Sigma_Z), \tag{2.5.10}$$

is defined by $u_0(\rho)=u(\sigma(\rho;0))$. Then the mapping $r'$ adjoint to $r$ makes

$$L(\Sigma_Z) \xrightarrow{r'} L(\Sigma_m) \tag{2.5.11}$$

with $r'(\rho)=\sigma(\rho;0)$.

Above we assumed that the support of $rK_m(\Sigma_m)$ is all of $Z$. Let us strengthen this to the "almost always" made assumption that the Banach subspace of $\mathscr{B}(\Sigma_Z)$ generated by the $rK_m(\Sigma_m)$ is all of $\mathscr{B}(\Sigma_Z)$, i.e. $rK_m(\Sigma_m)$ separates the set $L(\Sigma_Z)$. This assumption just means that two different registration effects of the states at time $t=0$ cannot have the same frequencies for all preparable ensembles. Only occasionally one abandons this assumption for computational reasons (see the end of this §2.5) or to discuss imprecise embeddings (see §3).

Due to this assumption, $r\mathscr{B}_m(\Sigma_m)$ is dense in $\mathscr{B}(\Sigma_Z)$. From $rK(\Sigma_m)=K(\Sigma_Z)$ then follows $r\tilde{K}_m(\Sigma_m)=K(\Sigma_Z)$; hence $\tilde{K}_m(\Sigma_m) \xrightarrow{r} K(\Sigma_Z)$ is surjective.

The set $K_m(\Sigma_m)$ determines the dynamics inasmuch as a $u \in K_m(\Sigma_m)$ determines the distribution over the trajectories. The assumption that the systems do not have memories therefore just means that each $u \in K_m(\Sigma_m)$ is already uniquely determined by the corresponding $u_0 = ru$, i.e. that the mapping $K_m(\Sigma_m) \xrightarrow{r} K(\Sigma_Z)$ is injective. Therefore the above assumption implies that $\tilde{K}_m(\Sigma_m) \xrightarrow{r} K(\Sigma_Z)$ is bijective.

Thus we give the assumption that the systems have "no memory" the final mathematical formulation:

The mapping $\tilde{K}_m(\Sigma_m) \xrightarrow{r} K(\Sigma_Z)$ is bijective.

According to §2.4, this is equivalent to asserting $r'\Sigma_Z = \Sigma_{mk}$. Hence considering $\Sigma_Z$ as a $\Sigma'$ from §2.4, we can adopt the whole theory from §2.4 with $\Sigma' = \Sigma_Z$ and with $r'\rho = \sigma(\rho; 0)$. But let us emphasize that we do *not* assume $\tilde{K}_m(\Sigma_m) = K_m(\Sigma_m)$ nor the existence of a mapping $T_{\check{a}}$ as in §2.4.

The mappings

$$A_\tau = r V_\tau^{(s)} \tilde{s} \tag{2.5.12}$$

defined in (2.4.21) represent mixture-morphisms of $K(\Sigma_Z)$. Since they form a semigroup, there is an operator $B$ defined densely in $K(\Sigma_Z)$ with

$$A_\tau = e^{B\tau}. \tag{2.5.13}$$

With $u$ from the domain of definition of $B$, for $u_\tau = A_\tau u$ we then obtain the differential equation

$$\frac{d}{d\tau} u_\tau = B u_\tau, \tag{2.5.14}$$

called the "master" equation. Since in applications only $B$ can be calculated (approximately), this equation (2.5.14) plays a large role; it is viewed as *the* dynamical equation for systems without memory. Below we shall learn in more detail, why (2.5.14) is so highly regarded.

The mapping $S'_r$ introduced in §2.4, when restricted to $\Sigma_Z$ coincides with $\chi$. Therefore, if $\chi$ has been guessed, then $S_r$ is known. As in the proof (in §2.4) that the mapping $\mathscr{B}'(\Sigma') \xrightarrow{s'r'} \mathscr{B}'_m(\Sigma_m)$ is injective, one shows that $S'_r$ injective is equivalent to $\chi(\sigma)$ being an extreme point of $S'_r L(\Sigma_z)$ for all $\sigma \in \Sigma_Z$. Hence the observable $\Sigma_Z \xrightarrow{\chi} L$ must be chosen so that

$$\chi(\Sigma_Z) \subset \partial_e S'_r L(\Sigma_Z)$$

holds; otherwise no $\hat{T}$ with $\hat{T}S'_r = 1$ can exist. If $S'_r$ is injective, then there is a mapping $\hat{T}$ with $\hat{T}S'_r = 1$, defined at least on $S'_r \mathscr{B}'(\Sigma_Z)$. How must $\hat{T}$ be introduced in a larger domain that comprises all of $L_m$? This depends (as shown in §2.4) on the partitioning into classes, conditioned by $K_m^*$, of the elements of $\mathscr{B}'$. Restricted to $S'_r \mathscr{B}'(\Sigma_Z)$, the mapping $l'$ must be injective. Hence $K_m^*$ must be so comprehensive that $K_m^*$ separates the elements of $S'_r \mathscr{B}'(\Sigma_Z)$. One must yet guess a set $K_m^*$ suitable in this sense, in order to define $\hat{T}$ according to §2.4. Thus the domain of definition of $\hat{T}$ is the set of all $y \in \mathscr{B}'$ for which there is a $z \in S'_r \mathscr{B}'(\Sigma_Z)$ with $l'y = l'z$; then $\hat{T}y = x$ holds with $z = S'_r x$. Since $K_m^*$ separates the elements of $S'_r \mathscr{B}'(\Sigma_Z)$ and $S'_r$ is injective, $\hat{T}$ is well defined. It is important that in general $\hat{T}$ cannot be defined on all of $\mathscr{B}'$; this we shall recognize below.

If $K_m^*$ is determined, then $A'_\tau$ and hence $A_\tau$ are by (2.4.26) determined with the aid of the Hamiltonian from $\mathscr{P}\mathscr{T}_{q\,\exp}$. It is not "automatically" guaranteed that these $A_\tau$ form a semigroup.

The following is a frequent choice for $K_m^*$: One assumes that there is an increasing sequence $\rho_\nu \in \mathscr{B}(Z)$ with $\bigcup_\nu \rho_\nu = Z$ and $\operatorname{tr} \chi(\rho_\nu) < \infty$. Let $\Sigma_Z^f$ be the set of all $\rho \in \Sigma_Z$

with tr $\chi(\rho)<\infty$. Then one chooses

$$K_m^* = \overline{\text{co}} \bigcup_{\rho \in \Sigma_Z^f} \frac{\chi(\rho)}{\text{tr}\,\chi(\rho)}. \tag{2.5.15}$$

Since $S_r'$ is injective, $K_m^*$ separates all elements of $S_r' \mathscr{B}'(\Sigma_Z)$ if $\mu(\chi(\rho), S_r' y)=0$ for all $\rho \in \Sigma_Z^f$ implies $y=0$. Equivalent to this is that $\langle S_r \chi(\rho), y\rangle = 0$ implies $y=0$, i.e. the $m_\rho = S_r \chi(\rho)[\text{tr}\,\chi(\rho)]^{-1}$ separate the elements of $\mathscr{B}'(\Sigma_Z)$. We have

$$\begin{aligned} m_\rho(\rho') &= [\text{tr}\,\chi(\rho)]^{-1} \langle S_r \chi(\rho), \rho'\rangle \\ &= [\text{tr}\,\chi(\rho)]^{-1} \mu(\chi(\rho), \chi(\rho')). \end{aligned} \tag{2.5.16}$$

If the $\{m_\rho \| \rho \in \Sigma_Z^f\}$ separate the elements of $\mathscr{B}'(\Sigma_Z)$, then it follows, conversely, that as well $S_r'$ is injective and also $K_m^*$ separates the elements of $S_r' \mathscr{B}'(\Sigma_Z)$. The elements of $K(\Sigma_Z)$ given in (2.5.16) must separate the elements of $L(\Sigma_Z)$ (hence those of $\mathscr{B}'(\Sigma_Z)$). To this important condition on the observable $\Sigma_Z \xrightarrow{\chi} L$ we shall return below.

Therefore, $\hat{T} y = x$ is defined for a $y \in \mathscr{B}'$ for which there is an $x \in \mathscr{B}'(\Sigma_Z)$ with

$$\mu(\chi(\rho), y) = \mu(\chi(\rho), S_r' x) \tag{2.5.17}$$

for all $\rho \in \Sigma_Z^f$.

From (2.4.32) follows

$$A_\tau' = \hat{T} U_\tau \chi(\rho) U_\tau^+ .$$

Since the $m_\rho$ from (2.5.16) separate the elements of $\mathscr{B}'(\Sigma_Z)$, we find $A_\tau' \sigma$ (and hence $A_\tau'$) uniquely determined by

$$\begin{aligned} \langle m_\rho, A_\tau' \sigma\rangle &= \langle m_\rho, \hat{T} U_\tau \chi(\sigma) U_\tau^+ \rangle \\ &= \langle m_\rho, \hat{T} \chi_\tau(\rho)\rangle. \end{aligned}$$

With the definition of $\hat{T}$ that follows from (2.5.17) and with (2.5.16), we obtain

$$\begin{aligned} [\text{tr}\,\chi(\rho)]^{-1} \mu(\chi(\rho), \chi_\tau(\rho)) &= [\text{tr}\,\chi(\rho)]^{-1} \mu(\chi(\rho), S_r' \hat{T} \chi_\tau(\rho)) \\ &= \langle m_\rho, \hat{T} \chi_\tau(\rho)\rangle = \langle m_\rho, A_\tau' \sigma\rangle = \langle A_\tau m_\rho, \sigma\rangle. \end{aligned}$$

Therefore, $A_\tau$ is uniquely determined by the equations

$$\begin{aligned} m_\rho(\sigma) &= [\text{tr}\,\chi(\rho)]^{-1} \mu(\chi(\rho), \chi(\sigma)), \\ A_\tau m_\rho(\sigma) &= [\text{tr}\,\chi(\rho)]^{-1} \mu(\chi(\rho), \chi_\tau(\rho)), \end{aligned} \tag{2.5.18}$$

which hold for all $m_\rho$.

In order to illustrate how the considered operators act, let us introduce the frequently used representation of $\mathscr{B}(\Sigma_Z)$ and $\mathscr{B}'(\Sigma_Z)$ by functions over $Z$.

To this end, we assume that $\chi$ ($\sigma$-additive relative to the $\sigma(\mathscr{B}', \mathscr{B})$-topology) is $\sigma$-additive relative to the $\sigma(\mathscr{B}, \mathscr{B}')$-topology for $\chi(\rho)$ as elements of $\mathscr{B}$ (for $\rho \in \Sigma_Z^f$).

Then the meausre

$$\bar{m}(\rho) = \text{tr}\,\chi(\rho) = \mu(\chi(\rho), 1) \tag{2.5.19}$$

is also $\sigma$-additive.

Under the above assumption, there is an increasing sequence $\rho_\nu \in \mathscr{B}(Z)$ with $\bigcup_\nu \rho_\nu = Z$ and $\bar{m}(\rho_\nu) < \infty$.

Each $y \in \mathscr{B}'(\Sigma_Z)$ can be represented by a measurable, essentially bounded, real function $y(z)$ over $Z$, so that

$$\langle m, y\rangle = \int y(z)\, dm(z) \tag{2.5.20}$$

holds for $m \in K(\Sigma_Z)$. For $y \in L(\Sigma_Z)$ one can assume $0 \leq y(z) \leq 1$.

According to the Radon-Nykodym theorem, every measure $m \in K(\Sigma_Z)$ can be represented by a positive, measurable function $\gamma(z)$ with

$$\int \gamma(z)\, d\bar{m}(z) = 1$$

and with (2.5.21)

$$\langle m, y\rangle = \int y(z)\, dm(z) = \int \gamma(z)\, y(z)\, d\bar{m}(z).$$

Using this well-known method of representing $K(\Sigma_Z)$ and $L(\Sigma_Z)$, we can reformulate the above considerations of $K_m^*$, $S_r'$, $\hat{T}$, $A_\tau$ as follows:

Since it was assumed that $\chi(\rho)$ is $\sigma$-additive (in the $\sigma(\mathscr{B}, \mathscr{B}')$-topology!), by the Radon-Nykodym theorem there is a measurable function $Z \xrightarrow{\omega} K$ with

$$S_r'\, \rho = \chi(\rho) = \int_\rho \omega(z)\, d\bar{m}(z). \tag{2.5.22}$$

For $y \in \mathscr{B}'(\Sigma_Z)$ this implies

$$S_r'\, y = \int y(z)\, \omega(z)\, d\bar{m}(z) \tag{2.5.23 a}$$

and

$$\langle S_r\, w, y\rangle = \mu(w, S_r'\, y) = \int y(z)\, \mu(w, \omega(z))\, d\bar{m}(z).$$

Since this yields

$$S_r\, w = \mu(w, \omega(z)), \tag{2.5.23 b}$$

we now have expressed $S_r\, w$ by the density function corresponding to it.

From (2.5.15) thus follows

$$K_m^* = \overline{\mathrm{co}} \bigcup_{z \in Z} \omega(z). \tag{2.5.24}$$

For $\rho, \rho' \in \Sigma_Z^f$ we obtain

$$\mu(\chi(\rho), \chi(\rho')) = \int_\rho d\bar{m}(z) \int_{\rho'} d\bar{m}(z')\, \mu(\omega(z), \omega(z')), \tag{2.5.25}$$

which by (2.5.16) gives

$$m_\rho(\rho') = \left[\int_\rho d\bar{m}(z)\right]^{-1} \int_\rho d\bar{m}(z') \int_{\rho'} d\bar{m}(z)\, \mu(\omega(z'), \omega(z)).$$

Therefore, to the measure $m_\rho$ there corresponds the density function $\gamma_\rho$ with

$$\gamma_\rho(z) = \left[\int_\rho d\bar{m}(z)\right]^{-1} \int_\rho d\bar{m}(z')\, \mu(\omega(z'), \omega(z)).$$

Hence all convex combinations of the measures $m_\rho$ are given by the density functions

$$\bar{\gamma}(z) = \int \gamma(z')\, d\bar{m}(z')\, \mu(\omega(z'), \omega(z)) \tag{2.5.26}$$

with $\gamma \in K(\Sigma_Z)$. The latter means in short that $\gamma(z)$ represents an element of $K(\Sigma_Z)$. Thus the set $S_r K_m^*$ is just the set of all $\bar{\gamma} \in K(\Sigma_Z)$ which by (2.5.26) can be calculated from the $\gamma \in K(\Sigma_Z)$. Hence (2.5.26) represents a mixture-morphism of $K(\Sigma_Z)$ onto $S_r K_m^* \subset K(\Sigma_Z)$, which describes a "smearing" of $\gamma$ to $\bar{\gamma}$ by the symmetric integral kernel $\mu(w(z'), w(z))$. In applications, one should think of $\mu(w(z'), w(z))$ as a function of $z'$, something like a Gauss distribution around $z$.

The assumption that the $m_\rho$ in (2.5.16) separate all elements of $\mathscr{B}'(\Sigma_Z)$ just means that

$$\int y(z)\,\bar{\gamma}(z)\,d\bar{m}(z)=0$$

implies $y=0$. This is to say,

$$\int \mu(w(z'), w(z))\,y(z)\,d\bar{m}(z)=0$$

implies $y=0$, i.e. the mapping $\mathscr{B}'(\Sigma_Z)\to\mathscr{B}'(\Sigma_Z)$ given by

$$\bar{y}(z)=\int \mu(w(z), w(z'))\,y(z')\,d\bar{m}(z') \tag{2.5.27}$$

is injective. This can be tested most simply in the form (2.5.27). If the integral kernel in applications forms a Gaussian, it is plausible that the mapping (2.5.27) is injective. Thus, according to the considerations immediately following (2.5.16), it is guaranteed that $\mathscr{B}'(\Sigma_Z)\to\mathscr{B}'$ and $S'_r L(\Sigma_Z)\xrightarrow{l'} L$ are injective. Hence (2.5.17) makes $\hat{T}$ well defined.

Let the smearing (2.5.26) be denoted briefly by

$$K(\Sigma_Z)\xrightarrow{Q} K(\Sigma_Z). \tag{2.5.28}$$

The mapping (2.5.27) is the corresponding dual

$$L(\Sigma_Z)\xrightarrow{Q'} L(\Sigma_Z). \tag{2.5.29}$$

One observes that $Q$ is not a projector, i.e. $Q^2 \neq Q$.

By (2.5.17) and (2.5.23a), $\hat{T}$ is defined by $\hat{T}y=x$ with

$$\begin{aligned}\mu(\chi(\rho), y)&=\langle \chi(\rho), S'_r x\rangle\\ &=\int \mu(\chi(\rho), w(z))\,x(z)\,d\bar{m}(z)\\ &=\int_\rho d\bar{m}(z')\int d\bar{m}(z)\,x(z)\,\mu(w(z'), w(z)).\end{aligned}$$

(2.5.22) implies

$$\mu(\chi(\rho), y)=\int_\rho \mu(w(z), y)\,d\bar{m}(z) \tag{2.5.30}$$

for arbitrary $y\in\mathscr{B}'$.

By $\tilde{y}(z)=\mu(w(z), y)$, to each $y\in\mathscr{B}'$ there is assigned a $\tilde{y}\in\mathscr{B}'(\Sigma_Z)$.

This mapping $\mathscr{B}'\to\mathscr{B}'(\Sigma_Z)$ is dual to the mixture-morphism $\mathscr{B}(\Sigma_Z)\xrightarrow{\bar{T}}\mathscr{B}$ defined by

$$\bar{T}m=\int \gamma(z)\,w(z)\,d\bar{m}(z), \tag{2.5.31}$$

where $\gamma$ is the density corresponding to $m$. There follows $\bar{T}K(\Sigma_Z)=K_m^*$. From (2.5.28), (2.5.26) and (2.5.23b) follows

$$Q=S_r\,\bar{T}, \tag{2.5.32}$$

thus $\tilde{y}=\bar{T}'\,y$. According to (2.5.17), $\hat{T}$ is determined by

$$\mu(\bar{T}m, y)=\mu(\bar{T}m, S_r'\,x) \qquad \text{for all } m\in K(\Sigma_Z). \tag{2.5.33}$$

Hence $\hat{T}y=x$ is defined for all $y$ for which there is an $x$ such that holds (2.5.33) and thus

$$\bar{T}'\,y=\bar{T}'\,S_r'\,x=Q'\,x. \tag{2.5.34}$$

Therefore the domain of definition of $\hat{T}$ is $\bar{T}'^{-1}\,Q'\,\mathscr{B}'(\Sigma_Z)$; and there we get $\hat{T}=Q'^{-1}\,\bar{T}'$. It is essential that the domain of definition of $\hat{T}$ not be all of $\mathscr{B}'$. This is so since $Q'\,L(\Sigma_Z)$ is not all of $L(\Sigma_Z)$. The "smearing" $Q'$ is injective but not surjective. The existence of a $T_{\bar{a}}$ with (2.4.47) would imply $\hat{T}=T_{\bar{a}}'$ (see §2.4); then we could take $\bar{T}=T_{\bar{a}}$ and thus $Q=1$. An embedding without smearing would be possible. According to §2.4 this is only possible if $\Sigma_Z$ is atomic.

$T$ has "nearly" the properties of $T_{\bar{a}}$. Instead of (2.4.47) we have (2.5.32), and $Q$ should "nearly" be 1. In fact, experience with macrosystems has not given any indication that only the ensembles in $Q\,K(\Sigma_Z)$ can be prepared (rather than all in $K(\Sigma_Z)$). In §3 we shall discuss a measure for the difference between $Q$ and 1.

The introduction by (2.5.31) of the map $\bar{T}$ with the property $\bar{T}K(\Sigma_Z)=K_m^*$ suggests to invert the way which provides $K_m^*$ by (2.5.15). Instead of (2.5.15) we may introduce a mixing morphism $B(\Sigma_Z)\xrightarrow{\bar{T}}B$ and define $K_m^*$ by $\bar{T}K(\Sigma_Z)$. Then a function $\bar{\omega}(z)$ can be defined in analogy to (2.5.31) by

$$\bar{T}m=\int \gamma(z)\,\bar{\omega}(z)\,d\bar{m}(z). \tag{2.5.35}$$

Here $\bar{\omega}(z)$ may differ from the $\omega(z)$ defined by (2.5.22)!

From (2.5.23b) follows the "smearing" operator $Q=S_r\,\bar{T}$ given by

$$Q\,m=\int \gamma(z')\,\mu(\bar{\omega}(z'), \omega(z)\,d\bar{m}(z') \tag{2.5.36}$$

rather than (2.5.26). In this way the symmetric kernel $\mu(\omega(z'), \omega(z))$ may in general be replaced by the asymmetric $\mu(\bar{\omega}(z'), \omega(z))$. Since it does not essentially change the formulas, this generalization shall not be carried out.

When to $m\in K(\Sigma_z)$ we assign the density function $\gamma(z)$, the defining equation (2.5.18) for $A_\tau$ reads

$$A_\tau\,Q m=\int \gamma(z')\,d\bar{m}(z')\,\mu(\omega(z'), U_\tau\,\omega(z)\,U_\tau^+). \tag{2.5.37}$$

We define $Q_\tau\,m$ by the right side of (2.5.37). $Q_\tau$ is given by the integral kernel $\mu(\omega(z'), U_\tau\,\omega(z)\,U_\tau^+)$ and we have $Q_0=Q$. Therefore, (2.5.37) can briefly be written

$$A_\tau Q=Q_\tau, \quad \text{i.e.} \quad A_\tau=Q_\tau\,Q^{-1}, \tag{2.5.38a}$$

where (2.5.37) makes

$$Q_\tau=S_r\,\mathscr{U}_\tau'\,\bar{T}. \tag{2.5.38b}$$

Hence, to begin with, $A_\tau$ is defined only on $QK(\Sigma_Z)$. But since $A_\tau$ must be a mixture-morphism of $K(\Sigma_Z)$ into itself, one must be able to extend the operator $Q_\tau\,Q^{-1}$ to the whole subspace of $\mathscr{B}$ generated by $Q\,K(\Sigma_Z)$, i.e. to all of $\mathscr{B}$. One mostly writes $A_\tau$ with an integral kernel $g_\tau(z', z)$ in the form

$$(A_\tau \gamma)(z) = \int \gamma(z')\, g_\tau(z', z)\, d\bar{m}(z'). \tag{2.5.39}$$

In the applications, one mostly assumes that $\gamma(z)$ is so "smooth" that $\gamma(z) \in QK(\Sigma_Z)$ holds. By (2.5.39) one may then replace $g_\tau(z', z)$ by $\mu(w(z'), U_\tau w(z) U_\tau^+)$. Therefore one often writes

$$g_\tau(z', z) \sim \mu(w(z'), U_\tau w(z) U_\tau^+). \tag{2.5.40}$$

Thus we see more clearly how, with $\Sigma_Z \xrightarrow{\chi} L$ given and $K_m^*$ chosen by (2.5.15), the dynamical operator $A_\tau$ in $Z$ is determined by the Hamiltonian in $\mathscr{P}\mathscr{T}_{q\,\mathrm{exp}}$. By (2.5.37) the $A_\tau$ determined by (2.5.37) and (2.5.38) are not guaranteed to satisfy the semigroup property (not even approximately). Rather, the semigroup property presents a condition on $\Sigma_Z \xrightarrow{\chi} L$.

In practice, one cannot calculate $A_\tau$ since $U_\tau$ is not calculable due to the complexity of the Hamiltonian. Therefore, one tries to calculate (more or less well) at least $B$ from (2.5.14), by trying an expansion

$$\mathrm{tr}(w(z'), U_\tau w(z) U_\tau^+) = \mathrm{tr}(w(z') w(z)) + \tau h(z', z) + \dots \tag{2.5.41}$$

With this, from (2.5.37) one obtains

$$BQm = \int \gamma(z')\, h(z', z)\, d\bar{m}(z').$$

But in the calculation of $h(z', z)$ one must not fall into the error of differentiating the left side of (2.5.41) formally with respect to $\tau$, thus equating $h(z', z)$ to $\mathrm{tr}(w(z')[iHw(z) - iw(z)H])$. Because of the smearing hidden in $\mu(w(z'), U_\tau w(z) U_\tau^+)$ (see the operator $Q$), the expression $\mathrm{tr}(w(z'), U_\tau w(z) U_\tau^+)$ does not change at all for microscopically small times. Hence, one must calculate the left side for microscopically large times in order to obtain "macroscopic" changes of (2.5.41). This holds even if one wishes to obtain on the right side only that term which increases linearly with $\tau$. Just this makes it difficult to determine the operator $B$ for the master equation (2.5.14).

The set $\Sigma_{mk}$ consists of all $\sigma(\rho; 0)$ with $\rho \in \Sigma_Z$. The set $V_\tau^{(s)} \Sigma_{mk}$ consists of all $\sigma(\rho, \tau)$. Therefore, $\bar{\Sigma}_{mk}$ is the complete Boolean subring of $\Sigma_m$ generated by the set $\{\sigma(\rho; \tau) | \rho \in \Sigma_Z, \tau \geq 0\}$. Having assumed the systems to be dynamically continuous, we get $\bar{\Sigma}_{mk} = \Sigma_m$. According to §2.4, the $A_\tau$ determine the mapping $\tilde{s}$ and hence $K_m(\Sigma_m) = \tilde{s} S_r K_m^*$, i.e. the dynamics of the system. One can choose $\Sigma_m \xrightarrow{S_r' \tilde{s}'} L$ as the macro-observable $\Sigma_m \xrightarrow{\tilde{F}_{b_0}} L$. Let us still express the $\tilde{s}'[\sigma(\rho_1; \tau_1) \cap \sigma(\rho_2; \tau_2) \cap \dots]$ calculated in §2.4 explicitly by the $g_\tau(z', z)$ from (2.5.39).

For example, (2.4.27) implies

$$\tilde{s}'[\sigma(\rho_1; \tau_1) \cap \sigma(\rho_2; \tau_2)] = \int_{\rho_1} g_{\tau_1}(z, z')\, d\bar{m}(z') \int_{\rho_2} g_{\tau_2 - \tau_1}(z', z'')\, d\bar{m}(z''); \tag{2.5.42}$$

hence the macro-observable $F_{\tilde{b}_0} = S_r' s'$ with (2.5.23 a) yields

$$\begin{aligned}(S_r' \tilde{s}')&[\sigma(\rho_1; \tau_1) \cap \sigma(\rho_2; \tau_2)] \\ &= \int w(z)\, d\bar{m}(z) \int_{\rho_1} g_\tau(z, z')\, d\bar{m}(z') \int_{\rho_2} g_{\tau_2 - \tau_1}(z', z'')\, d\bar{m}(z''). \end{aligned} \tag{2.5.43}$$

Thus it becomes understandable that often one does not worry about the observable $\Sigma_m \xrightarrow{F\bar{b}_0} L$, but only about the $\Sigma_Z \xrightarrow{\chi} L$ by which $\omega(z)$ is determined in (2.5.22).

For the probability $\langle u, \sigma(\rho_1; \tau_1) \cap \sigma(\rho_2, \tau_2)\rangle$ with $u \in K_m(\Sigma_m)$, which obeys $u = \tilde{s}m$ with $m \in K_m(\Sigma_Z) = S'_r K^*_m$, from (2.5.42) follows

$$\langle u, \sigma(\rho_1; \tau_1) \cap \sigma(\rho_2, \tau_2)\rangle$$
$$= \int \gamma(z)\, d\bar{m}(z) \int_{\rho_1} g_{\tau_1}(z, z')\, d\bar{m}(z') \int_{\rho_2} g_{\tau_2 - \tau_1}(z', z'')\, d\bar{m}(z''). \tag{2.5.44}$$

Here $\gamma(z)$ is the density function corresponding to $m$. Because of (2.5.44), one calls $g_\tau(z, z')$ the transition probability from $z$ to $z'$ in the time $\tau$. Then (2.5.44) says: The probability that the trajectory goes through the region $\rho_1 \subset Z$ at the time $\tau_1$ and through the region $\rho_2 \subset Z$ at the time $\tau_2$ is the sum of the probabilities

$$\gamma(z)\, d\bar{m}(z) \quad g_{\tau_1}(z, z')\, d\bar{m}(z') \quad g_{\tau_2 - \tau_1}(z', z'')\, d\bar{m}(z''). \tag{2.5.45}$$

Each of these is the product of the probability $\gamma(z)\, d\bar{m}(z)$ that the system has the state $z$ at time "zero", the probability $g_{\tau_1}(z, z')\, d\bar{m}(z')$ that the system goes over from $z$ to $z'$ within the time $\tau_1$, and the probability $g_{\tau_2 - \tau_1}(z', z'')\, d\bar{m}(z'')$ that the system then goes over to $z''$ within the time interval from $\tau_1$ to $\tau_2$.

The product form (2.5.45) is not a consequence of the probability concept, but rather a special structure for systems without memory, obtained from the embedding condition (2.5.4).

For systems without memory, we have thus shown that the embedding of $\mathscr{PT}_m$ into $\mathscr{PT}_{q\,\exp}$ leads to the master equation, where the transition probabilities are determined by the $g_\tau(z', z)$ from (2.5.39). Then $\tilde{S}_{\bar{b}_0}$ is also "essentially" determined since one can replace $S_{\bar{b}_0}$ by $S_r$. Finally, also $K_m(Y)$ and hence the support $S_m$ are determined by $K^*_m$ and $\tilde{S}_{\bar{b}_0}$.

This finishes the description of our example. Let us still establish the connection with the widespread method of *macroscopic cells*, since the representation of the observable $\Sigma_Z \xrightarrow{\chi} L$ in the form of a general (*not* decision) observable is not yet customary.

The description by macroscopic cells goes back to J. v. Neumann [25]. It arose from the notion of his days that *each*(!) observable is representable by a self-adjoint operator. Then it is identical with the structure which in VIII D 3.1 we called a scale observable. According to this idea, to the macroscopic measurements ought to correspond "macro-observables" as scale observables. Since all macroscopic measurements can be performed "simultaneously" (formerly said instead of "coexistently"), all spectral families of these macro-observables commute (so one thought*)), i.e. together generate a single decision observable in the sense of VII D 2.1.

One soon recognizes that the assumption of $\Sigma_m \xrightarrow{F\bar{b}_0} L$ as a decision observable is not tenable. Indeed J. v. Neumann did not at all mean a macroscopic observable to be our $\Sigma_m \xrightarrow{F\bar{b}_0} L$ but rather the observable we have called $\Sigma_Z \xrightarrow{\chi} L$. Since

*) The different registration methods $b_0 \in i\mathscr{R}_0$ (see §2.2) need *not* lead to observables coexistent in $\mathscr{PT}_{q\,\exp}$!

in applications $\Sigma_Z$ is not atomic, the assumption that $\Sigma_Z \xrightarrow{\chi} L$ is a decision observable also leads to contradictions with the embedding. It was the fact of the so-called imprecision of macroscopic measurements, intuitively seen very correctly by J. v. Neumann, that led him to introduce discrete decision observables as macroscopic observables. Here, we shall not follow the path of J. v. Neumann but rather go the way indicated in §2.4.

For this purpose, we somewhat mollify the requirement $r'\Sigma_Z = \Sigma_{mk}$ (equivalent to assuming "no memory"). We imagine a Boolean subring $\Sigma'$ of $\Sigma_Z$ such that $r'\Sigma' = \Sigma_{mk}$. We complete the weakening by the sharpened requirement of the existence of a mixture-morphism $T_{\tilde{a}}$, as introduced in §2.4 by (2.4.37). We can then use the derivations in §2.4 from (2.4.39) on, adding the special condition $\Sigma' \subset \Sigma_Z$. We must now observe that $r$ is the mapping of $\mathscr{B}(\Sigma_m)$ onto $\mathscr{B}(\Sigma')$.

From §2.4 follows that $\Sigma'$ must be atomic. Then the atoms $\rho_\nu$ of $\Sigma'$ must form a "cell partition" of $Z$, since $\bigcup_\nu \rho_\nu = Z$. Below we shall say more about a "reasonable" magnitude of the cells. According to §2.4 one can choose $K_m^* = T_{\tilde{a}} K(\Sigma')$. In §2.4 we denoted the support of $T_{\tilde{a}} K(\Sigma')$ by $\bar{e}$. The $T_{\tilde{a}} m_\nu = w_\nu$ had to satisfy the condition $w_\nu = e_\nu w e_\nu$, but were otherwise undetermined. Therefore, $K_m^*$ as the set of all convex combinations of the $w_\nu$ must be determined by a special choice of the $w_\nu$, analogous to that in (2.5.15). For this purpose we assume that for all atoms $\rho_\nu$ with $\chi(\rho_\nu) = e_\nu + f_\nu$ ($e_\nu \leq \bar{e}, f_\nu \leq \bar{e}^\perp$) we have $\operatorname{tr}(e_\nu) < \infty$. This $\operatorname{tr}(e_\nu)$ is just the dimension of the subspace $e_\nu \mathscr{H}$ of $\mathscr{H}$. Analogously to (2.5.16), we try to fix $K_m^*$ by

$$K_m^* = \overline{\operatorname{co}} \bigcup_\nu \frac{\chi(\rho_\nu)}{\operatorname{tr} \chi(\rho_\nu)}. \tag{2.5.46}$$

$[\operatorname{tr} \chi(\rho_\nu)]^{-1} \chi(\rho_\nu) \in K_m^*$ is possible only when $f_\nu = 0$, since we assumed

$$\text{support of } K_m^* = \bar{e}.$$

Hence let us put $f_\nu = 0$.

According to (2.4.51), we immediately find $K_m(\Sigma') = S_r K_m^* = K(\Sigma')$; thus we must have $K_m(\Sigma_m) = \tilde{K}_m(\Sigma_m) = \tilde{s} K(\Sigma')$. Since $A_\tau$ has the form given in (2.4.53), it makes

$$\begin{aligned} A_\tau m_\nu &= \sum_\mu [\operatorname{tr} e_\nu]^{-1} \mu(U_\tau^+ e_\nu U_\tau, e_\mu) m_\mu \\ &= \sum_\mu \lambda_\tau(\nu, \mu) m_\mu, \end{aligned} \tag{2.5.47}$$

where

$$\lambda_\tau(\nu, \mu) = [\operatorname{tr} e_\nu]^{-1} \mu(U_\tau^+ e_\nu U_\tau, e_\mu). \tag{2.5.48}$$

This $\lambda_\tau(\nu, \mu)$ is called the transition probability (within the time $\tau$) from the cell $\rho_\nu$ to the cell $\rho_\mu$. The $A_\tau$ form a semigroup if the

$$\tilde{\mathscr{U}}'_\tau = P_{\tilde{a}} \mathscr{U}'_\tau P_{\tilde{a}} \tag{2.5.49}$$

with $\mathscr{U}'_\tau w = U_\tau^+ w U_\tau$ form a semigroup (also see (2.4.32)), where the $\tilde{\mathscr{U}}'_\tau$ are given explicitly in (2.4.54).

The question of the (approximate) semigroup property of the $P_a \mathscr{U}'_\tau P_a$ has often been investigated in the literature.

The semigroup property of the $A_\tau$ can easily be formulated in terms of the $\lambda_t(\nu, \mu)$:

$$\lambda_{\tau_1+\tau_2}(\nu, \omega) = \sum_\mu \lambda_{\tau_1}(\nu, \mu)\, \lambda_{\tau_2}(\mu, \omega). \tag{2.5.50}$$

According to J. v. Neumann's idea, the "magnitude" of the $e_\nu$ ought to be described by the macroscopic imprecision of measurement. Here we can give this notion a precise meaning, since we not only know the $e_\nu$ but rather have started with the atoms $\rho_\nu$ of $\Sigma'$: A $\rho_\nu$ should be chosen as large as possible but still so small that the various $z \in \rho_\nu$ are practically not distinguishable by macroscopic measurements. The mapping $\chi$ introduced in (2.5.9) should include the macroscopic imprecision. Above, this imprecision entered best through the integral kernel $\mu(\omega(z), \omega(z'))$, which differs from zero for $z \neq z'$. In order to permit a comparison between the preceding theory with $\chi$ from (2.5.9) and the atomic description by $\Sigma'$, we imagine the cells $\rho_\nu$ to be so chosen that $[\mathrm{tr}(e_\nu)]^{-1} e_\nu$ is a good approximation to $\omega(z)$ for suitable $z \in \rho_\nu$. The $\rho_\nu$ have, say, the extension of the partition described by $\mu(\omega(w(z), \omega(z'))$ about a point $z \in \rho_\nu$.

That we have considered only $\Sigma'$ instead of $\Sigma_Z$ has the disadvantage that the dynamics of the system is not completely described by the $A_\tau$ in $K(\Sigma')$. In fact, the Boolean subring $\bar{\Sigma}_{mk}$ generated by all the $V_\tau^{(s)} \sigma_\nu$ with $\sigma_\nu = \sigma(\rho_\nu; 0)$ is not all of $\Sigma_m$.

In particular, not all $\sigma(\rho; 0)$ with $\rho \in \Sigma_Z$ are in $\bar{\Sigma}_{mk}$. Therefore, $\tilde{s}' \sigma(\rho; 0)$ is not uniquely determined. In order to calculate the form of $\sigma(\rho; 0)$ for a $\rho \notin \Sigma'$ also, we use

$$\rho = \bigcup_\nu \rho \cap \rho_\nu,$$

hence $\tilde{s}' \sigma(\rho; 0) = \sum_\nu \tilde{s}' \sigma(\rho \cap \rho_\nu; 0)$. Because of $\tilde{s}' \sigma(\rho_\nu \cap \rho; 0) \leq \tilde{s}' \sigma(\rho_\nu; 0) = \rho_\nu$, this gives

$$\tilde{s}' \sigma(\rho_\nu \cap \rho; 0) = \lambda_\nu(\rho \cap \rho_\nu)\, \rho_\nu. \tag{2.5.51}$$

The $\lambda_\nu(\rho)$ form a real, positive measure on the Boolean subring $[0, \rho_\nu]$ of $\Sigma_Z$. This measure is not fixed, but rather can be "chosen". From

$$\tilde{s}' \sigma(\rho; 0) = \sum_\nu \lambda_\nu(\rho \cap \rho_\nu)\, \rho_\nu \tag{2.5.52}$$

then follows

$$s'_r \tilde{s}' \sigma(\rho; 0) = \sum_\nu \lambda_\nu(\rho \cap \rho_\nu)\, e_\nu \tag{2.5.53}$$

for the macro-observable $S'_r \tilde{s}'$.

The right side of (2.5.53) corresponds to the effect that above we denoted by $\chi(\rho)$. For the $\rho \leq \rho_\nu$ ($\nu$ fixed), we have

$$S'_r \tilde{s}' \sigma(\rho; 0) = \lambda_\nu(\rho)\, e_\nu. \tag{2.5.54}$$

Even if the $\rho$ become smaller and smaller, the *type* of the effect $\chi(\rho) = \lambda e_\nu$ does not change any more, only the factor $\lambda$ decreases. On the other hand, every (arbitrarily small) $\rho \leq \rho_\nu$ yields

$$\| S'_r \tilde{s}' \sigma(\rho; 0) \|^{-1} S'_r \tilde{s}' \sigma(\rho; 0) = e_\nu.$$

In general, we expect the $\chi(\rho)$ to behave similarly; this statement will be formulated in §2.6.

Having prepared the formulas (2.4.47), (2.4.48), we leave it to the reader to calculate $\tilde{s}$ on $\bar{\Sigma}_{mk}$. This results in formulas analogous to (2.5.43), carried over to the "discrete" case. In applications, one should be content with $\Sigma'$ instead of the whole $\Sigma_Z$ and with $\bar{\Sigma}_{mk}$ instead of the whole $\Sigma_m$.

The method of discrete cells $\rho_\nu$ has the advantage of mathematical simplicity. In the choice of the cells (i.e. of $\Sigma'$ from $\Sigma_Z$), however, lies a physically odious arbitrariness.

## §2.6 Dynamically Determined Systems and Contracted State Spaces

Let us return to §2.4, i.e. at first admit systems with memories.

We call the dynamics preparation-determined if $\Sigma_{mk} = \Sigma_m$. Then $\Sigma_{mk}$ is naturally a Boolean ring, while $\tilde{s}'$ is the identical mapping $\mathscr{B}'(\Sigma_m) \xrightarrow{\tilde{s}'} \mathscr{B}'(\Sigma')$. Thus, $\tilde{s}$ is also the identical mapping $\mathscr{B}(\Sigma') \xrightarrow{\tilde{s}} \mathscr{B}(\Sigma_m)$, such that $\tilde{K}_m(\Sigma_m) = K(\Sigma_m)$. Identifying $\tilde{K}_m(\Sigma_m)$ with $\tilde{K}_m(\hat{S}_m)$, we then have $\bar{\tilde{K}}^\sigma_m(\hat{S}_m) = K(\hat{S}_m)$ and hence (see II §3.3) $\partial_e \bar{\tilde{K}}^\sigma(\hat{S}) = \partial_e K(\hat{S}_m)$, which is the set of all point measures on $\hat{S}_m$.

In this sense, each trajectory can be prepared "mentally" by the elements of $\tilde{K}_m(\Sigma_m)$. Of course, we can have $K_m(\Sigma_m) \neq \tilde{K}_m(\Sigma_m)$; but then one can interpret the elements of $K_m(\Sigma_m)$ in the sense that the individual trajectories can be prepared only with *finite* imprecision (see II §3.3). But also conversely: If $\delta_e \bar{\tilde{K}}^\sigma_m(\hat{S}_m)$ is the set of all point measures on $\hat{S}_m$, we have $\tilde{K}_m(\Sigma_m) = K(\Sigma_m)$ and hence $\Sigma_{mk} = \Sigma_m$.

Therefore, if the dynamics is preparation-determined, then in §2.4 we identify $\Sigma'$ with $\Sigma_m$. Then $\tilde{s}$ and $r$ are identity mappings of $\mathscr{B}(\Sigma_m)$ into itself. In particular this implies

$$A'_\tau = V^{(s)}_\tau$$

and hence

$$V^{(s)}_\tau = \hat{T} \mathscr{U}_\tau S'_{\tilde{b}_0}. \tag{2.6.1}$$

The mapping $L(\Sigma_m) \xrightarrow{S'_{\tilde{b}_0}} S'_{\tilde{b}_0} L(\Sigma_m)$ is bijective, and $\hat{T}$ is defined on all $y \in \mathscr{B}'$ for which there is an $x \in \mathscr{B}'(\Sigma_m)$ with $\mu(w, y) = \mu(w, S'_{\tilde{b}_0} x)$ for all $w \in K^*_m$. Then it gives $\hat{T} y = x$.

Initially it remains unclear how to find $S'_{\tilde{b}_0}$ and $K^*_m$. Therefore we sharpen our assumptions by adding that the systems have no memory.

This implies that the mapping $K(\Sigma_m) \xrightarrow{r} K(\Sigma_Z)$ is bijective. Then $r$ is an isomorphism $B(\Sigma_m) \xrightarrow{r} B(\Sigma_Z)$ of the Banach spaces; and $\Sigma_Z \xrightarrow{r'} \Sigma_m$ is an isomorphism of the Boolean rings. For the continuous dynamics considered in §2.5, this yields a bijective mapping $Z \xrightarrow{d} S_m \subset Y$ for which the $d$ extended to $\mathscr{B}(Z)$ coincides with $r'$. The mapping $d$ is equivalent to a continuous mapping

$$Z \times [0, \infty) \to Z, \tag{2.6.2}$$

which we write

$$Z \xrightarrow{d_\tau} Z \qquad \text{for } \tau \geq 0. \tag{2.6.3}$$

Then $y = dz$ holds with $y = z(\tau) = d_\tau z(0)$.

Conversely, if a continuous mapping (2.6.2) is given, it yields an injective mapping $Z \xrightarrow{d} Y$ with $S_m \overset{\text{def}}{=} dZ$ as a closed set. Then a bijective mapping $\mathscr{B}(Z) \xrightarrow{d} \mathscr{B}(S_m)$ is given by $\rho \in \mathscr{B}(Z)$ and $\rho \to \sigma = \{y \mid y = dz, z \in \rho\}$. Each $\sigma$-additive measure over $\mathscr{B}(Z)$ leads in this way to a $\sigma$-additive measure over $\mathscr{B}(S_m)$ and hence over $\mathscr{B}(Y)$. The choice of a $\sigma$-additive measure $u$ over $\mathscr{B}(Z)$, which is effective on all open sets, then leads to a Boolean ring $\Sigma_Z = \mathscr{B}(Z)/\mathscr{J}(Z)$ with $\mathscr{J}(Z)$ as the set of elements in $\mathscr{B}(Z)$ having $u$-measure zero. Hence all the totally continuous measures $u$ over $\mathscr{B}(Z)$ can be identified with the elements of $K(\Sigma_Z)$. Then via $d$ we can identify $K(\Sigma_Z)$ with $K(\Sigma_m)$.

If we require $r\tilde{K}_m(\Sigma_m) = K(\Sigma_Z)$ (see §2.5), this implies $\tilde{K}_m(\Sigma_m) = K(\Sigma_Z)$. Hence the systems have no memory and are preparation-determined.

Systems without memory, whose dynamics is preparation-determined, are because of (2.6.3) called "initialvalue-determined" systems. From $S_m = dZ$ follows

$$\sigma_m(d_\tau \rho; \tau) = \sigma(d_\tau \rho; \tau) \cap S_m = \sigma(\rho; 0) \cap S_m.$$

With $\mathfrak{s}'$ from §2.5, this is equivalent to

$$\mathfrak{s}' \sigma(d_\tau \rho; \tau) = \rho. \tag{2.6.4}$$

Therefore, systems without memory for which there is a continuous function (2.6.2) such that (2.6.4) is satisfied, are initialvalue-determined.

From (2.6.4) follows

$$A'_\tau d_\tau \rho = \rho, \quad \text{i.e.} \quad A'_\tau d_\tau = 1 \tag{2.6.5}$$

as a mapping of $\mathscr{B}'(\Sigma_Z)$ into itself; hence the $d_\tau$ also form a semigroup. Although (2.6.5) is very illustrative, it is difficult to evaluate this equation. Therefore let us reformulate it by the methods from §2.5.

To this end, we multiply (2.6.5) by $Q'$:

$$Q' A'_\tau d_\tau = Q' d_\tau = Q'. \tag{2.6.6}$$

We can express this either by

$$\int \mu(\omega(z), \omega(z'))\, y(z')\, d\bar{m}(z') = \int \mu(\omega(z), U_\tau \omega(d_\tau z')\, U_\tau^+)\, y(z')\, d\bar{m}(d_\tau z') \tag{2.6.7}$$

for all $y \in \mathscr{B}'(\Sigma_Z)$, or by

$$\mu(w(z), \chi(\rho)) = \mu(w(z), U_\tau(d_\tau \rho)\, U_\tau^+) \tag{2.6.8}$$

for all $\rho \in \Sigma_Z$.

From (2.6.8) follows

$$\tilde{\chi}(\rho) = U_\tau \chi(d_\tau \rho)\, U_\tau^+ = \chi(\rho) + \alpha(\rho), \tag{2.6.9}$$

where $\alpha(\rho)$ is "orthogonal" to all $w(z)$, i.e.

$$\mu(w(z), \alpha(\rho)) = 0.$$

Hence, also $\mu(\chi(\rho'), \alpha(\rho)) = 0$ holds for all $\rho'$. Therefore (2.6.9) implies

$$\operatorname{tr}(\tilde{\chi}(\rho)^2) \geq \operatorname{tr}(\chi(\rho)^2).$$

Because of $\tilde{\chi}(\rho) \leq \mathbf{1}$, we further find

$$\operatorname{tr}(\chi(\rho)^2) \leq \operatorname{tr}(\tilde{\chi}(\rho)^2) \leq \operatorname{tr}(\tilde{\chi}(\rho)).$$

Hence we get

$$\operatorname{tr}(\chi(\rho)^2) \leq \operatorname{tr}(\chi(d_\tau \rho)), \tag{2.6.10}$$

where the left side can be written

$$\operatorname{tr}(\chi(\rho))^2 = \int_\rho \int_\rho \mu(w(z), w(z'))\, d\bar{m}(z)\, d\bar{m}(z'). \tag{2.6.11}$$

From $\chi(Z) = 1$ follows

$$\int_Z \mu(w(z), w(z'))\, d\bar{m}(z') = 1.$$

For fixed $z$, the measure $\mu(w(z), w(z'))$ is essentially different from zero only in a "macroscopic" imprecision neighborhood of $z$. Hence we also have

$$\int_\rho \mu(w(z), w(z'))\, d\bar{m}(z') \approx 1, \tag{2.6.12}$$

if $\rho$ contains the macroscopic imprecision neighborhood of $z$. Now let $\rho$ be so large that there is a "small number" $n$ of imprecision neighborhoods (for the mathematical formulation and estimation of such imprecision sets, see §3 and [44]). To within a factor in the order of magnitude $n$, we then get

$$\int_\rho \int_\rho \mu(w(z), w(z'))\, d\bar{m}(z)\, d\bar{m}(z') \sim \int_\rho d\bar{m}(z) = \operatorname{tr}(\chi(\rho)).$$

Since $\operatorname{tr}(\chi(\rho))$ is a "very large" number, let us take the logarithm. Then (2.6.10) yields

$$\log \operatorname{tr} \chi(\rho) \leq \log \operatorname{tr} \chi(d_\tau \rho) \tag{2.6.13}$$

if $\rho$ contains "some" imprecision neighborhoods. It may appear unsatisfactory that (2.6.13) is not an "exact" mathematical inequality. But it just is essential in the embedding problem that one replaces exact equations and inequalities by others having "very small" errors. Error estimates are always possible, as shown in §3 and [44]. Such estimates then justify that in practice one may deal with equations resp. inequalities as in (2.6.13).

The left side of (2.6.13) contains a measure for the macroscopic imprecision, since $\rho$ in (2.6.13) contains "some" imprecision sets. The macroscopic imprecisions described by

$$\chi(\rho)=\int_\rho \omega(z)\,d\bar{m}(z),$$

however, can *not* be guessed by the "correspondence" principle. This principle can at most yield "reference points" for guessing the observable $\Sigma_Z \xrightarrow{\chi} L$. Therefore, for the macroscopic imprecision one must know more about macroscopic measurements. Perhaps the embedding conditions could be used to find the smallest possible imprecision sets still compatible with the embedding (see §3 and [44]). Of course, one can increase the imprecision without thereby violating the embedding conditions (also see the contracted state spaces at the end of this §2.6). If one increases the imprecision, however, the dynamics can lose the "no memory" characteristic (again see below).

The measure of macroscopic imprecision appearing on the left side of (2.6.13) can be given a somewhat different form. Since $\rho$ is large for "some" imprecision sets, by (2.6.12) we have

$$\mu(\omega(z),\chi(\rho))\approx 1$$

and hence $\|\chi(\rho)\|\approx 1$. On the other hand, in case $\rho$ is not "too large", one may put

$$\begin{aligned}\chi(\rho)&=\int_\rho \omega(z)\,d\bar{m}(z)\approx\omega(z)\int_\rho d\bar{m}(z)\\&=\omega(z)\operatorname{tr}\chi(\rho)\end{aligned}\tag{2.6.14}$$

up to a factor the order of magnitude $n$. Hence $1\approx\|\chi(\rho)\|=\omega(z)\operatorname{tr}\chi(\rho)$ permits us to put

$$\log\operatorname{tr}\chi(\rho)=-\log\|\omega(z)\|\tag{2.6.15}$$

(here $\|\omega(z)\|$ is the norm of $\omega(z)$ as an element of $\mathscr{B}'$!). In the form (2.6.15), the state $z$ about which the macroscopic imprecision is considered comes more clearly to the fore.

The measure of imprecision introduced in (2.6.15) is intimately connected with the concept of entropy. This concept is not always introduced in a uniform way. As two possible definitions of "entropy at the state $z$", let us here consider

$$s_1(z)=-\log\|\omega(z)\|\tag{2.6.16}$$

and

$$s_2(z) = -\operatorname{tr}(\omega(z) \log \omega(z)). \qquad (2.6.17)$$

Which of these expressions one uses, depends on which one can handle better mathematically. In the case (2.5.46), i.e. with $\omega(z) = \operatorname{tr}(e_v)^{-1} e_v$, we have

$$s_1(z) = s_2(z) = \log \operatorname{tr}(e_v) \quad \text{for } z \in \rho_v.$$

With $\chi(\rho)$ from (2.6.14), the difference between $s_1$ and $s_2$ is of the order of magnitude

$$\frac{-1}{\operatorname{tr} \chi(\rho)} \operatorname{tr}(\chi(\rho) \log \chi(\rho)).$$

Hence $s_1$ and $s_2$ are "physically" equal if

$$\frac{-1}{\operatorname{tr} \chi(\rho)} \operatorname{tr}(\chi(\rho) \log \chi(\rho)) \ll -\log \operatorname{tr} \chi(\rho).$$

This happens when $\rho$ does not contain too few imprecision sets.

From (2.6.13) follows

$$s_1(z) \leq \log \operatorname{tr}(\chi(d_\tau \rho)), \qquad (2.6.18)$$

where $\rho$ is a neighborhood of $z$ with "some" imprecision sets. If $d_\tau \rho$ were essentially smaller than a macroscopic imprecision set about $d_\tau z$, then (2.6.18) could be violated. Therefore, $d_\tau \rho$ cannot be arbitrarily small. But if $d_\tau \rho$ contains only "some" imprecision sets about $d_\tau z$, then the right side of (2.6.18) equals the entropy $s_1(d_\tau z)$ and we obtain

$$s_1(z) \leq s_1(d_\tau z). \qquad (2.6.19)$$

This is the well-known theorem on the growth of entropy.

On the other hand, if $d_\tau \rho$ is "very large", i.e. $d_\tau \rho$ contains a large number $N$ of imprecision sets, we can have $\log \operatorname{tr} \chi(d_\tau \rho) > s_1(d_\tau z)$, namely if $\log N$ is not small relative to $s_1(d_\tau z)$. In such a case, $d_\tau \rho$ can become very large in spite of $\rho$ being small; this is called the instability of the dynamics. Therefore, the entropy theorem is assured only for an initialvalue-determined and *stable* dynamics. It is often not sufficiently emphasized that for unstable dynamics the entropy theorem can be violated. In this context, let us recall that a lowering of the macroscopic imprecision (by admission of finer measurement methods) can also raise the deterministic aspect of the dynamics, since fluctuation phenomena can be made accessible to refined measurement methods.

Since the $d_\tau$ form a semigroup, (2.6.19) implies the continuous growth of the entropy for each stable trajectory.

Frequent objections are raised to the assertion that the entropy theorem is compatible with $\mathscr{PT}_{q\,\text{exp}}$. In our words, one suspects that a $\mathscr{PT}_m$ with a deterministic dynamics cannot be embedded in $\mathscr{PT}_{q\,\text{exp}}$. We have already handled the recurrence objection in §2.3. More serious objections rest on the fact that the motion reversal transformation $C$ (described in IX §2) exists in $\mathscr{PT}_{q\,\text{exp}}$. The relation given in IX

(2.21) through (2.23) has the consequence that in $\mathcal{PT}_{q\,\text{exp}}$ for each "motion" there is the motion running in the "reverse" direction, which is not possible in most of the theories $\mathcal{PT}_m$ (irreversibility in $\mathcal{PT}_m$!). This objection is not impressive since the embedding just clarifies that not all of $K$ is realizable. But $\tilde{K}_m$ (unlike $K$) is not invariant under $C$:

$$C\tilde{K}_m C \neq \tilde{K}_m!$$

More important is the following objection.

Let a motion reversal $Z \xrightarrow{\tilde{C}} Z$ with $\tilde{C}^2 = \tilde{C}$ be defined in $Z$. We write briefly $\tilde{C}(z) = z'$. Intuitively, $z'$ is that state which differs from $z$ by reversed velocity fields (e.g. in hydrodynamics). With $C$ as in $\mathcal{PT}_{q\,\text{exp}}$, we obtain

$$C\chi(\rho)\,C = \chi(\tilde{C}\rho) = \chi(\rho'). \tag{2.6.20}$$

A motion from $z_1$ to $z_2 = d_\tau z_1$ is called reversible if $d_\tau z_2' = z_1'$. In many $\mathcal{PT}_m$ most motions are irreversible. Is this compatible with the embedding in $\mathcal{PT}_{q\,\text{exp}}$?

Choosing $\rho$ in (2.6.12) from the magnitudes of "some" imprecision sets about $z$ (hence writing $\rho(z)$), one obtains

$$\operatorname{tr}(\omega(z)\,\chi(\rho(z))) \approx 1. \tag{2.6.21}$$

From (2.6.8) then follows

$$\operatorname{tr}(\omega(z)\,U_\tau\,\chi(d_\tau\,\rho(z))\,U_\tau^+) \approx 1. \tag{2.6.22}$$

Let us consider the five states $z_1, z_2 = d_\tau z_1, z_1' = \tilde{C} z_1, z_2' = \tilde{C} z_2$ and $d_\tau z_2'$. Proceeding from $\rho(z_1)$ we put $\rho(z_2) = d_\tau \rho(z_1)$, $\rho(z_1') = \tilde{C}\rho(z_1)$, $\rho(z_2') = \tilde{C}\rho(z_2)$. From (2.6.22) follows

$$\operatorname{tr}(U_\tau^+\,\omega(z_2')\,U_\tau\,\chi\,(d_\tau\,\rho(z_2'))) \approx 1 \tag{2.6.23}$$

and also

$$\operatorname{tr}(\omega(z_1)\,U_\tau\,\chi(\rho(z_2))\,U_\tau^+) \approx 1.$$

With $\operatorname{tr}(CAC) = \operatorname{tr}(A)$ this implies

$$\operatorname{tr}(\omega(z_1')\,U_\tau^+\,\chi(\rho(z_2'))\,U_\tau) \approx 1. \tag{2.6.24}$$

Only when the entropy increases noticeably in the motion from $z_1$ to $z_2 = d_\tau z_1$, it should allegedly not be compatible with the embedding. Assuming $s_1(z_1) \lneqq s_1(d_\tau z_1) = s_1(z_2)$, with $\|w(z)\| = \|Cw(z)\,C\| = \|w(\tilde{C}z)\|$ we obtain

$$s_1(z_1') \lneqq s_1(z_2) = s_1(z_2') \leq s_1(d_\tau z_2').$$

For this reason, $d_\tau z_2'$ significantly differs from $z_1$, i.e. the motion from $z_1$ to $z_2 = d_\tau z_1$ is essentially irreversible. Therefore, with $z_1'$ and $d_\tau z_2'$ sufficiently far removed from one another, we get

$$\rho(z_1') \cap d_\tau\,\rho(z_2') = \emptyset. \tag{2.6.25}$$

This implies

$$1 \geq \chi(\rho(z_1') \cup d_\tau \rho(z_2')) = \chi(\rho(z_1')) + \chi(d_\tau \rho(z_2')),$$

hence

$$\operatorname{tr}(w \chi(\rho(z_1'))) + \operatorname{tr}(w \chi(d_\tau(z_2'))) \leq 1 \tag{2.6.26}$$

for all $w \in K$. Putting $w = U_\tau^+ \mathscr{w}(z_2') U_\tau$, with (2.6.23) we find

$$\operatorname{tr}(U_\tau^+ \mathscr{w}(z_2') U_\tau \chi(\rho(z_2'))) \approx 0. \tag{2.6.27}$$

The two equations (2.6.24) and (2.6.27) in fact contradict each other: Using the approximation $\chi(\rho(z_1')) \approx \mathscr{w}(z_1') \operatorname{tr} \chi(\rho(z_1'))$, from (2.6.27) (with $\approx 0$ replaced by $=0$) we get $\operatorname{tr}(U_\tau^+ \mathscr{w}(z_2') U_\tau \mathscr{w}(z_1')) = 0$. If in (2.6.24) the approximation $\chi(\rho(z_2')) = \mathscr{w}(z_2') \operatorname{tr} \chi(\rho(z_2'))$ were used one would on the left side obtain 0, in contradiction to $\approx 1$. Of course, the fallacy is due to not having performed a tidy error estimate. Under certain conditions we shall see that (2.6.27) can indeed be compatible with (2.6.24). Instead of (2.6.27) we write more precisely

$$\operatorname{tr}(\chi(\rho(z_1') U_\tau^+ \mathscr{w}(z_2') U_\tau)) = \varepsilon. \tag{2.6.28}$$

where $\varepsilon$ is so small that the probability $\varepsilon$ cannot "physically" be distinguished from the probability 0. Thus we must "physically exclude" the effect $\chi(\rho(z_1'))$ in the ensemble $U_\tau^+ \mathscr{w}(z_2') U_\tau$ (also see [3] §11).

From (2.6.28), i.e. from

$$\int_{\rho(z_1')} \operatorname{tr}(\mathscr{w}(z) U_\tau^+ \mathscr{w}(z_2') U_\tau) \, d\bar{m}(z) = \varepsilon,$$

one quite certainly cannot conclude

$$\operatorname{tr}(\mathscr{w}(z) U_\tau^+ \mathscr{w}(z_2') U_\tau) = 0.$$

We obtain

$$\int_{\rho(z_1')} \operatorname{tr}(\mathscr{w}(z) U_\tau^+ \mathscr{w}(z_2') U_\tau) \, d\bar{m}(z) \leq \| \mathscr{w}(z_2') \| \int_{\rho(z_1')} d\bar{m}(z) = \| \mathscr{w}(z_2') \| \operatorname{tr} \chi(\rho(z_1')).$$

Therefore, with $\log \operatorname{tr} \chi(\rho(z_1')) = s_1(z_1')$ and $-\log \| \mathscr{w}(z_2') \| = s_1(z_2')$ we find

$$\operatorname{tr}(\chi(\rho(z_1')) U_\tau^+ \mathscr{w}(z_2') U_\tau) \leq e^{s_1(z_1') - s_1(z_2')}. \tag{2.6.29}$$

With $s_1(z_2') = (1+\alpha) s_1(z_1')$, from (2.6.29) one obtains

$$\operatorname{tr}(\chi(\rho(z_1')) U_\tau^+ \mathscr{w}(z_2') U_\tau) \leq e^{-\alpha s_1(z_1')}. \tag{2.6.30}$$

If $\alpha$ is not small relative to 1, then $\alpha \sigma_1(z_1')$ is large relative to 1, so that (2.6.30) as a probability cannot be distinguished from zero (estimates for $\operatorname{tr} \chi(\rho)$, i.e. the dimension of the macroscopic cells, are given in §3.2 and [44]).

If (2.6.30) is very small relative to 1, then despite this (2.6.24) *can* be satisfied. In fact, although the integrand in

$$\int_{\rho(z_2')} \operatorname{tr}(\omega(z_1')\, U_\tau^+\, \omega(z)\, U_\tau)\, d\bar{m}(z)$$

is smaller than $\|\omega(z')\| = e^{-s_1(z_1')}$, we find

$$\int_{\rho(z_2')} d\bar{m}(z) = \operatorname{tr} \chi(\rho(z_2'))\, e^{-s_1(z_2')}.$$

Instead of a contradiction, we have obtained the important embedding relation (2.6.29), in which we can again replace $z_1', z_2'$ by $z_1, z_2 = d_\tau z_1$. Hence, if the probability

$$e^{s_1(z_1) - s_1(d_\tau z_1)} \tag{2.6.31}$$

is practically zero, then the embedding of a macroscopically determined and irreversible dynamics is possible.

The appearance of an entropy in a formula for estimating probabilities as in (2.6.29) is the true meaning of the often mystified designation of $e^{s_1(z)}$ as a "thermodynamic" probability.

For systems *with* memory but with preparation-determined dynamics (mentioned at the beginning of this §2.6), till now no investigation as systematic as for systems without memory has been possible. Certainly, one would again have to guess the observable $\Sigma_Z \xrightarrow{\chi} L$. But a set $K_m^*$ with $\tilde{S}_{\tilde{b}_0} K_m^*$ dense in $K_m(\Sigma_m)$ will be essentially larger than the set (2.5.15). In concrete cases, various systems with memory have been investigated. Many magnetic processes in macrosystems belong to this domain. Perhaps one will attain a more systematic theory on the basis of known examples.

Also for systems without memory, the above discussion shows that the choice of the observable $\Sigma_Z \xrightarrow{\chi} L$ (for given state space $Z$) is physically not at all trivial. In fact, this observable actually contains a physically as important quantity as the entropy for non-equilibrium states. The observable $\Sigma_Z \xrightarrow{\chi} L$ and the whole macroscopic observable $\Sigma_m \xrightarrow{F\tilde{b}_0} L$ inform us about macroscopic measurement possibilities. These possibilities pose a deep physical problem; it concerns the interactions of the macrosystem with its surroundings.

But already the choice of $Z$ expresses what is macroscopically measurable. There is no theory which tells us how to choose $Z$, i.e. how to define it mathematically and to interpret it physically. Up to now, it has been left to the physicist's intuition to guess (on the basis of systems such as rarefied gases) the state space $Z$ (e.g. the set of Boltzmann partitions $f(\boldsymbol{r}, \boldsymbol{v})$). Yet it is easy to go over to "contracted" state spaces; e.g. from the space $Z$ of Boltzmann partitions to the space $Z^{(r)}$ of hydrodynamic magnitudes. With $m$ the mass of the atoms, this $Z^{(r)}$ contains

$$\begin{aligned} \mu(\boldsymbol{r}) &= m \int f(\boldsymbol{r}, \boldsymbol{v})\, d^3 \boldsymbol{v}, \\ \boldsymbol{u}(\boldsymbol{r}) &= \frac{m}{\mu(\boldsymbol{r})} \int \boldsymbol{v} f(\boldsymbol{r}, \boldsymbol{v})\, d^3 \boldsymbol{v}, \\ T(\boldsymbol{r}) &= \frac{m}{\mu(\boldsymbol{r})} \int \boldsymbol{v}^2 f(\boldsymbol{r}, \boldsymbol{v})\, d^3 \boldsymbol{v}. \end{aligned} \tag{2.6.32}$$

Therefore, the transition from a state space $Z$ to a contracted space $Z^{(r)}$ consists in a mapping $Z \xrightarrow{f} Z^{(r)}$, such as by (2.6.32). This mapping is physically meaningful if it is uniformly continuous with respect to the uniform structures of physical imprecision. Thus, we assume that a uniform structure of physical imprecision is given not only in $Z$, but also in $Z^{(r)}$. Let $Z_p$ resp. $Z_p^{(r)}$ be the corresponding uniform spaces. Hence we assume $Z_p \xrightarrow{f} Z_p^{(r)}$ uniformly continuous. In $Z_p$, this mapping $f$ defines a second uniform structure $pr$ as the initial uniform structure corresponding to $f$. This $pr$ is coarser than $p$ and of course need not separate the points of $Z$ ($Z_p$ and $Z_p^{(r)}$ are assumed as spaces with separating uniform structures). The mapping $f$ can be extended to the completions $\hat{Z}_p \xrightarrow{f} \hat{Z}_p^{(r)}$. Since $\hat{Z}_p$ is compact, $f(\hat{Z}_p)$ is a compact subset of $\hat{Z}_p^{(r)}$. One can identify $f(\hat{Z}_p)$ with the separated completion $\hat{Z}_{pr}$. Therefore, the introduction of a contracted state space $Z_p^{(r)}$ is physically equivalent to that of a uniform structure $pr$ in $Z$ that is coarser than $p$.

The embedding condition is satisfied for the description in the contracted state space $Z^{(r)}$ as soon as it is for the description in $Z$. While this is conceptually clear, let us briefly indicate the mathematical proof.

To each trajectory from $Y = C(\theta, Z)$, the mapping $\theta \to Z_p \xrightarrow{f} Z_p^{(r)}$ assignes a trajectory from $Y^{(r)} = C(\theta, Z^{(r)})$ (for simplicity we assume $f$ surjective). Let this mapping also be called $f$; then $Y_p \xrightarrow{f} Y_p^{(r)}$ is uniformly continuous so that it can be extended as a mapping $\hat{Y} \xrightarrow{f} \hat{Y}^{(r)}$ such that $f(\hat{Y}) = \hat{Y}^{(r)}$.

Then $\sigma \in \mathscr{B}(\hat{Y}^{(r)})$ and $\sigma \to f^{-1}(\sigma) \in \mathscr{B}(\hat{Y})$ define a mapping $\mathscr{B}(\hat{Y}^{(r)}) \to \mathscr{B}(\hat{Y})$ by which a mapping $K(\Sigma_m) \xrightarrow{g} K(\Sigma_m^{(r)})$ is generated in the canonical way. It is easy to see $V_\tau' g = g V_\tau'$. Then (2.5.4) with $\tilde{S}_{b_0}^{(r)} = g \tilde{S}_{b_0}$ implies that the embedding condition holds for the "contracted" description:

$$V_\tau' \tilde{S}_{b_0}^{(r)} w = \tilde{S}_{b_0}^{(r)} \mathscr{U}_\tau' w. \tag{2.6.33}$$

However, one must observe the following: The fact that the dynamics in $Z$ is "without memory" by no means implies that the dynamics in $Z^{(r)}$ is also without memory. With $r$ from §2.5 and a corresponding $\bar{r}$ for the contracted description, there is a $\bar{g}$ such that the diagram

$$\begin{array}{ccc} K(\Sigma_m^{(r)}) & \xrightarrow{\bar{r}} & K(\Sigma_Z^{(r)}) \\ \uparrow{\scriptstyle g} & & \uparrow{\scriptstyle \bar{g}} \\ K(\Sigma_m) & \xrightarrow{r} & K(\Sigma_Z) \end{array} \tag{2.6.34}$$

is commutative. If the dynamics in $Z$ is without memory (if the mapping $\tilde{s}$ introduced in §2.5 exists), it does not yet follow that there is an analogous mapping $\bar{\tilde{s}}$ relative to $\bar{r}$.

In many practical cases (also in the contracted hydrodynamic description indicated by (2.6.32); see [1] XV §10.6) for $\bar{g}$ there is a mixture-morphism $\bar{h}$ with $\bar{g}\bar{h} = 1$ on $K(\Sigma_Z^{(r)})$. However, $\bar{h}\bar{g}$ as a projector on $K(\Sigma_Z)$ can *differ* from 1. Then a mapping $K(\Sigma_Z^{(r)}) \xrightarrow{\bar{\tilde{s}}} K(\Sigma_m^{(r)})$ is defined by $\bar{\tilde{s}} = g\tilde{s}\bar{h}$, with $\bar{r}\bar{\tilde{s}} = \bar{r}g\tilde{s}\bar{h} = \bar{g}r\tilde{s}\bar{h} = \bar{g}\bar{h} = 1$. Therefore a dynamics without memory again holds in $Z^{(r)}$ *if* we restrict ourselves to ensembles from $\bar{h}K(\Sigma_Z^{(r)})$. That is, we restrict the preparation procedures so that only such initial ensembles $u \in K(\Sigma_Z)$ are allowed for which also $u \in hK(\Sigma_Z^{(r)})$ holds (hence $\bar{h}\bar{g}u = u$).

Let us mention that there are many other contraction methods to describe very complex macrosystems, as e.g. computers or brains. One tries to find "global" descriptions, where all details are omitted which are inessential for the "interesting" parts. All these contractions can be described by contractions of the Boolean ring $\Sigma_m$ to subrings. Such contractions are also meaningful if the dynamics is not deterministic or even not without memory. In such contractions the time scale can partly disappear (becoming one of the uninteresting details). The selection of the subrings depends on the complex structure of the state space $Z$. For instance, $Z$ can be a product $\prod_i Z_i$ of many "parts" $Z_i$. Such contractions can be highly significant in practice. But they obviously do not alter the embedding problem as discussed in §2. Therefore let us not bring any example.

## §2.7 Disturbances by Measurements

In the literature discussing the measuring process (in particular in quantum mechanics), we frequently find the opinion that classical physics neglects the disturbances by measurements whereas in quantum mechanics these disturbances are substantial and cannot be eliminated. Such an idea fails to see the core of the measuring process. Any measurement, whether on micro- or macrosystems, is a disturbance. Also there are measurements of macrosystems which considerably disturb these systems. For a measurement it is only essential that it yields conclusions about the measured systems as it was *before* the measurement. What happens to the systems *after* the measurement is unimportant also in macrophysics. The sand bag mentioned in II §3.1 is a suitable measurement method for a bullet.

Measurement methods (more correctly registrations) may be called *hard* if they essentially disturb the systems, otherwise *soft*. Any registration method in quantum mechanics thus is a hard method. Yet the objectivating description of classical physics does not depend on whether there are soft registration methods or not.

A soft registration method $b_0$ can be defined more precisely by the condition (2.3.12). Of course also any $b \in \mathcal{R}(\tilde{b}_0)$ will be called a soft registration procedure. In this sense the conclusions of (2.3.12), described in §2.3–§2.6, presuppose the existence of soft registration methods.

That $\tilde{b}_0$ is a soft registration method in $\mathcal{R}_{0m}$ does not imply that $i\tilde{b}_0$ is a soft method in the theory $\mathcal{PT}_{q\,\mathrm{exp}}$. The disturbances by the registrations of the method $i\tilde{b}_0$ (as by all registrations in quantum theory) are so violent that they considerably change the imagined measurement results of certain hypothetical observables in $\mathcal{PT}_{q\,\mathrm{exp}}$. The embedding condition (2.3.12) for $\tilde{b}_0$ just asserts that these disturbances by $i\tilde{b}_0$ can be neglected for the trajectories.

If the dynamics of the systems is preparation-determined, (2.3.12) implies that every individual trajectory is undisturbed by the ongoing measurement according to $\tilde{b}_0$. This implies a stability of the trajectories under small measurement disturbances (see §2.6). If the dynamics is only statistical it is imaginable that the measurement disturbances change individual trajectories, but without changing the statistics determined by the preparation. It is only "imaginable" and not detectable that the trajectories of individual systems are changed considerably by the measurement (individual trajectories cannot be reproduced by preparation).

If a soft registration method with (2.3.12) is possible corresponding to a state space $Z$, then all the more it exists corresponding to a contracted state space (see §2.6). Experience suggests the presumption that most macrosystems in our laboratories allow a soft registration method for their trajectories if we select a not too fine state space. To say it in another way:

It seems very difficult and expensive to construct macrosystems with such state spaces that *only* hard registration methods remain. Therefore we have attached great importance to the structure analysis of the inferences from (2.3.12). For this reason we will continue to presuppose (2.3.12) in XI.

Is it possible to recognize from the macrotheory $\mathscr{P}\mathscr{T}_m$ whether there are *only* hard registration methods? Of course not: The existence of a $\tilde{b}_0$ with (2.3.12) is a certain hypothesis (in the sense of XIII §4.3 and [3] §10) in $\mathscr{P}\mathscr{T}_m$. Therefore it must be regarded as physically possible if we have no severe objections to $\mathscr{P}\mathscr{T}_m$ as a g.G.-closed theory (see again XIII §4.3 and [3] §10). But now we have a theoretical possibility (namely by embedding) to investigate whether $\mathscr{P}\mathscr{T}_m$ is closed and to find a more comprehensive theory if necessary.

The possibility consists in investigating the following question. Assuming (2.3.12), examine the embedding according to §2.3 and 2.4. If the embedding should turn out as impossible under the condition (2.3.12), we can try to replace (2.3.12) by a weaker condition. For instance, we may demand that for all possible but small intervals $\tau_1 \leq \ldots \leq \tau_2$ there are registration methods $\tilde{b}_0$ which satisfy

$$\overline{\mathrm{co}}^\sigma \, \psi_{s\tilde{b}_0} \, \Sigma_{\tilde{b}_0} = L(\Sigma_{m;\,\tau_1 \leq \ldots \leq \tau_2}). \tag{2.7.1}$$

With $L(\hat{Y}, \geq \tau)$ and $L(\hat{Y}, \leq \tau)$ as in II (4.1.3) and II (4.1.4), the right side of (2.7.1) is defined by $L(\hat{Y}, \leq \tau_2) \cap L(\hat{Y}, \geq \tau_1)$ as $L(\Sigma_m)$ is by $L(\hat{Y})$.

Macrosystems for which high precision embedding with (2.3.12) is impossible, shall be called measurement-sensible. For such systems one would get a more comprehensive theory than $\mathscr{P}\mathscr{T}_m$ by detecting the limitations of macroscopic measurements. For instance, we could estimate the intervals $\tau_2 - \tau_1$ for which an embedding with the condition (2.7.1) is possible, or we could estimate the $L^i_m(\Sigma_m)$ we defined after (2.4.7). In this sense, the investigation of measurement-sensible systems requires investigating at first the possibility of soft registration methods for embedding.

This discussion of the problems posed by measurement-sensible systems makes it convincing that a theory of such systems is still unknown.

All the investigations of macrosystems preceding in this §2 presumed that the systems are "isolated" between preparation and registration. This isolation need not be exact, as we have discussed in §2.2. But in many applications we are concerned with non-isolated systems, also called "open" systems. The dynamics of these systems is essentially influenced by the environment. There are many methods to approximate the influence of the surroundings. The method of embedding can in principle not solve the problem of open systems since quantum mechanics (and therefore $\mathscr{P}\mathscr{T}_{q\,\exp}$) presupposes that the microsystems are isolated between preparation and registration. This we have established by the introduction of the "normative" axioms in III

§5.1. Only in the special case of microsystems in "external fields" an approximate method is developed in quantum mechanics (see [2] VIII §6). Therefore it is small wonder that also for macrosystems and the embedding in $\mathcal{PT}_{q\,\text{exp}}$ we should like to describe the surroundings by this method of external fields, e.g. introducing "boundary" conditions.

This method of external fields is useless if the systems change the environment so that there is a feedback of the surroundings on the system. In this general case there only remains to consider a larger system which includes part of the environment and can be considered as isolated. This is the reason for not treating open systems in this book.

In closing this §2 let us emphasize again that the structure analysis of the relation between $\mathcal{PT}_m$ and $\mathcal{PT}_{q\,\text{exp}}$ is not a "proof" of their compatibility. Rather it presents an organizing principle for the whole field usually called "statistical" mechanics. In this sense, not all problems have been solved but received a novel formulation. Such equestions in novel formulation are:

How can we find the state spaces and the corresponding macro-observables? How can a soft ongoing measurement be described in detail? How can we verify the embedding conditions? How does the dynamics of concrete macrosystems such as gases, fluids, solids look like? How could the many hitherto existing "fundamental investigations" of statistical mechanics be organized under the principle of embedding?

Regarding the last question, some readers may have the opinion that much of the work done in the foundations of statistical mechanics contradicts the concepts presented here. This is not true if we divest these approaches of the added philosophy. Then these approaches in fact present proofs or ways to possible proofs for parts of the structure analysis of §2 (for special systems). Such a synopsis of the achieved results in statistical mechanics would require a new book.

Let us conclude with the following remarks. This §2 has in a certain way presented a mathematically correct basis for the developments in [1] XV. There the mathematical considerations about measure theory had been indicated only in outline (the Boolean ring $\Xi$ in [1] XV corresponds to the Boolean ring $\Sigma_m$ introduced in §2.3).

## §3 Examples for Approximate Embedding

This section will bring examples for approximate embeddings. They prove that the embedding conditions in §2 can be satisfied approximately. We have discussed physically imprecise embedding in XIII §3 and [3] (for the general definition of imprecision sets see also [40]). The examples are therefore also an illustration of these general discussions.

We will not discuss the most general problem of §2 but only special cases. One of these is given by (2.4.47), (2.4.48), where $S_r$ and $T_{\tilde{a}}$ are mixture-morphisms $\mathcal{B} \xrightarrow{S_r} \mathcal{B}(\Sigma')$ and $\mathcal{B}(\Sigma') \xrightarrow{T_{\tilde{a}}} \mathcal{B}$. In addition we will identify $\Sigma'$ with $\Sigma_Z$ as in §2.5. The assumption $\Sigma' = \Sigma_Z$ is not essential since one can show that it is possible to introduce a finer state space $Z_f$ which makes $\Sigma' = \Sigma_{Z_f}$ and $Z$ a contracted state space (in the sense of §2.6) relative to $Z_f$.

In this section we omit the labels $r$ and $\tilde{a}$. Then (2.4.47) and (2.4.48) take the form

$$ST = \mathbf{1}, \tag{3.1}$$

$$A_\tau = S\,\mathscr{U}'_\tau\, T \tag{3.2}$$

for the mixture-morphism $\mathscr{B} \xrightarrow{S} \mathscr{B}(\Sigma_Z)$ and $\mathscr{B}(\Sigma_Z) \xrightarrow{T} \mathscr{B}$, where $A_0 = \mathbf{1}$ and $\{A_\tau\}$ is a contracting semigroup for $\tau \geq 0$.

In §2.4 we have also seen that (3.1) cannot be fulfilled unless $\Sigma_Z$ is atomic. In most applications, $Z_p$ ($Z$ with the uniform structure of physical imprecision; see II §1) is not a discrete space, therefore $\Sigma_Z$ is not atomic. We will demonstrate that nevertheless (3.1) and (3.2) can be fulfilled imprecisely, so that

$$ST \sim \mathbf{1}, \tag{3.3}$$

$$A_\tau \sim S\,\mathscr{U}'_\tau\, T. \tag{3.4}$$

We must explain in detail the meaning of the sign $\sim$. We will do this in §3.1.

As a second case we take $T$ in (3.3), (3.4) as the mixture-morphism $T$ of (2.5.31) or (2.5.35) and $S$ as before. Then $Q = ST$ is the smearing operator (2.5.32), given by the kernel $\mu(\omega(z), \omega(z'))$ or $\mu(\bar{\omega}(z), \omega(z'))$. In this interpretation, condition (3.3) becomes

$$Q \sim \mathbf{1} \tag{3.5}$$

and does not mean an imprecise embedding. The embedding described in §2.5 is exact for the time zero. The difference between $Q$ and $\mathbf{1}$ then describes the fact that not all ensembles in $K(\Sigma_Z)$ can be prepared but only those in $QK(\Sigma_Z)$. We will see in §3.2 that for macrosystems in finite space regions the Banach space $\mathscr{B}_m$ will be finite dimensional. Then $QK(\Sigma_Z)$ cannot span all of $\mathscr{B}(\Sigma_Z)$. Therefore the assumption $r'\,\Sigma_{mk} = \Sigma_Z$ in §2.5 is only an idealization ($r'\,\Sigma_{mk}$ can only approximately equal $\Sigma_Z$). In what sense this must be understood will be described briefly in §3.1.

In §3.2 we will see that it is very difficult (sometimes impossible) to detect the difference between $QK(\Sigma_Z)$ and $K(\Sigma_Z)$ by experiments. Nevertheless this difference is important for the problem of the reality of atoms as parts of a macrosystem (see XIII §4.8).

With $Q_\tau$ in (2.5.38b), the relation (3.4) becomes

$$A_\tau \sim Q_\tau. \tag{3.6}$$

This tells us that the $\{Q_\tau\}$ approximately form the semigroup $\{A_\tau\}$. We have used this approximation already in (2.5.40).

An estimate of the imprecision between $A_\tau$ and $Q_\tau Q^{-1}$, i.e. of an approxmation to (2.5.38a), has not been given. The main problem is, that $Q_\tau K(\Sigma_Z)$ will *not* be a subset of $Q\,K(\Sigma_Z)$ (then $\{Q_\tau Q^{-1}\}$ would be a semigroup). Hence $Q_\tau Q^{-1}$ must be extended to all of $K(\Sigma_Z)$ (see the remarks after (2.5.38)). One must expect that a semigroup $\{A_\tau\}$ is defined on all of $K(\Sigma_Z)$, where the $A_\tau$ act approximately on $Q\,K(\Sigma_Z)$ as the $Q_\tau Q^{-1}$ do. This will be demonstrated in §3.3.

The following sections §3.1 to §3.3 are excerpts from [44].

## § 3.1 Imprecision Sets

In II § 1 we introduced a uniform structure of physical imprecision in $Z$. The state space $Z$ with this structure was denoted by $Z_p$, the completion of this $Z_p$ by $\hat{Z}$. This $\hat{Z}$ is a compact and metrizable space. The Boolean ring $\mathscr{B}(\hat{Z})/\mathscr{J}(\hat{Z})$, where $\mathscr{J}(\hat{Z})$ contains no open set of $\hat{Z}$, was called $\Sigma_Z$ (see § 2.5). Therefore we can identify $C(\hat{Z})$ with a subspace $\mathscr{D}(\hat{Z})$ of $\mathscr{B}'(\Sigma_Z)$ (see the analogous consideration of $\mathscr{B}(\hat{S}_m)/\mathscr{J}(\hat{S}_m)$ in § 2.3). $C(\hat{Z})$ is normseparable since $\hat{Z}$ is compact and metrizable and therefore separable. $K(\Sigma_Z)$ is normseparable according to its construction. Therefore $\mathscr{B}(\Sigma_Z)$, $\mathscr{D}(\Sigma_Z)$ is a special case of the general structure $\mathscr{B}$, $\mathscr{D}$ in IV, namely that cf a "classical system" in the sense of VII § 5.3.

We abbreviate $L(\Sigma_Z)=[0,1]$ in $\mathscr{B}'(\Sigma_Z)$ and $L(\hat{Z})=L(\Sigma_Z)\cap\mathscr{D}(\Sigma_Z)=[0,1]$ in $C(\hat{Z})$. We write the bilinear forms of the dualities $\mathscr{B}(\Sigma_Z)$, $\mathscr{B}'(\Sigma_Z)$ and $C'(\hat{Z})$, $C(\hat{Z})$ as $\langle \dots, \dots \rangle$, that of the duality $\mathscr{B}$, $\mathscr{B}'$ as $\mu(\dots, \dots)$.

As discussed in IV, the uniform structure of physical imprecision on $K(\Sigma_Z)$ is given by the topology $\sigma(K(\Sigma_Z), \mathscr{D}(\Sigma_Z))=\sigma(K(\Sigma_Z), L(\hat{Z}))$. With $\bar{K}^\sigma(\Sigma_Z)$ as the closure of $K(\Sigma_Z)$ in $\mathscr{D}'(\Sigma_Z)$, the set $\partial_e \bar{K}^\sigma(\Sigma_Z)$ can be identified with $\hat{Z}$. The topology $\sigma(\partial_e \bar{K}^\sigma(\Sigma_Z), \mathscr{D}(\Sigma_Z))$ is identical with that of $\hat{Z}$ (this follows since $\mathscr{D}'(\Sigma_Z)$ can be identified with $C'(\hat{Z})$; see also [7] V § 8.1).

The essential condition for the embedding was the invariance of the probability. For an imprecise embedding we do not use the exact relation $\langle m,f\rangle=\alpha$ ($m\in K(\Sigma_Z)$, $f\in L(\Sigma_Z)$) for the probablity. Rather a smeared relation $\tilde{\mu}$ defined with imprecision sets as elements of the uniform structure of physical imprecision (see also XIII § 1) is employed. Let $U$ be an imprecision set, i.e. a vicinity belonging to $\sigma(K(\Sigma_Z), \mathscr{D}(\Sigma_Z))$. We define

$$\tilde{\mu}(m,f;\alpha)=\{(m,f;\alpha)\,|\text{ there is a } m'\in K(\Sigma_Z) \text{ with } \langle m',f\rangle=\alpha \text{ and } (m,m')\in U\}.$$

Instead of $\langle STm,f\rangle=\langle m,f\rangle$ let us only postulate $\langle m,f\rangle=\alpha\Rightarrow\tilde{\mu}(STm,f;\alpha)$ and [not $\tilde{\mu}(m,f;\alpha)]\Rightarrow\langle STm,f\rangle\neq\alpha$ (see [3] § 8 and XIII § 3). We can combine this by writing the postulate (3.3), resp. (3.5) in the form

$$(STm, m)=(Qm, m)\in U \qquad \text{for all } m\in K(\Sigma_Z). \tag{3.1.1}$$

The selection of the imprecision set $U$ is a physical problem. We hope that such a $U$ can be chosen that it is *physically* (not in the mathematical theory!) impossible to distinguish by measurements between elements $(m, m')\in U$.

$\sigma(K(\Sigma_Z), \mathscr{D}(\Sigma_Z))$ is metrizable. Therefore we can select a metric $d$, which generates the same uniform structure as $\sigma(K(\Sigma_Z), \mathscr{D}(\Sigma_Z))$. If we take $d$ to select an imprecision set by

$$U=\{(m,m')\,|\,d(m,m')<\varepsilon\}, \tag{3.1.2}$$

the metric gets a physical significance which the more general uniform structure cannot describe.

Obviously we cannot give any physical arguments for the selection of a metric as long as we do not specify the state space $Z$. Therefore we can discuss only generally the physical significance of a metric as an enrichment of the uniform structure.

**D 3.1.1** Let $\Lambda$ be a subset of $L(\hat{Z})=L(\Sigma_Z)\cap\mathscr{D}(\Sigma_Z)$. By $d_\Lambda(z,z')=\sup_{f\in\Lambda}|f(z)-f(z')|$ a metric is defined if $\Lambda$ separates points, i.e. if $f(z)=f(z')$ for all $f\in\Lambda$ implies $z=z'$.

We have $d_\Lambda \leq 1$. For $\Lambda = L(\hat{Z}) = [0,1]$, in $C(\hat{Z})$ we have $d_{L(\hat{Z})}(z,z') = \{0$ for $z=z'$ and 1 for $z \neq z'\}$. According to the Stone-Weierstrass-theorem ([7] V §8.1) $\Lambda$ separates points if $la\,\Lambda$ is normdense in $L(\hat{Z})$ where $la$ is defined as in II §3.1.

Since $\Lambda$ separates points, the initial uniform structures on $\hat{Z}$ generated by $\Lambda$ and $L(\hat{Z})$ are equal to that of the compact space $\hat{Z}$.

**T 3.1.1** The uniform structure of $\hat{Z}$ and that generated by the metric $d_\Lambda$ are equal if and only if $\Lambda$ is norm-precompact. For every metric $d \leq 1$ which generates the uniform structure of $\hat{Z}$, there is a set $\Lambda \subset L(\hat{Z})$ with $d = d_\Lambda$. For instance $\Lambda$ may be

$$\Lambda_m = \{f \mid f \in L(\hat{Z}) \text{ and } |f(z) - f(z')| \leq d(z,z')\} \tag{3.1.3}$$

($|f(z) - f(z')| \leq d(z,z')$ is known as the Lipschitz condition).

*Proof.* Immediately follows $d_{\Lambda_m} \leq d$ with $\Lambda_m$ in (3.1.3). If $d \leq 1$ generates the uniform structure of $\hat{Z}$, we get $f(z) = d(z, z_0) \in L(\hat{Z})$ and $|f(z) - f(z')| = |d(z,z_0) - d(z',z_0)| \leq d(z,z')$, therefore $d(z,z_0) \in \Lambda_m$ and $d_{\Lambda_m}(z,z_0) \geq d(z,z_0)$. Since $z_0$ is arbitrary, there follows $d_{\Lambda_m} \geq d$. Let $d_\Lambda$ be a metric generating the uniform structure of $\hat{Z}$. Then $f \to f(z)$ defines a mapping $\Lambda \xrightarrow{z} [0,1] \subset \mathbf{R}$. The initial uniform structure generated by all these mappings $z \in \hat{Z}$ is precompact. We will show that on $\Lambda$ this uniform structure coincides with that generated by the norm.

Obviously, the norm uniform structure is finer than the initial stucture for the mappings $z$. Therefore it remains to be shown that in $\{(f_1, f_2) \mid \|f_1 - f_2\| \leq \varepsilon\}$ there is a vicinity

$$\{(f_1, f_2) \mid |f_1(z_i) - f_2(z_i)| < \delta \; (i = 1, \ldots, n)\}. \tag{3.1.4}$$

Since $\hat{Z}$ is compact, there are $n$ elements $z_1, \ldots, z_n$ with $d_\Lambda(z, z_i) < \frac{\varepsilon}{4}$ for all $z \in \hat{Z}$ and a suitable $z_i$. We have

$$\begin{aligned} |f_1(z) - f_2(z)| &\leq |f_1(z) - f_1(z_i)| + |f_1(z_i) - f_2(z_i)| + |f_2(z_i) - f_2(z)| \\ &\leq 2 d_\Lambda(z, z_i) + |f_1(z_i) - f_2(z_i)| \\ &\leq \frac{\varepsilon}{2} + |f_1(z_i) - f_2(z_i)| < \frac{\varepsilon}{2} + \frac{\varepsilon}{2} = \varepsilon \end{aligned} \tag{3.1.5}$$

if in (3.1.4) we choose $\delta < \frac{\varepsilon}{2}$.

That $d_\Lambda$ for a norm-precompact set $\Lambda$ is a metric for $\hat{Z}$ can be proved very similarly by a relation analogous to (3.1.5):

$$\begin{aligned} |f(z) - f(z')| &\leq |f(z) - f_i(z)| + |f_i(z) - f_i(z')| + |f_i(z') - f(z')| \\ &\leq 2\|f - f_i\| + |f_i(z) - f_i(z')|. \quad \square \end{aligned}$$

T 3.1.1 immediately shows that $\Lambda_m$ is the greatest set with $d = d_\Lambda$. One easily proves that $\Lambda_m$ is norm closed (and therefore norm compact); it is a lattice and obeys $\Lambda_m = \mathbf{1} - \Lambda_m$. We have $\Lambda_m \neq L(\hat{Z})$ since $L(\hat{Z})$ is not norm compact. This is not a contradiction to $la\,\Lambda_m$ norm dense in $L(\hat{Z})$, since we made $la\,\Lambda \overset{\text{def}}{=} F \cap L(\hat{Z})$ with $F$ the smallest subspace of $C(\hat{Z})$ containing $\Lambda$ which contains $|f(z)|$ whenever it contains $f(z)$ (see II §3.1).

By T 3.1.1 the physical selection of a metric is equivalent to the selection of a norm precompact set $\Lambda \subset L(\hat{Z})$.

In II § 3.1 we have discussed the physical interpretation of the elements $f$ of $L(\hat{Y})$ by introducing the probability $\lambda_{\text{Meas}}$. This discussion can be taken over if we replace $\hat{Y}$ by $\hat{Z}$. Therefore we may be brief, saying only that $f(z)$ is the probability (of an indication on a measuring device) for the state $z \in \hat{Z}$.

The set $\Lambda$ (with $la\,\Lambda$ norm-dense in $L(\hat{Z})$) can be interpreted in the same way as the set $\psi_m(\phi)$ in II § 3.1 (for which we presumed $la\,\psi_m(\phi)$ norm dense in $L(\hat{Y})$). Therefore we will interpret $\Lambda$ as the set of all "state effects" which can occur in devices for the measurement of the state $z \in \hat{Z}$. That $\Lambda$ itself is not norm-dense in $L(\hat{Z})$ but norm precompact, is an additional physical structure beyond the idealization of taking all of $L(\hat{Z})$ as possible "state effects". We mentioned in II § 3.1 that the presumption of $\psi_m(\phi)$ norm-dense in $L(\hat{Y})$ is perhaps too strong. We could on the contrary presume that $\psi_m(\phi)$ is norm precompact. The consequences for the embedding problem have not been investigated.

If the set $\Lambda$ is given, then two states $z_1, z_2$ cannot be distinguished if the difference of their probabilities obeys $|f(z_1)-f(z_2)|<\varepsilon$ for all $f \in \Lambda$, with $\varepsilon$ so small that a repetition of more than $\varepsilon^{-1}$ experiments is in principle impossible. In this manner we have found an $\varepsilon$ for an imprecision set $\{(z_1, z_2) \mid d_\Lambda(z_1, z_2)<\varepsilon\}$ in $\hat{Z}$. The metric and the equivalent set $\Lambda$ cannot be deduced from $\hat{Z}$, but form an additional structure describing the measuring possibilities. These measuring possibilities are very essentially connected with the embedding problem; hence we may perhaps get conditions for $\Lambda$ from the postulate that the embedding be possible.

If we distinguish points of $\hat{Z}$ by an imprecision, we may replace $\Lambda$ by a finite subset $\Lambda' \subset \Lambda$. Since $\Lambda$ is norm compact, there is such a $\Lambda'$ that for every $f \in \Lambda$ there is an $f' \in \Lambda'$ with $\|f-f'\|<\frac{\varepsilon}{2}$. We have

$$\begin{aligned} |f(z)-f(z')| &\le |f(z)-f'(z)|+|f'(z)-f'(z')|+|f'(z')-f(z')| \\ &\le \varepsilon+|f'(z)-f'(z')| \end{aligned}$$

and therefore

$$d_\Lambda(z, z') \le \varepsilon + \sup_{f' \in \Lambda'} |f'(z)-f'(z')|.$$

The $f' \in \Lambda'$ define an initial uniform structure in $Z$, which is weaker than that of $\Lambda$. This uniform structure defined by $\Lambda'$ does not necessarily separate points of $Z$. But the completion in this uniform structure is a new state space $Z'$ which with non-zero imprecision cannot be distinguished from $\hat{Z}$. This $Z'$ is finite dimensional since we can take the values of the $f' \in \Lambda'$ as coordinates; then $d_{\Lambda'}$ is a metric in $Z'$.

In this sense, every state space $Z$ may be replaced with non-zero imprecision by a finite dimensional state space $Z'$. If $Z'$ is suitably chosen, there is no possibility to distinguish physically between $Z$ und $Z'$. The original $Z$ can be viewed as a mathematical idealization of $Z'$. Such an idealization $Z$ is sometimes more practical than a realistic $Z'$!

For the embedding we need a metric $d$ for the $\sigma(K(\Sigma_Z), \mathscr{D}(\Sigma_Z))$-topology to define the imprecision set (3.1.2). It would be physically senseless, in $\hat{Z}$ and in $K(\Sigma_Z)$ to take two metrics which have no physical connection. On the contrary, it seems meaningful to take the same set $\Lambda$, in order to test the elements of $K(\Sigma_Z)$. This can be done as shown by the next definition and the subsequent theorem.

**D 3.1.2** Take $\Lambda$ as in D 3.1.1. Then

$$d_\Lambda^K(m, m') = \sup_{f \in \Lambda} |\langle m, f\rangle - \langle m', f\rangle|$$

defines a metric if $\Lambda$ separates the elements of $K(\Sigma_Z)$.

$\Lambda$ separates the elements of $K(\Sigma_Z)$ if and only if the linear span of $\Lambda \cup \{\mathbf{1}\}$ is $\sigma(\mathscr{B}'(\Sigma_Z), \mathscr{B}(\Sigma_Z))$-dense in $\mathscr{B}'(\Sigma_Z)$ or norm dense in $\mathscr{D}(\Sigma_Z) = C(\hat{Z})$. Since $\Lambda$ separates the elements of $K(\Sigma_Z)$, the topologies $\sigma(K(\Sigma_Z), \Lambda)$ and $\sigma(K(\Sigma_Z), \mathscr{D}(\Sigma_Z))$ are equal (on $K(\Sigma_Z)$).

**T 3.1.2** $d_\Lambda^K$ is a metric of $\sigma(K(\Sigma_Z), \mathscr{D}(\Sigma_Z))$ if and only if $\Lambda$ is norm precompact. The set

$$\Lambda_m^K = \{f \mid f \in L(\hat{Z}) \text{ and } |\langle m, f\rangle - \langle m', f\rangle \leq d_\Lambda^K(m, m')\}$$

is the greatest $\Lambda' \subset L(\hat{Z})$ with $d_\Lambda^K = d_{\Lambda'}^K$.

The proof proceeds in the same way as for T 3.1.1. The metric $d_\Lambda^K$ can be defined on $\bar{K}^\sigma(\Sigma_Z)$, and the set $\partial_e \bar{K}^\sigma(\Sigma_Z)$ can be identified with $\hat{Z}$. Hence $d_\Lambda^K$ on $\partial_e \bar{K}^\sigma(\Sigma_Z)$ is identical with $d_\Lambda$ on $\hat{Z}$. But $d_{\Lambda_1} = d_{\Lambda_2}$ does not imply $d_{\Lambda_1}^K = d_{\Lambda_2}^K$ on $K(\Sigma_Z)$. In general we have only $\Lambda_m^K \subset \Lambda_m$, where $\Lambda_m$ is given in (3.1.3) with $d = d_\Lambda$.

Therefore we select only such metrics $d_\Lambda$, $d_\Lambda^K$ on $\hat{Z}$ resp. $K(\Sigma_Z)$ which have $\Lambda = \Lambda_m$.

With (3.1.2) the condition (3.1.1) for an imprecise embedding (or alternatively for the approximation of $Q$ by $\mathbf{1}$) then becomes

$$d_\Lambda(STm, m) = d_\Lambda(Qm, m) < \varepsilon \quad \text{for all } m \in K(\Sigma_Z) \tag{3.1.6a}$$

or equivalently

$$|\langle m, f\rangle - \langle STm, f\rangle| = |\langle m, f\rangle - \langle Qm; f\rangle| < \varepsilon$$
$$\text{for all } m \in K(\Sigma_Z) \text{ and all } f \in \Lambda. \tag{3.1.6b}$$

Similar considerations can be performed to introduce a metric $d$ for $\sigma(L(\Sigma_Z), K(\Sigma_Z))$ by selecting a suitable subset $K_d$ of $K(\Sigma_Z)$:

$$d(k_1, k_2) = \sup_{m \in K_d} |\langle m, k_1 - k_2\rangle|.$$

This $K_d \subset K(\Sigma_Z)$ must be norm compact. For an imprecision $d(k_1, k_2) < \varepsilon$, it therefore suffices to use finitely many elements of $K_d$. We must expect that in this sense the elements of $\Sigma_Z$ ($\Sigma_Z$ as $\partial_e L(\Sigma_Z)$) can only be approximated with high (but finite) precision by the elements of $r' \Sigma_{mk}$ (see the remarks after (3.5)). In the next section we will not go into details of estimating such an imprecision lying between $r' \Sigma_{mk}$ and $\Sigma_Z$.

## § 3.2 Embedding at Time Zero

This section is to prove that an imprecise embedding in the sense of (3.1.6) is possible. We also estimate the accuracy of the embedding by relations for $\Lambda$, $\varepsilon$ and the Hilbert space. With the exact embedding at time zero (of § 2.5), we estimate the magnitude of the "smearing" by the operator $Q$.

**T 3.2.1** Let $\Lambda$ be a norm precompact subset of $L(\Sigma_Z)$ (e.g. a norm precompact subset of $L(\hat{Z})$) and choose $\varepsilon>0$. Then there are a finite dimensional Hilbert space $\mathscr{H}$ and two mixture-morphisms $\mathscr{B}(\mathscr{H})\xrightarrow{S}\mathscr{B}(\Sigma_Z)$, $\mathscr{B}(\Sigma_Z)\xrightarrow{T}\mathscr{B}(\mathscr{H})$ such that

$$|\langle m,f\rangle-\langle STm,f\rangle|<\varepsilon \quad \text{for all } m\in K(\Sigma_Z) \text{ and all } f\in\Lambda. \tag{3.2.1}$$

The condition for $\Lambda$ cannot be weakened.

*Proof.* It suffices to consider the case of finite $\Lambda$: Suppose the theorem has been proved for finite $\Lambda_1$. Since $\Lambda$ is norm precompact there is a finite set $\Lambda_1$, such that $\|f-f_\nu\|<\varepsilon/4$ for all $f\in\Lambda$ and suitable $f_\nu\in\Lambda_1$. Then there are mixture-morphisms $S$, $T$ such that $|\langle m, f_\nu\rangle-\langle STm, f_\nu\rangle|<\varepsilon/2$ for all $m\in K(\Sigma_Z)$ and all $f_\nu\in\Lambda_1$. Then

$$\begin{aligned}|\langle m,f\rangle-\langle STm,f\rangle|&\le|\langle m,f\rangle-\langle m,f_\nu\rangle|+|\langle m,f_\nu\rangle-\langle STm,f_\nu\rangle|\\&\quad+|\langle STm,f_\nu\rangle-\langle STm,f\rangle|\le 2\,\|f-f_\nu\|+\varepsilon/2\le\varepsilon.\end{aligned}$$

Let $\Lambda$ be finite and $\varepsilon>0$. Each $f_\nu\in\Lambda_1$ may be approximated in the norm by a step functional $\sum_n \alpha_n^\nu l_{\sigma_n}$. Here $\sigma_n$ is a partition of the unit element $e$ in $\Sigma$, obeying $\langle m, l_{\sigma_n}\rangle=m(\sigma)$ (see the spectral theorem [2] IV Th. 2.1.15). We can take the common partition $e=\bigvee_{n=1}^{N}\sigma_n$ for all $f_\nu\in\Lambda_1$ such that $\left\|f_\nu-\sum_{n=1}^{N}\alpha_n^\nu l_{\sigma_n}\right\|<\varepsilon/2$.

Let $\mathscr{H}$ be a Hilbert space of dimension $N$. Let $\varphi_n$ be a complete orthonormal system in $\mathscr{H}$.

We set ($m\in K(\Sigma_Z)$, $w\in K(\mathscr{H})$)

$$Tm=\sum_{n=1}^{N} m(\sigma_n)\,P_{\varphi_n},\qquad Sw=\sum_{n=1}^{N}\mu(w,P_{\varphi_n})\,m_n$$

with $m_n(\sigma)=m_0(\sigma_n)^{-1}\,m_0(\sigma\wedge\sigma_n)$, where $m_0\in K(\Sigma_Z)$ is effective. Then

$$STm=\sum_{n=1}^{N} m(\sigma_n)\,m_n$$

and therefore

$$(STm)\,\sigma_n=m(\sigma_n)$$

and

$$\left\langle m-STm,\ \sum_{n=1}^{N}\alpha_n^\nu l_{\sigma_n}\right\rangle=0.$$

Thus follows

$$|\langle m-STm,f_\nu\rangle|\le 2\left\|f_\nu-\sum_{n=1}^{N}\alpha_n^\nu l_{\sigma_n}\right\|<\varepsilon.$$

Let $\Lambda$ be a set such that for $\varepsilon>0$ there are $S$, $T$ with (3.2.1) and a finite dimensional Hilbert space. Then we get

$$\|f-T' S' f\|<\varepsilon \quad \text{for all } f\in\Lambda,$$

where $T'$ is a norm continuous mapping of $L(\mathscr{H})$ in $L(\Sigma_Z)$. Since $\mathscr{H}$ is finite dimensional, $L(\mathscr{H})$ is norm compact. Therefore $T' L(\mathscr{H})$ is norm compact. Thus there are finitely many $g_\nu \in T' L(\mathscr{H})$ with $\|T' S' f-g_\nu\|<\varepsilon$ and therefore $\|f-g_\nu\|<2\varepsilon$, i.e. finitely many balls of radius $2\varepsilon$ which cover $\Lambda$. Since $\varepsilon$ was arbitrary, $\Lambda$ is precompact. □

Theorem T 3.2.1 is neither a solution of the embedding problem as described in the previous sections nor an estimate of the smearing by $Q$. This theorem does not take into account essential physical constraints (for instance the time development according to (3.2) and the fact that in $Z$ a macroscopic energy is defined which is a constant of motion). Rather T 3.2.1 demonstrates that the impossibility of (3.1) for non-atomic Boolean rings is no severe objection, because "in physical approximation" $\Sigma_Z$ can be replaced by an atomic ring (a partition of the unit $e\in\Sigma_Z$). In this sense, T 3.2.1 is a first justification of the method of "cell partitions" of $Z$ which describes the imprecision sets in §2.5.

The definition of $T$ in the proof of T 3.2.1 rests on a mapping of a cell in $Z$ on a projection operator $P_\varphi$. We saw in §2.5 that this is extremely unrealistic. A cell must be related to a projection $E$ with high dimension; otherwise an embedding of the dynamics would be impossible. In the proof of T 3.2.1 we can replace the $P_{\varphi_n}$ by such higher dimensional projectors $E_n$; only the dimension of the Hilbert space must exceed the number $N$ of the $\sigma_n$. This is a first hint that the dimensions will play a decisive role in the embedding problem.

In this section let us consider the constraint that a macroscopic energy is defined. Afterwards also the time development will be taken into account.

We presume that a macroscopic energy is defined as a continuous function $E(z)$ in $Z$. Then an interval $E_1\leq E(z)\leq E_2$ defines a closed subset of $Z$ which we call "energy shell". $\Delta E=E_2-E_1$ is called the width of the shell.

The constraint for the embedding is the following: Let $e(\lambda)$ be the spectral family of the Hamiltonian. We presume that there is a partial observable (of the macroscopic observable) which represents the macroscopic energy. This partial observable may be given by a family of effects $f(\lambda)$ (increasing with $\lambda$) for which

$$\|(e(\lambda)-f(\lambda-\delta))_-\|<\varepsilon$$

and

$$\|(f(\lambda+\delta)-e(\lambda))_-\|<\varepsilon.$$

Here $\varepsilon$ cannot be distinguished physically from 0 when $\delta>0$ is a positive number characterizing the macroscopic imprecision of energy measurements.

Then we have $f(E_2+2\delta)-f(E_1-2\delta)\gtrsim e(E_2+\delta)-e(E_1-\delta)\gtrsim f(E_2)-f(E_1)$. Therefore we may replace the yet unknown $f(\lambda)$ by the $e(\lambda)$ in the following way: There are two values $E_1'\leq E_1$ and $E_2'\geq E_2$ (i.e. $E_1'\sim E_1-\delta$ and $E_2'\sim E_2+\delta$) such that $S'$ maps the subring of the Borel sets into the energy shell of $Z$ in $\mathscr{B}'((e(E_2')-e(E_1'))\mathscr{H})$. Likewise, $T$ maps the measures with supports in this energy shell of $Z$ into $\mathscr{B}((e(E_2')-e(E_1'))\mathscr{H})$. Here $(e(E_2')-e(E_1'))\mathscr{H}$ is called an energy shell in $\mathscr{H}$.

To simplify the notation we denote a fixed energy shell again by $Z$. Then $\hat{Z}$ is the completion of this $Z$. In the same way we denote $(e(E_2')-e(E_1'))\mathscr{H}$ again by $\mathscr{H}$. Then we have $\mathscr{B}(\Sigma_Z)\xrightarrow{T}\mathscr{B}(\mathscr{H})$ and $\mathscr{B}(\mathscr{H})\xrightarrow{S}\mathscr{B}(\Sigma_Z)$ with $Z$ and $\mathscr{H}$ as energy shells.

Since in the applications the systems are enclosed in a finite volume, the energy shell $\mathscr{H}$ has a finite dimension. T 3.2.1 makes us expect that $\Delta E$ (i.e. the macroscopic imprecision of the energy) is not too small, since the energy shell $\mathscr{H}$ must be of very high dimension in order to make an embedding possible. In this section we want to get relations between the imprecision in $Z$ and the dimension of $\mathscr{H}$.

In the following, $Z$ and $\mathscr{H}$ are always energy shells.

**D 3.2.1** Let $\mathscr{B}$ be a base-normed Banach space. Two mixture-morphisms $\mathscr{B}\xrightarrow{S}\mathscr{B}(\Sigma_Z)$ and $\mathscr{B}(\Sigma_Z)\xrightarrow{T}\mathscr{B}$ are called an embedding with an $\varepsilon$-approximation, with an $\varepsilon$-smearing if

$$|\langle m,f\rangle-\langle STm,f\rangle|=|\langle m,f\rangle-\langle Qm,f\rangle|<\varepsilon \quad \text{for all } m\in K(\Sigma_Z) \text{ and } f\in\Lambda.$$

This is equivalent to

$$d(m,STm)=d(m,Qm)<\varepsilon \quad \text{for all } m\in K(\Sigma_Z).$$

In the following we omit the labels on $d_\Lambda$ and $d_\Lambda^K$.

**D 3.2.2** With $S$, $T$ as in D 3.2.1, we define

$$\Delta(\hat{Z},\mathscr{B})=\inf_S\inf_T\sup_{m\in K(\Sigma_Z)}\sup_{f\in\Lambda}|\langle m,f\rangle-\langle STm,f\rangle|.$$

$\Delta(\hat{Z},\mathscr{B})$ is the smallest $\varepsilon$ for which an $\varepsilon$-approximation, resp. an $\varepsilon$-smearing of the embedding is possible.

The following theorem is a first step in the calculation of $\Delta(\hat{Z},\mathscr{B})$.

**T 3.2.2** Let $\mathscr{D}$ be a $\sigma(\mathscr{B}',\mathscr{B})$-dense subspace of $\mathscr{B}'$ with $\mathbf{1}\in\mathscr{D}$ (see e.g. $\mathscr{D}$ in IV §4). Then $\bar{K}^\sigma$ (the $\sigma(\mathscr{D}',\mathscr{D})$-closure of $K$ in $\mathscr{D}'$) is the base of $\mathscr{D}'$, while $K(\hat{Z})=\bar{K}^\sigma(\Sigma_Z)$ is the base of $C'(\hat{Z})=\mathscr{D}'(\Sigma_Z)$. For any positive linear map $C(\hat{Z})\xrightarrow{R}\mathscr{D}$ with $R\mathbf{1}=\mathbf{1}$ and $z\in\hat{Z}$ we define $\eta(R,z)=\sup\{\lambda\,|\,\lambda\in\mathbf{R}, R(d(z,\cdot))\geq\lambda\mathbf{1}\}$ and $\eta(R)=\sup_{z\in\hat{Z}}\eta(R,z)$. Here $d(z',\cdot)$ denotes $d(z',z)$ as function of $z$ with $z'$ fixed; similarly we employ $\eta(R,\cdot)$.

Then

(i) $\eta(R,\cdot)\in\Lambda$,

(ii) $\eta(R)=\inf_T\sup_{m\in K(\Sigma_Z)}\sup_{f\in\Lambda}|\langle m,f\rangle-\mu(Tm,Rf)|$.

Here the infimum can be taken over any of the sets:

(a) all mixture-morphisms $\mathscr{B}(\Sigma_Z) \xrightarrow{T} \mathscr{D}'$,

(b) all mixture-morphisms $\mathscr{B}(\Sigma_Z) \xrightarrow{T} \mathscr{B}$,

(c) all maps $C'(\hat{Z}) \to \operatorname{lin} K$ of the form

$$T(u) = \sum_{i=1}^{N} \langle u, f_i \rangle \, w_i$$

with $f_i \in C(\hat{Z})$, $f_i \geq 0$ and $\sum_{i=1}^{N} f_i = \mathbf{1}$ and $w_i \in K$; in this case $\sup_m$ can be taken over all $m \in K(\hat{Z})$. The last $T$ is also a special mixture-morphism $\mathscr{B}(\Sigma_Z) \xrightarrow{T} \mathscr{B}$.

(iii) For any positive linear map $C(\hat{Z}) \xrightarrow{R} \mathscr{D}$ with $R\mathbf{1} = \mathbf{1}$ and any $\varepsilon > 0$, there is another such map $\mathscr{B}'(\Sigma_Z) \xrightarrow{\tilde{R}} \mathscr{D}$ satisfying $\| Rf - \tilde{R}f \| < \varepsilon$ for all $f \in \Lambda$.

(iv) $\Delta(\hat{Z}, \mathscr{B}) = \inf_R \inf_T \sup_{m \in K(\Sigma_Z)} \sup_{f \in \lambda} |\langle m, f \rangle - \mu(Tm, Rf)|$, where $R$ is any positive linear map $\mathscr{B}'(\Sigma_Z) \xrightarrow{R} \mathscr{B}'$ with $R\mathbf{1} = \mathbf{1}$. This $\Delta(\hat{Z}, \mathscr{B})$ does not depend on the set $K(\Sigma_Z)$ as a norm separable and $\sigma(C'(\hat{Z}), C(\hat{Z}))$-dense subset of the base $K(\hat{Z})$ of $C'(\hat{Z})$ (therefore our notation in D 3.2.2 was meaningful).

*Proof.* (i) $R(d(z, \cdot) - d(z', \cdot)) \leq R(d(z, z')\,\mathbf{1}) \leq d(z, z')\,\mathbf{1}$; hence $Rd(z', \cdot) \geq Rd(z', \cdot) \geq Rd(z, \cdot) - d(z, z')\,\mathbf{1} \geq (\eta(R, z) - d(z, z'))\,\mathbf{1}$. Consequently $\eta(R, z') \geq \eta(R, z) - d(z, z')$ and by symmetry $|\eta(R, z) - \eta(R, z')| \leq d(z, z')$.

(ii) Let $J_\alpha$ ($\alpha = a, b, c$) be the infimum over $T$ in the set $\alpha$. We begin by showing $\eta(R) \leq J_a$. Let $z_0 \in \hat{Z}$ be a point where $\eta(R, \cdot)$ takes its maximum $\eta(\mathbf{R})$. Define $f_0 = d(z_0, \cdot)$ and choose $m_\varepsilon \in K(\Sigma_Z)$ such that $\langle m_\varepsilon, f_0 \rangle \leq \varepsilon$. This is possible since $m_\varepsilon(\sigma) = m_0(\sigma_0^{-1})\, m_0(\sigma_0 \wedge \sigma)$, with $m_0$ an effective measure and $\sigma_0$ corresponding to the open set $\{z \mid d(z_0, z) < \varepsilon\}$, is such an $m_\varepsilon$. Then

$$\begin{aligned} J_a &= \inf_{T \in (a)} \sup_{m \in K(\Sigma_Z)} \sup_{f \in \Lambda} |\langle m, f \rangle - \mu(Tm, Rf)| \\ &\geq \inf_{T \in (a)} |\langle m_\varepsilon, f_0 \rangle - \mu(Tm_\varepsilon, Rf_0)| \geq \inf_{w \in \bar{K}^\sigma} (\mu(w, Rd(z_0, \cdot)) - \varepsilon) \\ &= \eta(R, z_0) - \varepsilon = \eta(R) - \varepsilon. \end{aligned}$$

By trivial inclusions we get $\eta(R) \leq J_a \leq J_b \leq J_c$, so that only $J_c \leq \eta(R)$ remains to be proven. Let $\varepsilon > 0$ and pick a set $\{z_i\} \subset Z$ ($i = 1, \ldots, N$) such that for every $z \in \hat{Z}$ there is a $z_i$ with $d(z, z_i) \leq \varepsilon/2$.

With $g_i$ as the positive part of $\varepsilon - d(z_i, \cdot)$ we have $\sum_{i=1}^{N} g_i(z) \geq \varepsilon/2 > 0$. Consequently the functions $f_i = g_i \left( \sum_{k=1}^{N} g_k \right)^{-1}$ are continuous and satisfy $f_i \geq 0$, $\sum_{i=1}^{N} f_i = \mathbf{1}$, $f_i(z) \neq 0 \Rightarrow d(z_i, z) \leq \varepsilon$. By the definition of $\eta(R, z_i)$, and because $K$ is $\sigma(\mathscr{D}', \mathscr{D})$-dense in $\bar{K}^\sigma$, we can find elements $w_i \in K$ such that $\mu(w_i, Rd(z_i, \cdot)) \leq \eta(R, z_i) + \varepsilon$.

Defining $S$ as in (ii) with (c), for any $u \in \bar{K}^\sigma(\Sigma_Z) = K(\hat{Z})$ and $f \in \Lambda$ we have

$$\begin{aligned}
|\langle u, f\rangle - \mu(Tu, Rf)| &= \left|\left\langle u, f - \sum_{i=1}^{N} f_i\, \mu(w_i, Rf)\right\rangle\right| \\
&\leq \sup_{z \in \hat{Z}} \left| \sum_{i=1}^{N} f_i(z)[f(z) - \mu(w_i, Rf)]\right| \\
&\leq \sup_{z \in \hat{Z}} \sum_{i=1}^{N} f_i(z)\, |\mu(w_i, R(f(z)\mathbf{1} - f))| \\
&\leq \sup_{z \in \hat{Z}} \sum_{i=1}^{N} f_i(z)\, \mu(w_i, R(|f(z)\mathbf{1} - f|)) \\
&\leq \sup_{z \in \hat{Z}} \sum_{i=1}^{N} f_i(z)\, \mu(w_i, R d(z, \cdot)) \\
&\leq \sup_{z \in \hat{Z}} \sum_{i=1}^{N} f_i(z)[\mu(w_i, R d(z_i, \cdot)) + d(z, z_i)] \\
&\leq \sup_{z \in \hat{Z}} \sum_{i=1}^{N} f_i(z)\, \eta(R, z_i) + \varepsilon + \sup_{z \in \hat{Z}} \sum_{i=1}^{N} f_i(z)\, d(z, z_i) \\
&\leq \eta(R) + 2\varepsilon
\end{aligned}$$

since $d(z, z_i) \leq \varepsilon$ for $f_i(z) \neq 0$ and $\eta(R, z_i) \leq \sup_{z \in \hat{Z}} \eta(R, z)$.

(iii) It suffices to construct a map $\mathscr{B}'(\Sigma_Z) \xrightarrow{P} C(\hat{Z}) = \mathscr{D}(\Sigma_Z)$ with $\| Pf - f \| < \varepsilon$ for all $f \in \Lambda$. Then $\tilde{R} = RP$ has the desired property. For the construction of $P$ we use the $f_i$ from the proof of (ii) with (c). Let $\sigma_i$ be the support of $f_i$, and $m_{0\sigma_i}$ the measure $m_0(\sigma_i)^{-1}\, m_0(\sigma \wedge \sigma_i)$ for an effective $m_0$. Then

$$Pf = \sum_i f_i \langle m_{0\sigma_i}, f\rangle \in C(\hat{Z})$$

and for $f \in \Lambda$:

$$\begin{aligned}
\| Pf - f \| &= \sup_{z \in \hat{Z}} \left|\sum_i f_i(z)[\langle m_{0\sigma_i}, f\rangle - f(z)]\right. \\
&\leq \sup_z \sum_i f_i(z) \langle m_{0\sigma_i}, |f - f(z)\, 1|\rangle \\
&\leq \sup_z \sum_i f_i(z) \langle m_{0\sigma_i}, d(z, \cdot)\rangle \\
&\leq \sup_z \sum_i f_i(z) [m_{0\sigma_i}, d(z_i, \cdot)\rangle + d(z, z_i)] \leq 2\varepsilon
\end{aligned}$$

since $d(z_i, z) \leq \varepsilon$ for $z \in \sigma_i$ ($z$ in the support of $f_i$).

(iv) First we shall prove that for any $R$ there is an $S$ with $\| S' f - Rf \| < \varepsilon$ for all $f \in \Lambda$. According to the proof of T 3.2.1 it suffices to prove this for finitely many step functionals $f_\nu = \sum_n \alpha_n^\nu\, l_{\sigma_n}$. We define

$$Sw = \sum_n \mu(w, R l_{\sigma_n})\, m_{0\sigma_n}$$

with $m_{0\sigma_n}$ as above. Then we get

$$S' f = \sum_n R l_{\sigma_n} \langle m_{0\sigma_n}, f \rangle,$$

which for $f = f_\nu$ implies $S' f_\nu = R f_\nu$.

For $\mathscr{D} = \mathscr{B}'$, (iii) and (ii) imply that $\sup_R \inf_T \sup_m \{\ldots\}$ can be taken either over all positive linear maps $\mathscr{B}'(\Sigma_Z) \xrightarrow{R} \mathscr{B}'$ with $R\mathbf{1} = \mathbf{1}$ and all mixture-morphisms $\mathscr{B}(\Sigma_z) \xrightarrow{T} \mathscr{B}$ and $m \in K(\Sigma_Z)$ or over all positive linear maps $C(\hat{Z}) \xrightarrow{R} \mathscr{B}'$ with $R\mathbf{1} = \mathbf{1}$ and all $T$ of (ii) (c) and all $m \in K(\hat{Z})$. The last objects do not depend on $K(\Sigma_Z)$. $\square$

For $\hat{Z}$ with the metric $d$ and an $\varepsilon > 0$, we define the covering number $N_c(\varepsilon)$, the packing number $N_p(\varepsilon)$, and the Lipschitz partitioning number $N_L(\varepsilon)$. The covering number $N_c(\varepsilon)$ is the smallest number of balls of radius $\varepsilon$ which cover $\hat{Z}$ i.e.

$$N_c(\varepsilon) = \min\{N \mid \text{there are } N \text{ points } z_1, \ldots, z_N \text{ such that for every } z \in \hat{Z} \text{ there is a } z_i \text{ with } d(z, z_i) < \varepsilon\}.$$

The packing number is the biggest number of balls of radius $\varepsilon$ which can be packed into $\hat{Z}$ without overlaps, i.e.

$$N_p(\varepsilon) = \max\{N \mid \text{there are } N \text{ points } z_1, \ldots, z_N \text{ with } d(z_i, z_j) \geq 2\varepsilon \text{ for } i \neq j\}.$$

$N_L(\varepsilon)$ is defined by

$$N_L(\varepsilon) = \max\left\{N \mid \text{there are } N \text{ elements } f_1, \ldots, f_N \in L(\hat{Z}) \text{ with } \|f_i\| = 1, \sum_{i=1}^N f_i \leq 1 \text{ and } \varepsilon f_i \in \Lambda\right\}$$

with $\Lambda = \Lambda_m$ and $\Lambda_m$ in (3.13).

**T 3.2.3**

(i) $N_c(2\varepsilon) \leq N_p(\varepsilon) \leq N_c(\varepsilon)$.

(ii) $N_L(2\varepsilon) \leq N_p(\varepsilon) \leq N_L(\varepsilon)$.

(iii) For $\alpha = c, p, L$, the numbers $N_\alpha(\varepsilon)$ increase if $\varepsilon$ decreases and $\lim_{\varepsilon \to 0} N_\alpha(\varepsilon) = \operatorname{card} \hat{Z}$, which equals $N$ if $\hat{Z}$ has $N$ points and $\infty$ if $\hat{Z}$ has infinitely many points.

*Proof.* (i) Choose $z_1, \ldots, z_N$ as in the definition of $N_p(\varepsilon)$. Since $N = N_p(\varepsilon)$ is maximal, one cannot find any $z' \in \hat{Z}$ with $d(z', z_i) \geq 2\varepsilon$ for all $i$. That is, for every $z$ there is a $z_i$ with $d(z, z_i) < 2\varepsilon$ which is to say that the balls of radius $2\varepsilon$ around the $z_i$ cover $\hat{Z}$.

Suppose $N_p(\varepsilon) > N_c(\varepsilon)$. Choose $z_1, \ldots, z_{N_p}$ and $y_1, \ldots, y_{N_c}$ as in the definition of $N_p(\varepsilon)$ resp. $N_c(\varepsilon)$. Then to each $z_i$ there corresponds at least one $y_k$ with $d(z_i, y_k) < \varepsilon$. Since $N_p > N_c$, this correspondence cannot be injective, hence one can find $y_k, z_i, z_j$ $(i \neq j)$ such that $d(z_i, y_k) < \varepsilon$ and $d(y_k, z_j) < \varepsilon$. Thus $d(z_i, z_j) < 2\varepsilon$ in contrast to the definition of $N_p(\varepsilon)$.

(ii) Choose $f_1, \ldots, f_{N_L}$ as in the definition of $N_L(2\varepsilon)$. Since $\|f_i\| = 1$ and $\hat{Z}$ is compact, there are points $z_1, \ldots, z_{N_L}$ with $f_i(z_i) = 1$, hence $f_i(z_j) = \delta_{ij}$. Since $2\varepsilon f_i \in \Lambda$:

$$2\varepsilon = 2\varepsilon\,|f_i(z_i) - f_i(z_j)| \leq d(z_i, z_j) \qquad \text{for all } i \neq j.$$

Choose $z_1, \ldots, z_{N_p}$ as in the definition of $N_p(\varepsilon)$. Then the functions

$$f_i(z) \overset{\text{def}}{=} \tfrac{1}{2}\,|1-\varepsilon^{-1}\,d(z, z_i)| + \tfrac{1}{2}(1-\varepsilon^{-1}\,d(z, z_i))$$

have disjoint supports and satisfy the definition of $N_L(\varepsilon)$.

(iii) It is trivial that $N_\alpha(\varepsilon)$ increases with decreasing $\varepsilon$. Moreover $N_p(\varepsilon)$ is obviously bounded by card $\hat{Z}$. On the other hand, suppose that $\hat{Z}$ contains at least $N$ different points $z_1, \ldots, z_N$ and put $\varepsilon = \frac{1}{2} \min_{i \neq j} d(z_i, z_j)$. Then $N_p(\varepsilon) \geq N$. □

Let $\mathscr{H}_D$ denote a Hilbert space of the finite dimension $D$.

**T 3.2.4** If $D \geq N_c(\varepsilon)$ and $\delta > 0$, then

$$\delta\left(1-\frac{D}{N_L(\delta)}\right)_+ \leq \Delta(\hat{Z}, \mathscr{B}(\mathscr{H}_D)) \leq \varepsilon,$$

where $(\ldots)_+$ is the positive part.

*Proof* for the lower bound: Let $C(\hat{Z}) \xrightarrow{R} \mathscr{B}'(\mathscr{H}_D)$ as before. Let $N = N_L(\delta)$ and $f_1, \ldots, f_N$ as in the definition of $N_L(\delta)$ and $z_1, \ldots, z_N$ such that $f_i(z_j) = \delta_{ij}$. Since $\delta f_i \in \Lambda$, we have $f_i \geq (1-\delta^{-1}\,d(z_i, \cdot))_+$. Hence, with the notation of T 3.2.2:

$$\begin{aligned} D &= \operatorname{tr}(R(\mathbf{1})) \geq \operatorname{tr}(R \sum_i f_i) \\ &\geq \sum_i \operatorname{tr}(R(1-\delta^{-1}\,d(z_i, \cdot))_+) \\ &\geq \sum_i \| R((1-\delta^{-1}\,d(z_i, \cdot))_+) \| \\ &\geq \sum_i \sup_{w \in K} \mu(w, 1-\delta^{-1}\,R d(z_i, \cdot)) \\ &= \sum_i (1-\delta^{-1}\,\eta(R, z_i)) \geq N(1-\delta^{-1}\,\eta(R)). \end{aligned}$$

Hence $\Delta(\hat{Z}, \mathscr{B}(\mathscr{H}_D)) = \inf_R \eta(R) \geq \delta\left(1-\frac{D}{N}\right)$.

*Proof* for the upper bound: Let $N = N_c(\varepsilon) \leq D$ and $z_1, \ldots, z_N$ as in the definition of $N_c(\varepsilon)$. Define $R$ by $Rf = \sum_{i=1}^{N} f(z_i)\, P_i$, where $\{P_i\}$ is a family of orthogonal projections with $\sum_{i=1}^{N} P_i = \mathbf{1}$. Thus $z \in \hat{Z}$ gives

$$\eta(R, z) = \sup\left\{\lambda \,\middle|\, \sum_{i=1}^{N} d(z, z_i)\, P_i \geq \lambda\, \mathbf{1}\right\} = \min d(z, z_i) \leq \varepsilon$$

and hence

$$\Delta(\hat{Z}, \mathscr{B}(\mathscr{H}_D)) = \inf_{\tilde{R}} \eta(\tilde{R}) \leq \eta(R) = \sup_z \eta(R, z) \leq \varepsilon. \quad \square$$

The upper bound in this theorem can be interpreted as an existence theorem: If $N_c(\varepsilon) \leq D$, then an embedding of $Z$ into $\mathscr{B}(\mathscr{H}_D)$ with an $\varepsilon$-approximation, resp. with an $\varepsilon$-smearing exists. How should we interpret the lower bound?

It is fundamental for the structure of physical theories that infinite sets are mathematical idealizations which have no analogue in the real world. These idealizations must be "revoked" by imprecision sets (see XIII §2.5 and [3] §§6 and 9; and [40]). The introduction of a state space $Z$ where $Z_p$ is precompact, is an example for such an idealization. It must be "revoked" by introducing imprecision sets as elements of the uniform structure $p$ (see §3.1 and II §1). But at the beginning we have no physical arguments other than experience, how to select useful imprecision sets. Now we have a theoretical method to find such imprecision sets; in particular we found a non-zero (!) lower bound for the imprecision: The state space $\hat{Z}$ has only then a physical meaning when the imprecision is greater than given by the lower bound in T 3.2.4. Therefore we can denote that lower bound as a fundamental "macroscopic imprecision" or a "thermodynamic imprecision relation" as done in [1] XV §8.

In the sense of the smearing operator $Q$ of the embedding from §2.5, we can describe the same situation as follows.

$\hat{Z}$ is well defined, but we can neither measure nor prepare the states with arbitrary precision. We cannot measure the states more precisely, if $\varepsilon$ is so small that a repetition of $\varepsilon^{-1}$ experiments is impossible (as described in §3.1). It is impossible to prepare the states more precisely since only ensembles in $QK(\Sigma_Z)$ can be prepared.

In order to understand the magnitude of the bounds of T 3.2.4 more clearly, it may be useful to restate T 3.2.4 in terms of the inverse function $D \to \Delta(\hat{Z}, \mathscr{B}(\mathscr{H}_D))$: For $\varepsilon > 0$, let $D(\varepsilon)$ be the smallest dimension for which an $\varepsilon$-embedding of $\hat{Z}$ into $\mathscr{B}(\mathscr{H}_D)$ exists. Then T 3.2.4 yields

$$\tfrac{1}{2} N_L(2\varepsilon) \leq \sup_{\delta} \left(1 - \frac{\varepsilon}{\delta}\right) N_L(\delta) \leq D(\varepsilon) \leq N_c(\varepsilon).$$

Thus the behavior of $D(\varepsilon)$ is closely related to the behavior of the numbers $N_L$, $N_p$, $N_c$ studied in T 3.2.3.

**T 3.2.5** Let $\hat{Z}$ be a bounded region $\mathscr{V}$ in $\mathbf{R}^n$ of total volume $V$ and $d$ the metric in $\mathbf{R}^n$. Then asymptotically for large integers $D$ we get:

$$C_-(n) \left(\frac{V}{DV_n}\right)^{1/n} \leq \Delta(\hat{Z}, \mathscr{B}(\mathscr{H}_D)) \leq C_+(n) \left(\frac{V}{DV_n}\right)^{1/n},$$

where $V_n$ denotes the volume of the unit ball in $\mathbf{R}^n$, and $C_\pm(n)$ are constants depending only on $n$, satisfying $\frac{1}{2} \leq C_- \leq C_+ \leq 2$.

*Proof.* The calculation of $N_\alpha(\varepsilon)$ for $\varepsilon \to 0$ and $\alpha = p, L, c$ is reduced to determining the numbers

$$v_\alpha(n) = \lim_{\varepsilon \to 0} \frac{N_\alpha(\varepsilon) V_n \varepsilon^n}{V} \qquad (\alpha = p, L, c).$$

$v_p(n)$ and $v_c(n)$ have been studied extensively (see [51]). (The value of $v_p(n)$ is desired in Hilbert's 18th problem and still unknown for $n \geq 3$.) Of the estimates compiled in [51], we shall only use

$$v_p \geq \frac{\pi}{2\sqrt{3}}, \frac{\pi}{3\sqrt{2}}, \frac{\pi^2}{16}$$

and

$$v_c \leq \frac{2\pi}{2\sqrt{3}}, \frac{5\sqrt{5}\,\pi}{24}, \frac{3\pi^2}{5\sqrt{5}} \qquad \text{for } n=2,3,4;$$

and the asymptotic estimates

$$v_p(n) \geq 2^{1-n}; \qquad \frac{n}{e\sqrt{e}} \leq v_c(n) \leq n(\log n + \log\log n + 5). \tag{3.2.2}$$

The lower bounds for $v_p$ become bounds for $v_L$ by virtue of

$$v_L(n) \geq 2^{n/2}\, v_p(n). \tag{3.2.3}$$

This is proved as follows: Let $\{x_i\} \subset \mathbf{R}^n$ be a packing of $\varepsilon$-balls (i.e. $|x_i - x_j| \geq 2\varepsilon$); then the functions $f_i(x) = \left(1 - \frac{1}{\varepsilon\sqrt{2}}\,|x_i - x|\right)_+$ (+ means the positive part) satisfy $\varepsilon\sqrt{2}\, f_i \in \Lambda$ and $f_i \geq 0$. If they also satisfy $\sum_i f_i \leq 1$, we have $N_L(\sqrt{2}\,\varepsilon) \geq N_p(\varepsilon)$ by definition of $N_L$; thus the result follows. Define $g_i(x) = \left(1 - \frac{1}{2\varepsilon^2}\,|x - x_i|^2\right)_+$. Then $f_i \leq g_i$, and it suffices to prove $\sum_i g_i \leq 1$. The latter holds by the following elementary results:

Let $i'$ and $j'$ run over those $i$ which for fixed $x$ make $\left(1 - \frac{1}{2\varepsilon^2}\,|x - x_i|^2\right)_+ \neq 0$. From $|x_i - x_j|^2 = (x_i - x)^2 + (x_j - x)^2 - 2(x_i - x)\cdot(x_j - x)$ and $|x_i - x_j|^2 \geq 2\varepsilon$ for $i \neq j$ follows $(x_i - x)^2 + (x_j - x)^2 \geq 2(x_i - x)\cdot(x_j - x) + 4\varepsilon^2(1 - \delta_{ij})$. Summing this over $i'$ and $j'$ (let this be $N$ indices) we have

$$2N \sum_{i'} (x_{i'} - x)^2 \geq 2\Big(\sum_{i'} (x_{i'} - x)\Big)\cdot\Big(\sum_{j'} (x_{j'} - x)\Big) + 4\varepsilon^2 N(N-1) \geq 4\varepsilon^2 N(N-1),$$

consequently

$$\sum_i \left(1 - \frac{1}{2\varepsilon^2}\,|x_i - x|^2\right) \leq N - (N-1) = 1.$$

For the lower bound of the gap we need the bound $v_L(n) \leq 1 + n$. Choose $f_1, \ldots, f_N$ as in the definition of $N_L(\varepsilon)$ and assume $f_i(x_i) = 1$. Then we get $f_i(x) \geq (1 - \varepsilon^{-1}|x - x_i|)_+$, which with $\mathscr{V}_\varepsilon$ as the $\varepsilon$-neighborhood of $\mathscr{V}$ ($\mu$ = volume, i.e. $\mu(\mathscr{V}) = V$) yields

$$\mu(\mathscr{V}_\varepsilon) \geq \int_{\mathscr{V}_\varepsilon} d^n x \sum_i f_i \geq \sum_i \int_{\mathbf{R}^n} d^n x\,(1 - \varepsilon^{-1}|x - x_i|)_+$$

$$= N_L(\varepsilon)\,\frac{V_n\,\varepsilon^n}{1+n} \approx \frac{v_L(n)}{1+n}\,\mu(\mathscr{V}). \tag{3.2.4}$$

Given $v_L(n)$ and $v_c(n)$, the best bounds for $D(\varepsilon)$ take the form

$$\sup_\delta\left(1-\frac{\varepsilon}{\delta}\right)N_L(\delta)\approx\frac{V}{V_n\,\varepsilon^n}\sup_\delta\left(1-\frac{\varepsilon}{\delta}\right)\left(\frac{\varepsilon}{\delta}\right)^n v_L(n)=\frac{V}{V_n\,\varepsilon^n}v_-\leq D(\varepsilon)\leq\frac{V}{V_n\,\varepsilon^n}v_+$$

with $v_-=\dfrac{v_L(n)}{(1+n)\left(1+\frac{1}{n}\right)^n}$ and $v_+=v_c$. Taking inverse functions gives bounds of the desired type with

$$C_\pm(n)=(v_\pm(n))^{1/n}.$$

Combining these with the bounds in (3.2.2), (3.2.3), (3.2.4) yields the estimates stated in the theorem. □

At first an estimate of $D(\varepsilon)$ up to a small factor independent of $\varepsilon$ may appear to be fantastically accurate since in the applications $\log D$ is of the order of the particle number and thus very large. In this sense, the estimates only prove that an embedding is possible with fantastic precision. There are situations, however, in which the dimension of $\mathscr{H}$ is of an order for which the above estimates are relevant.

For example, in a theory of the Brownian motion the Hilbert space is of the form $\mathscr{H}_{\text{particle}}\times\mathscr{H}_{\text{bath}}$. The "particle" is "macroscopic" and $\mathscr{H}_{\text{particle}}$ is only the Hilbert space of its center of mass. The inner structure of the particle is not measured and shall not change macroscopically. The macroscopic measurements are in this sense only the measurements of the position of the particle. The embedding should therefore only proceed into $\mathscr{B}(\mathscr{H}_{\text{particle}})$. Since the particle is not "free", the energy of its motion is neither a macroscopic observable nor a constant of the motion. As finite dimensional subspace of $\mathscr{H}_{\text{particle}}$, we therefore must not take an energy shell but rather the space where the energy is smaller than a fixed constant.

As state space $Z$ we can take a bounded region of $\mathbf{R}^3$ (with the Euclidean metric). The Hamiltonian of the particle (i.e. of the center of mass motion) is $H=\dfrac{1}{2m}\boldsymbol{P}^2$, with $m$ the mass of the particle. As Hilbert space we take the subspace $\mathscr{H}_E$ where the eigenvalues of $H$ are smaller than $E$. The de Broglie wavelength of a free particle with energy $E$ is $\lambda_E=2\pi(2mE)^{-1/2}$.

For sufficiently large $E$ we get

$$0.23\leq\frac{\Delta(\hat{Z},\mathscr{B}(\mathscr{H}_E))}{\lambda_E}\leq 0.44. \tag{3.2.5}$$

We may derive (3.2.5) for the general case of $\mathbf{R}^n$ with $H=\dfrac{1}{2m}\Delta_n$; then (3.2.5) takes the general form

$$\Delta(\hat{Z},\mathscr{B}(\mathscr{H}_E))\sim C\frac{n\lambda_E}{2\pi e}\quad\text{with}\quad\frac{1}{\sqrt{2}}\leq C\leq 1.$$

This result merely combines T 3.2.5 with Weyl's asymptotic formula $N(\eta)\sim V\dfrac{V_n}{(2\pi)^n}\eta^{n/2}$ for the number of eigenvalues of the Laplacian below $\eta$ (see e.g. [52]

XIII, Theorem 78). Thus T 3.2.5 implies

$$\Delta(\hat{Z}, \mathscr{B}(\mathscr{H}_E)) \sim C\left(\frac{V}{V_n \dim \mathscr{H}_E}\right)^{1/n} \sim C V_n^{-2/n} \lambda_E \quad \text{with} \quad C \in [C_+(n), C_-(n)].$$

The energy $E$ has the magnitude of the temperature. If $m$ is high enough (greatly exceeding the mass of an atom), the bounds in (3.2.5) are very small.

From the theory of Brownian motion, we know that the bath makes the statistics of the trajectories of the form: "without memory". This is not true for a "free" particle. Nevertheless an embedding at "time zero" with the same bounds (3.2.5) is possible also for a free particle. The embedding at time zero cannot show whether the dynamics is without memory.

In §3.4 we will give an explicit form of the morphisms $T$ and $S$ for an embedding of a free particle, described in the 6-dimensional $\mu$-space (position-momentum space) as state space including the dynamics. This embedding provides a dynamics without memory. We shall also see that the embedding in the 3-dimensional position space as "reduced" state space (see §2.6) cannot yield a dynamics without memory. Since in §3.4 we shall describe the embedding in the $\mu$-space in an explicit form, for this case let us here not evaluate the bounds according to T 3.2.5.

The elements of the macroscopic state space $Z$ are frequently "fields", i.e. there are two spaces $\mathbf{R}^{n_1}$ and $\mathbf{R}^{n_2}$, and $Z = C(\mathbf{R}^{n_1}, \mathbf{R}^{n_2})$. Such a space is for instance that of the fields $(\boldsymbol{u}(\boldsymbol{r}), \rho(\boldsymbol{r}), T(\boldsymbol{r}))$ for hydrodynamics, where $\boldsymbol{u}$ is the velocity, $\rho$ the mass density and $T$ the temperature. Here the fields form a function $\mathbf{R}^3 \to \mathbf{R}^5$. Another example is the state space of the Boltzmann distribution functions $f(\boldsymbol{p}, \boldsymbol{q})$ which form functions $\mathbf{R}^6 \to \mathbf{R}$. These distribution functions can also be represented mathematically as measures (see below and §3.5).

We will consider as state space $Z$ only this last example; i.e. the base $K(X)$ of $C'(X)$ with the metric

$$d(m_1, m_2) = \sup_{f \in \Lambda(X)} |\langle m_1 - m_2, f \rangle|.$$

Here $X$ may be a compact space with the metric $d$. Then $K(X)$ is compact (see T 3.1.2 and the remarks after T 3.1.2), i.e. $Z = \hat{Z}$.

**T 3.2.6** Let $X$, $d$ and $K(X)$ be as above. Also choose $\varepsilon > 0$ and arbitrary integers $M$ and $N$. Then we find

(i) Let $X_M$ be the special space (for $X$) of $M$ points with the discrete metric and $S(M) \overset{\text{def}}{=} K(X_M)$ the $(M-1)$-dimensional simplex.

Then

$$\log N_p(S(M), \varepsilon) \geq (M-1)\left\{\log \frac{1}{\varepsilon} - \log 8\right\},$$

$$\log N_c(S(M), \varepsilon) \leq (M-1)\left\{\log \frac{1}{\varepsilon} + 1 - \log 4 + 4\varepsilon\right\}.$$

(ii)

$$\log N_p(K(X), \varepsilon_1 \varepsilon_2) \geq \log N_p(S(N_p(X, \varepsilon_1)), \varepsilon_2),$$
$$\log N_c(K(X), \varepsilon_1 + \varepsilon_2) \leq \log N_c(S(N_c(X, \varepsilon_1)), \varepsilon_2).$$

*Proof.* (i) We can identify $S(M)$ with $\{x \mid x \in \mathbf{R}^M,\ x_i \geq 0,\ \sum_i x_i = 1\}$. Then the metric is given by

$$d(x, y) = \max\{|\sum_i (x_i - y_i) f_i| \mid 0 \leq f_i \leq 1\} = \tfrac{1}{2} \sum_i |x_i - y_i|$$

(since $\sum_i (x_i - y_i) = 0$). The $\varepsilon$-ball around $x \in S(M)$ can be written $(x + \varepsilon DS(M))$ with $DS(M) = \{x - y \mid x, y \in S(M)\}$. Let $\mu$ be the Lebesgue measure in the hyperplane $\sum_i x_i = 1$ and suppose that the $\varepsilon$-balls at $\{x_j\}_{j=1}^N$ cover $S(M)$. Then

$$N \varepsilon^{M-1} \mu(DS(M)) \geq \mu(S(M)).$$

Since the relation

$$\mu(DS(M)) = \binom{2(M-1)}{M-1} \mu(S(M))$$

characterizes the simplex (Theorem 6.2 in [51]), we have

$$N_c(S(M), \varepsilon) \geq \varepsilon^{1-M} \frac{[(M-1!]^2}{(2M-2)!}$$

and

$$\log N_p(S(M), \varepsilon) \geq \log N_c(S(M), 2\varepsilon) \geq (M-1) \log \frac{1}{2\varepsilon} - (M-1) \log 4 + \frac{1}{2} \log(\pi M)$$

by Stirling's formula.

Let $K$ be an integer. We shall compute the largest distance $\delta(K)$ of points $\{p_i \in S(M)\,(i = 1, \ldots, M)\}$ to the points $\left\{\frac{K_i}{K} \middle| (i = 1, \ldots, M\right\}$ with integers $K_i$ and $\sum_i K_i = K$:

First find an $M$-tuple $\{K_i^0\}$ with $\frac{K_i^0}{K} \leq p_i \leq \frac{K_i^0 + 1}{K}$. Set $r_i = K p_i - K_i^0$ and $R = \sum_i r_i$, then $R$ is an integer with $0 \leq R \leq M$ and $K = \sum_i K_i^0 + R$. For any subset $\sigma \subset \{1, \ldots, M\}$ of cardinality $R$, the $M$-tuple $\{K_i^\sigma\}$ with $K_i^\sigma = K_i^0$ for $i \notin \sigma$ and $K_i^\sigma = K_i^0 + 1$ for $i \in \sigma$ is normalized to $K$. Moreover,

$$p = \{p_i\} \quad \text{and} \quad \frac{K^\sigma}{K} = \left\{\frac{K_i^\sigma}{K}\right\}$$

yield

$$d\left(p, \frac{K^\sigma}{K}\right) = \frac{1}{2} \sum_i \left| p_i - \frac{K_i^\sigma}{K} \right| = \frac{1}{2K} \left(\sum_{i \in \sigma} r_i - \sum_{i \notin \sigma} (1 - r_i)\right) = \frac{1}{K} \sum_{i \notin \sigma} r_i.$$

By a proper choice of $\sigma$, this $d$ can be made smaller than

$$\frac{1}{K}(M - R) \frac{R}{M} = \frac{M}{K}\left(1 - \frac{R}{M}\right) \frac{R}{M} \leq \frac{M}{4K}.$$

The number of $M$-tuples $\{K_i\}$ with $\sum_{i=1}^{M} K_i = K$ is equal to the number of combinations of the order $K$ of $M$ elements, i.e. $\binom{K + M - 1}{K}$. Hence

$$\log N_c\left(S(M), \frac{M}{4K}\right) < \log \binom{K + M + 1}{K} \leq (M-1)\left\{\log \frac{K}{M-1} + 1 + \frac{M-1}{K}\right\}.$$

ii) Let $x_1, \ldots, x_N \in X$ with $N = N_p(X, \varepsilon_1)$ be an $\varepsilon_1$-packing and $F \subset S(N)$ the set of centers of an $\varepsilon_2$-packing in $S(N)$. With $|F|$ as the cardinality of $F$ we have $|F| = N_p(S(N), \varepsilon_2)$. For $p \in F$ define $\mu_p \in K(X)$ by $\langle \mu_p, f \rangle = \sum_i p_i f(x_i)$. We show $d(\mu_p, \mu_{p'}) \geq 2\varepsilon_1 \varepsilon_2$ for all $p, p' \in F$ with $p \neq p'$. Since $F$ is a packing, there are $\alpha_i \in [0, 1]$ with $\sum_i (p_i - p'_i)\,\alpha_i \geq 2\varepsilon_2$. Let $f = \frac{1}{2} + \sum_i (\alpha_i - \frac{1}{2})(\varepsilon_1 - d(x_i, \cdot))_+$. Then $f \in \Lambda X$ and

$$\langle \mu_p - \mu_{p'}, f \rangle = \sum_{i,j} (p_i - p'_i)(\alpha_j - \tfrac{1}{2})(\varepsilon_1 - d(x_i, x_j))_+ = \varepsilon_1 \sum_{i,j} (p_i - p'_i)(\alpha_j - \tfrac{1}{2})\,\delta_{ij} \geq 2\varepsilon_1 \varepsilon_2 .$$

Now let $x_1, \ldots, x_N \in X$ with $N = N_c(X, \varepsilon_1)$ be an $\varepsilon_1$-covering and $F \subset S(N)$ the set of centers of an $\varepsilon_2$ covering of $S(N)$ with $|F| = N_c(S(N), \varepsilon_2)$. We must show that for every $\mu^* \in K(X)$ there is a $p \in F$ such that $|\langle \mu^* - \mu_p, f \rangle| \leq \varepsilon_1 + \varepsilon_2$ for all $f \in \Lambda(X)$.

Consider a partition $\{\sigma_i\}$ of $X$ with $x_i \in \sigma_i$ and $d(x, x_i) \leq \varepsilon$ for all $x \in \sigma_i$ and set $p_i^* = \langle \mu^*, \sigma_i \rangle = \mu^*(\sigma_i)$. Then

$$\langle \mu^* - \mu_{p^*}, f \rangle = \sum_i \int_{\sigma_i} (f(x) - f(x_i))\, d\mu^*(x) \leq \varepsilon_1 .$$

For $p^*$ we can find $p \in F$ with $d(p^*, p) \leq \varepsilon_2$, thus

$$\langle \mu_{p^*} - \mu_p, f \rangle = \sum_i (p_i^* - p_i)\, f(x_i) \leq \varepsilon_2 . \quad \square$$

Combining the last two results with T 3.2.4, for embeddings of $K(X)$ we obtain the estimates:

**T 3.2.7** Let $X$, $d$ be as before, let

$$N_p(X, \varepsilon) \geq A_p\, \varepsilon^{-n} \quad \text{and} \quad N_c(X, \varepsilon) \leq A_c\, \varepsilon^{-n},$$

and let $\mathcal{H}_D$ be a Hilbert space of dimension $D$ with $\log D \gg A_p$. Then we have

$$c_- \left( \frac{A_p}{\log D} \right)^{1/n} \leq \Delta(K(X), \mathcal{B}(\mathcal{H}_D)) \leq c_+ \left( \frac{A_c}{\log D} \right)^{1/n}$$

with $c_- = \frac{1}{8}(ne)^{-1/n}$ and $c_+ = 1{,}001 \left( 7 + \frac{1}{n} \log \frac{\log D}{A_c} \right)^{1/n}$.

*Proof* for the lower bound:

$$\log N_p(K(X), \varepsilon) \geq \log N_p \left( S(N_p(X, 8t\varepsilon)), \frac{1}{8t} \right) \geq (N_p(X, 8t\varepsilon) - 1) \log t$$

$$\geq A_p (8\varepsilon t)^{-n} \log t - \log t \qquad \text{for } 1 \leq 8t \leq \varepsilon^{-1}.$$

We may set $\log t = \frac{1}{n}$ to obtain

$$\log N_p(K(X), \varepsilon) \geq \frac{A_p}{8^n\, ne}\, \varepsilon^{-n}$$

where the small term $(-\log t)$ has been neglected. With the abbreviation

$$\bar{\varepsilon} = \left( \frac{A_p}{\log D \cdot 8^n\, ne} \right)^{1/n} \qquad \text{we have } \log N_p \geq \left( \frac{\bar{\varepsilon}}{\varepsilon} \right)^n \log D$$

hence T 3.2.4 yields

$$\Delta(K(X), \mathscr{B}(\mathscr{H}_D)) \geq \delta(1-\exp\{\log D - \log N_p(K(X), \delta)\})_+$$
$$\geq \delta\left\{1-\exp \log D\left(1-\left(\frac{\bar{\varepsilon}}{\delta}\right)^n\right)\right\}_+ .$$

This expression vanishes for $\delta \geq \bar{\varepsilon}$ and has a steep maximum for $\delta$ just below $\bar{\varepsilon}$. Setting $\left(\frac{\bar{\varepsilon}}{\varepsilon}\right)^n = 1 + \frac{\log\log D}{\log D}$ we obtain

$$\Delta(K(X), \mathscr{B}(\mathscr{H}_D)) \geq \bar{\varepsilon}\left(1+\frac{\log\log D}{\log D}\right)^{-1/n}\left(1-\frac{1}{\log D}\right).$$

Since $\log D$ was assumed large, we may set

$$\Delta(K(X), \mathscr{B}(\mathscr{H}_D)) \geq \bar{\varepsilon}.$$

*Proof* for the upper bound: Assume $\varepsilon$ so small that no $\varepsilon$-embedding of $K(X)$ into $\mathscr{H}_D$ exists. Then T 3.2.4 with $0<t<1$ implies

$$\log D \leq \log N_c(K(X), \varepsilon) \leq \log N_c(S(N_c(X,(1-t)\varepsilon), t\varepsilon)$$
$$\leq N_c(X,(1-t)\varepsilon)\left\{\log\frac{1}{t\varepsilon}+1-\log 4\right\} \leq A_c(1-t)^{-n}\varepsilon^{-n}\frac{1}{n}\log\left(\frac{e}{4t+\varepsilon}\right)^n.$$

Hence

$$\xi\log\xi \geq B \quad \text{with} \quad \xi = \left(\frac{e}{4t\varepsilon}\right)^n \quad \text{and} \quad B = \frac{n\log D}{A_c}\left(\frac{e(1-t)}{4t}\right)^n.$$

For sufficiently large $\log D$ we have $B \geq e$, hence $\xi \geq \frac{B}{\log B}$ because $\xi \to \xi\log\xi$ is monotonic. Solving this inequality for $\varepsilon$ yields

$$\varepsilon \leq \left(\frac{A_c}{\log D}\right)^{1/n}\frac{1}{1-t}\left\{\frac{1}{n}\log\frac{\log D}{A_c}+\frac{\log n}{n}+\log\frac{e(1-t)}{4t}\right\}^{1/n}$$
$$\leq \left(\frac{A_c}{\log D}\right)^{1/n} 1.001\left\{7+\frac{1}{n}\log\frac{\log D}{A_c}\right\}^{1/n} \qquad \text{for } t = 9\cdot 10^{-4}$$

and with $\frac{\log n}{n} \leq \frac{1}{e}$. Since $\Delta(\ldots)$ is the sup of all $\varepsilon$ for which no embedding exists, it must satisfy the same inequality. □

The most interesting example with a state space of the form $K(X)$ is Boltzmann's distribution. A Boltzmann distribution function $f(\boldsymbol{p}, \boldsymbol{r})$ shall by

$$N(\mathscr{V}) = \int_{\mathscr{V}} f(\boldsymbol{p}, \boldsymbol{r})\, d^3\boldsymbol{p}\, d^3\boldsymbol{r} \tag{3.2.6}$$

describe the "approximate number of particles" with $(\boldsymbol{p}, \boldsymbol{r})$ in the region $\mathscr{V}$ of the $\mu$-space. This can be given a better form: For a "smooth" function $g(\boldsymbol{p}, \boldsymbol{r})$,

$$\gamma = \int f(\boldsymbol{p}, \boldsymbol{r})\, g(\boldsymbol{p}, \boldsymbol{r})\, d^3\boldsymbol{p}\, d^3\boldsymbol{r}$$

for all practical purposes equals $\sum_{i=1}^{N} g(\boldsymbol{p}_i, \boldsymbol{r}_i)$, if the $N$ particles are distributed in the $\mu$-space as described by $f(\boldsymbol{p}, \boldsymbol{r})$ (see also §3.5). In this sense we may identify the Boltzmann distribution functions with the measures in $K(X)$, except that these measures are multiplied by $N$. Here $X$ must be a compactification of the $\mu$-space relative to a suitable metric.

The physically interesting gas of many atoms is enclosed in a finite region $X_1$ in space. Thus we may take $X = X_1 \times X_2$, where $X_1$ is compact with the usual Euclidean metric $|\boldsymbol{r}_1 - \boldsymbol{r}_2|$. To make the momentum space $X_2$ precompact, we there must choose another metric than the Euclidean one. With $\boldsymbol{e} = \boldsymbol{p}/|\boldsymbol{p}|$ one can introduce

$$\boldsymbol{p}' = \boldsymbol{e}\, \mathrm{tg}^{-1} \frac{|\boldsymbol{p}|}{p_m}$$

and (for instance) as metric in $X_2$ define

$$d(\boldsymbol{p}_1, \boldsymbol{p}_2) = |\boldsymbol{p}'_1 - \boldsymbol{p}'_2|.$$

Let us take another form, which is not so elegant but simpler to use. In $X_2$ we take a finite region $|\boldsymbol{p}| \leq p_{\max}$ and put

$$d(\boldsymbol{p}_1, \boldsymbol{r}_1; \boldsymbol{p}_2, \boldsymbol{r}_2) = \max\left\{\frac{|\boldsymbol{p}_1 - \boldsymbol{p}_2|}{p_m}, \frac{|\boldsymbol{r}_1 - \boldsymbol{r}_2|}{r_m}\right\}.$$

Here we might (for instance) choose $p_m = p_{\max}$ and $r_m$ as the diameter of the finite region $X_1$ in space. But we shall choose other fixed values for $p_m$ and $r_m$, such that an imprecision $d(\ldots) \leq \varepsilon$ does neither depend on $X_1$ nor on the temperature (i.e. on the energy) of the gas.

The set

$$\Lambda(X) = \{\varphi : X \to [0, 1] \mid |\varphi(x) - \varphi(x')| \leq d(x, x')\}$$

(similarly defined as (3.1.3)) induces a metric in $Z = K(X)$; it is again denoted by $d$ and makes $Z$ compact.

Provided the use of the asymptotic formulas for $N_\alpha(X_i, \varepsilon)$ is justified, i.e. if $(\varepsilon r_m)^3 \ll V$ and $\varepsilon p_m \ll p_{\max}$ (with $V$ the volume of the gas), we have

$$N_p(X, \varepsilon) \geq N_p(X_1, \varepsilon)\, N_p(X_2, \varepsilon) \geq \frac{3 v_p (3)^2}{4\pi} \left(\frac{p_{\max}}{p_m}\right)^3 \frac{V}{r_m^3} \varepsilon^{-6},$$

$$N_c(X, \varepsilon) \leq N_c(X_1, \varepsilon)\, N_c(X_2, \varepsilon) \leq \frac{3 v_c (3)^2}{4\pi} \left(\frac{p_{\max}}{p_m}\right)^3 \frac{V}{r_m^3} \varepsilon^{-6}.$$

We expect that the kinetic theory of an ideal gas of $N$ particles can be embedded into a quantum theory over the Hilbert space $\mathscr{H}$, if we restrict the particles to the finite region $X_1$ in space (by a suitable choice of the Hamiltonian) and restrict the energy to an interval (characterized by a temperature $T$). If $T$ is not too small, then $\log D$ ($D$ the dimension of $\mathscr{H}$) coincides with the thermostatic entropy given by the well known equation

$$\log D = N\left(\frac{5}{2} + \log\left\{\frac{V (2\pi)^3}{N} (2\pi m T)^{3/2}\right\}\right). \tag{3.2.7}$$

Here $T$ is measured in energy units, i.e. the Boltzmann constant is taken equal to one.

Let us apply T 3.2.7 to the above estimates for $N_\alpha(X, \varepsilon)$. The conditions $\varepsilon p_m \ll p_{\max}$ and $(\varepsilon r_m)^3 \ll V$ on which they depend can be checked in retrospect by setting $\varepsilon = \Delta(K(X), \mathscr{B}(\mathscr{H}))$. In order to control the term $\log \frac{\log D}{\log A_c}$ in the $c_+$ of T 3.2.7, $\frac{r_m p_m}{p_{\max}}$ must be bounded. For instance we may assume $p_m \leq p_{\max}$ and a fixed length $r_m$. Then we have

$$\Delta(K(X), \mathscr{B}(\mathscr{H}_D)) = C \left(\frac{p_{\max}}{p_m}\right)^{1/2} \left(\frac{V}{r_m^3 \log D}\right)^{1/6} \tag{3.2.8}$$

with $0.056 \leq C \leq 1.001 \left(7 + \frac{1}{6} \log \frac{\log D}{A_c}\right)^{1/6}$ and

$$A_c = \frac{3 v_c (3)^2}{4\pi} \left(\frac{p_{\max}}{p_m}\right)^3 \frac{V}{r_m^3}.$$

Taking the example of Helium at 0° C and atmospheric pressure, from (3.2.7) we get

$$\log D \approx 4.014 \cdot 10^{20} \frac{V}{\mathrm{cm}^3}. \tag{3.2.9}$$

Then we have

$$\frac{\log D}{A_c} = \left[\frac{3 v_c (3)^2}{4\pi} \left(\frac{p_{\max}}{p_m}\right)^3\right]^{-1} \left(\frac{r_m}{\mathrm{cm}}\right)^3 4.014 \cdot 10^{20}$$
$$\leq \left(\frac{3 v_c (3)^2}{4\pi}\right)^{-1} \left(\frac{r_m}{\mathrm{cm}}\right)^3 4.014 \cdot 10^{20}.$$

Taking $r_m \leq 10^{18}$ cm (obvious for physical reasons) we obtain

$$0.056 \leq C \leq 1.63.$$

Taking $\log D$ from (3.2.9) in (3.2.8), we thus get

$$\Delta(K(X), \mathscr{B}(\mathscr{H}_D)) = C \left(\frac{p_{\max}}{p_m}\right) \left(\frac{\mathrm{cm}}{r_m}\right)^{1/2} 3.68 \cdot 10^{-4}.$$

Putting (3.2.7) into (3.2.8), we see the very important fact that the bounds for $\Delta(\ldots)$ do not depend on the volume $V$ but only on the density $N/V$. This should be so if $\Delta(\ldots)$ specifies a macroscopic "imprecision" or "smearing".

If we put $\Delta(\ldots) = \varepsilon$, then (3.2.8) can also be read

$$(\varepsilon p_m)(\varepsilon r_m) \geq (0.056)^2 \, p_{\max} \left(\frac{V}{\log D}\right)^{1/3}.$$

Taking the values from above (for Helium with $N/N_0$ as the density relative to the density at atmospheric pressure) we obtain

$$(\varepsilon p_m)(\varepsilon r_m) \geq 1.1 \left(\frac{p_{\max}}{p_T}\right) \left(\frac{N}{N_0}\right)^{-1/3}$$

with $p_T = (3mT)^{1/2}$. If $p_{\max} \sim p_T$ and $N \sim N_0$, we have

$$(\varepsilon p_m)(\varepsilon r_m) \gtrsim 1.$$

This shows that the imprecision has the order of "Heisenberg's uncertainty relation", i.e. the imprecision is very small relative to the "macroscopic scales" $p_m$ and $r_m$.

$\Delta(\ldots)$ was defined as the smallest $\varepsilon$ for which an $\varepsilon$-embedding exists. In this definition no constraints were imposed (except on the energy). The actual possibilities for measurements on macroscopic systems, however, are not arbitrary. On the contrary, these possibilities are determined by the real interactions of the macrosystems with their surroundings. To formulate this problem of measurement possibilities seems to imply that it cannot be solved. This may be true, if one desires a general survey over "all" measurement possibilities. When one is content with less, it is possible to give special macroscopic observables for special systems. In this context one must stress that there can be more than one macroscopic observable (i.e. more than one $S$) for the same state space $Z$, since in general there is not only one measuring method for the states in $Z$. Therefore one must not be shocked that there is a certain arbitrariness in defining $S$, an arbitrariness within the imprecision of the description of the system in the state space $Z$.

To guess special macroscopic observables for special systems, one often succeeds by the intuition that macroscopic measurements of position, momentum, current density etc. resemble very imprecise measurements of the corresponding quantum mechanical observables. For example the streaming velocity of a fluid at a position $\boldsymbol{r}$ should have something to do with the quantum mechanical momentum of the atoms in a region around $\boldsymbol{r}$.

It was the ingenious discovery of Boltzmann, to introduce the "Boltzmann distribution function" as macroscopic observable. From this, we have only adopted the form $Z = K(X)$ of the state space, not yet using the significance of (3.2.6). This significance, however, will determine $S$ and $T$ within an arbitrariness given by the physical imprecision. Therefore the imprecision resp. the smearing by embedding can be greater than the above bounds. For the kinetic theory, in §3.5 we shall try to give such $S$ and $T$ which describe the meaning of the "distribution function".

## §3.3 The Embedding of the Time Evolution

As already mentioned at the beginning of §3, we have not succeeded to estimate the approximation to (2.5.38a), i.e. the difference between $A_\tau Q$ and $Q_\tau$. The main problem of such an approximation is that $A_\tau$ must be a halfgroup. We have given some general remarks to this problem in §2.3.

Our aim in this section is only to demonstrate that an approximation to (3.6) is not much worse than the approximation (3.3) discussed extensively in §3.2. Then the difference between $A_\tau Q$ and $Q_\tau$ cannot be worse than the errors in (3.6) and (3.3).

In II §1 we have given physical imprecision sets for trajectories, taking into account the imprecision in time. We can avoid this complication of time imprecision by presuming that the $A_\tau$ are "sufficiently" continuous. Therefore let us first contemplate continuity conditions for the time evolution operators $A_\tau$.

The quantum mechanical "time of evolution operator" $\mathcal{U}_\tau$ in (3.4) is given by the Hamiltonian in $\mathcal{P}\mathcal{T}_{q\,\mathrm{exp}}$. This Hamiltonian is a structure of nature and not at physicists' disposal. Nevertheless let us first prove that an imprecise embedding of the time evolution $A_\tau$ is possible for "suitable" Hamiltonians. Obviously, such an embedding for "suitable" Hamiltonians only proves that one cannot raise a "fundamental" objection to the embedding procedure. The real problem, to give an explicit form of $S$ and $T$ *and* to deduce from the "real" Hamiltonian the time evolution operator $A_\tau$ with its halfgroup property, is no yet soluble. Only more or less good approximations to the master equation operator $B$ in (2.5.14) have been deduced in special cases (see also §3.5).

For a dynamics without memory, in §2.5 we have presumed that the systems are dynamically continuous (II D3.4.1). We shall sharpen this to the assumption that $\langle A_\tau m, f\rangle$ is continuous in $\tau$ ($\tau \geq 0$) for all $m \in K(\Sigma_Z)$ and all $f \in L(\hat{Z})$ (see §3.1). In addition we presume that $A'_\tau$ transforms $L(\hat{Z})$ into itself, or equivalently that $A_\tau$ is $\sigma(\mathcal{B}(\Sigma_Z), \mathcal{D}(\Sigma_Z))$-continuous (for fixed $\tau$). Then $\langle m, A'_\tau f\rangle$ is continuous for each pair in $\mathcal{B}(\Sigma_Z) \times \mathcal{D}(\Sigma_Z)$.

**T 3.3.1** If $\tau \to \langle u, A'_\tau f\rangle$ is continuous for each pair $(u, f) \in \mathcal{B}(\Sigma_Z) \times \mathcal{D}(\Sigma_Z)$, then the mapping $\tau \to A_\tau u$ is norm-continuous for each $u \in \mathcal{B}(\Sigma_Z)$. If $\tau \to A_\tau u$ is norm-continuous for each $u \in \mathcal{B}(\Sigma_Z)$, then $\tau \to A'_\tau f$ is $\sigma(\mathcal{B}'(\Sigma_Z), \mathcal{B}(\Sigma_Z))$-continuous for all $f \in \mathcal{B}'(\Sigma_Z)$. When $\tau \to \langle u, A'_\tau f\rangle$ is continuous for all $(u, f) \in \mathcal{B}(\Sigma_Z) \times \mathcal{D}(\Sigma_Z)$, then it is continuous for all $(u, f) \in \mathcal{B}(\Sigma_Z) \times \mathcal{B}'(\Sigma_Z)$. If $\tau \to \langle u, A'_\tau f\rangle$ is continuous for all $(u, f) \in \mathcal{B}(\Sigma_Z) \times \mathcal{D}(\Sigma_Z)$, then $\tau \to A'_\tau f$ is norm-continuous for all $f \in \mathcal{D}(\Sigma_Z)$. If $\tau \to A'_\tau f$ is norm-continuous for all $f \in \mathcal{D}(\Sigma_Z)$, then $\tau \to A_\tau u$ is $\sigma(\mathcal{D}'(\Sigma_Z), \mathcal{D}(\Sigma_Z))$-continuous for all $u \in \mathcal{D}'(\Sigma_Z)$. If $\tau \to \langle u, A'_\tau f\rangle$ is continuous for all $(u, f) \in \mathcal{B}(\Sigma_Z) \times \mathcal{D}(\Sigma_Z)$, then it is continuous for all $(u, f) \in \mathcal{D}'(\Sigma_Z) \times \mathcal{D}(\Sigma_Z)$.

The proof will not be given, since it is similar to that of the theorems VI Th. 1.2.1 and 1.2.6 in [2].

The assumption that $(\tau, f) \to A'_\tau f$ is continuous is sufficient to prove the following theorem, which is an extension of T 3.2.1.

**T 3.3.2** Let $\Lambda_m$ be as in § 3.1, $\theta$ an interval $[0, \mathrm{T}]$ and $\varepsilon > 0$. Then there are a finite dimensional Hilbert space $\mathcal{H}$, mixture morphisms $\mathcal{B}(\mathcal{H}) \xrightarrow{S} \mathcal{B}(\Sigma_Z)$, $\mathcal{B}(\Sigma_Z) \xrightarrow{T} \mathcal{B}(\mathcal{H})$, and a continuous unitary group $U_\tau$ on $\mathcal{H}$, such that

$$|\langle A_\tau m, f\rangle - \langle S[U_\tau^+ (Tm) U_\tau], f\rangle| \leq \varepsilon$$

for all $m \in K(\Sigma_Z)$, all $f \in \Lambda_m$ and all $\tau \in \theta$.

The proof is very similar to that of T 3.2.1. In essence one must replace the set $\Lambda_m$ by $\Lambda_\theta = \bigcup_{\tau \in \theta} A'_\tau \Lambda_m$. We have $\Lambda_m \subset \Lambda_\theta$, while $\Lambda_\theta$ is as norm compact as $\Lambda_m$ since $\Lambda_\theta$ is the continuous image of the compact set $\theta \times \Lambda_m$. We do not give the whole proof here (see [44]), since below we will prove T 3.3.5 which makes a more precise statement about $\varepsilon$, $\theta$ and the dimension of $\mathcal{H}$. To this purpose we define

**D 3.3.1**

$$d_\theta(z,z') = \sup_{f\in\Lambda_m} \sup_{\tau\in\theta} |(A'_\tau f)(z) - (A'_\tau f)(z')|,$$

$$m(\tau) = \sup_{f\in\Lambda_m} \sup_{z\in Z} \sup_{\substack{\tau_1,\tau_2\in\theta\\ |\tau_1-\tau_2|\le\tau}} |(A'_{\tau_1} f)(z) - (A'_{\tau_2} f)(z)|.$$

**T 3.3.3** $d_\theta$ is a metric in $\hat{Z}$, uniformly equivalent to $d$ and $d_\theta \ge d$.

*Proof.* $d_\theta \ge d$ is obvious. The uniformity induced by $d_\theta$ is equivalent to the initial uniformity induced by the set $\Lambda_\theta$ as mapping $\hat{Z} \to \mathbf{R}$, since $\Lambda_\theta$ is norm compact. Since each $A'_\tau f$ is uniformly continuous on $\hat{Z}$ with $d$ as metric, $d$ and $d_\theta$ are uniformly equivalent. □

**T 3.3.4** $\lim_{\tau\to 0} m(\tau) = 0$.

*Proof.* We define a mapping $C(\hat{Z}) \times \hat{Z} \xrightarrow{\tilde{A}} C(\theta)$ by $[\tilde{A}(f,z)](\tau) = (A'_\tau f)(z)$; this $\tilde{A}$ is continuous in the sense

$$\|\tilde{A}(f,z) - \tilde{A}(f',z')\| \le \|f - f'\| + d_\theta(z,z').$$

Hence $\tilde{A}(\Lambda_m \times \hat{Z}) \subset C(\theta)$ is compact and thus its elements are equicontinuous by the Arzéla-Ascoli theorem. □

**T 3.3.5** Let $\hat{Z}$, $d$, $A_\tau$ and $\theta$ be given as above. Let $D_1$, $D_2$ be integers and $\varepsilon_1, \varepsilon_2 > 0$ such that

(1) $D_1$ balls of $d_\theta$-radius $\varepsilon_1$ cover $\hat{Z}$ and

(2) $(1+\pi)\, m\left(\frac{\mathbf{T}}{\pi} D_2^{-1/2}\right) \le \varepsilon_2$ (with $m(\tau)$ as in D 3.3.1).

Then there are a Hilbert space $\mathscr{H}$ with dimension $D_1 \cdot D_2$, a continuous unitary group $\{U_\tau\}$ on $\mathscr{H}$, and mixture-morphisms $\mathscr{B}(\mathscr{H}) \xrightarrow{S} \mathscr{B}(\Sigma_Z)$, $\mathscr{B}(\Sigma_Z) \xrightarrow{T} \mathscr{B}(\mathscr{H})$, such that

$$|\langle A_\tau m, f\rangle - \langle S[U_\tau^+ (Tm)\, U_\tau], f\rangle| \le 2\varepsilon_1 + \varepsilon_2$$

for all $m \in K(\Sigma_Z)$, all $f \in \Lambda_m$ and all $\tau \in \theta$.

*Proof.* Let $z_1, \dots, z_{D_1} \in \hat{Z}$ be centers of the covering and let $\{\sigma_i\}$ $(i = 1, \dots, D_1)$ be a partition of $\hat{Z}$ such that $d_\theta(z, z_i) \le \varepsilon_1$ for $z \in \sigma_i$. Then for all $m \in K(\Sigma_Z), f \in \Lambda_m$, $t \in \theta$ we get

$$|\langle m, A'_\tau f\rangle - \sum_i m(\sigma_i)(A'_\tau f)(z_i)| = |\sum_i m(\sigma_i)[\langle m_{\sigma_i}, A'_\tau f\rangle - (A'_\tau f)(z_i)]|$$

$$\le \sum_i m(\sigma_i)\, |\langle m_{\sigma_i}, d_\theta(\cdot, z_i)\rangle| \le \varepsilon_1 \sum_i m(\sigma_i) = \varepsilon_1,$$

where $m_{\sigma_i}(\sigma) = m(\sigma_i)^{-1}\, m(\sigma_i \cap \sigma)$.

We shall now construct a Hilbert space $\mathscr{H}_2$ with dimension $D_2$, a unitary group $\tilde{U}_\tau$, a $W_0 \in K(\mathscr{H}_2)$, and a map $C(\theta) \xrightarrow{\tilde{S}} \mathscr{B}(\mathscr{H}_2)$, such that $\varphi \in C(\theta)$ and $\tau \in \theta$ make

$$|\varphi(\tau)-\operatorname{tr}(W_0\,\tilde{U}_\tau(\tilde{S}\varphi)\,\tilde{U}_\tau^+)|\leq(1+\pi)\,m\left(\varphi,\frac{\mathbf{T}}{\pi}D_2^{-1}\right),$$

where $m(\varphi,\tau)=\sup\{|\varphi(\tau_1)-\varphi(\tau_2)|\,|\,\tau_1,\tau_2\in\theta,|\tau_1-\tau_2|<\tau\}$.

Consider the function $\tilde{\varphi}$ defined in $(-\mathbf{T},+\mathbf{T})$ by

$$\tilde{\varphi}(t)=\varphi(|t|)\qquad\text{for }|t|\leq\mathbf{T}.$$

Let $|v\rangle$ $(v=1,\ldots,D_2)$ be a basis in $\mathscr{H}_2$. We define $\tilde{U}_\tau$ by

$$\tilde{U}_\tau|v\rangle=e^{\pi i v\frac{\tau}{\mathbf{T}}}|v\rangle$$

and $W_0=P_\psi$ with $\langle\psi|v\rangle=D_2^{-1/2}$. Defining

$$\tilde{S}\varphi=\frac{D_2}{2\mathbf{T}}\int_{-\mathbf{T}}^{+\mathbf{T}}\tilde{\varphi}(t)\,\tilde{U}_t^+\,P_\psi\,\tilde{U}_\tau\,dt,$$

we find $\tilde{S}\varphi\geq0$ for $\varphi\geq0$ and

$$\langle v|\tilde{S}\,\mathbf{1})\,\mu\rangle=\frac{D_2}{2\mathbf{T}}\int_{-\mathbf{T}}^{+\mathbf{T}}e^{\pi i v\frac{t}{\mathbf{T}}}D_2^{-1}\,dt=\delta_{v\mu},$$

i.e. $\tilde{S}\,\mathbf{1}=\mathbf{1}$. We have

$$\begin{aligned}\operatorname{tr}(W_0\,\tilde{U}_\tau(\tilde{S}\varphi)\,\tilde{U}_\tau^+)&=\frac{D_2}{2\mathbf{T}}\int_{-\mathbf{T}}^{+\mathbf{T}}\tilde{\varphi}(\tau')\operatorname{tr}(P_\psi\,\tilde{U}_\tau\,\tilde{U}_{\tau'}^+\,P_\psi\,\tilde{U}_{\tau'}\,\tilde{U}_\tau^+)\,d\tau'\\&=(\tilde{\varphi}\circ k)(\tau)=\frac{1}{2\mathbf{T}}\int_{-\mathbf{T}}^{+\mathbf{T}}\tilde{\varphi}(\tau')\,k(\tau-\tau')\,d\tau',\end{aligned}$$

where $\circ$ denotes the convolution and $k$ the kernel

$$\begin{aligned}k(t)&=D_2\operatorname{tr}(P_\psi\,\tilde{U}_t\,P_\psi\,\tilde{U}_t^+)=D_2\,|\langle\psi,\tilde{U}_t\,\psi\rangle|^2\\&=D_2^{-1}\left|\sum_{v=1}^{D_2}e^{\pi i v\frac{t}{\mathbf{T}}}\right|^2=D_2^{-1}\left(\frac{\sin\frac{\pi}{2}D_2\frac{t}{\mathbf{T}}}{\sin\frac{\pi}{2}\frac{t}{\mathbf{T}}}\right)^2.\end{aligned}$$

This is called a Fejér-kernel, for which the estimate

$$\|\tilde{\varphi}-\tilde{\varphi}\circ k\|\leq(1+\pi)\,m\left(\varphi,\frac{\mathbf{T}}{\pi}D_2^{-1/2}\right)$$

can be found in [53] (section 2.4 and 2.5.1).

Let $\mathscr{H}_1$ be a Hilbert space with dimension $D_1$, and $P_i$ $(i=1,\ldots,D_1)$ a family of one-dimensional orthogonal projections. Let $\mathscr{H}=\mathscr{H}_1\times\mathscr{H}_2$, $U_\tau=1\times\tilde{U}_\tau$ and

$$Tm=\sum_{i=1}^{D_1}m(\sigma_i)\,P_i\times P_\psi,$$

$$Sw=\frac{D_2}{2\mathbf{T}}\sum_{i=1}^{D_1}\int_0^{\mathbf{T}}A_t\,m_{0\sigma_i}\operatorname{tr}(w[P_i\times(\tilde{U}_t^+\,P_\psi\,\tilde{U}_t+\tilde{U}_t\,P_\psi\,\tilde{U}_t^+)])\,dt,$$

where $m_{0\sigma_i}(\sigma)=m_0(\sigma_i)^{-1}\,m_0(\sigma_i\cap\sigma)$ for an effective measure $m_0$.

Then we get

$$
\begin{aligned}
|\langle A_\tau m, f\rangle - \langle S[U_\tau^+ (Tm)\, U_\tau]; f\rangle|
&\leq |\langle m, A'_\tau f\rangle - \sum_i m(\sigma_i)(A'_\tau f)(z_i)| \\
&\quad + \left|\sum_i m(\sigma_i)(A'_\tau f)(z_i) - \frac{D_2}{2\mathbf{T}} \int_{-\mathbf{T}}^{+\mathbf{T}} \sum_i m(\sigma_i)\langle m_{0\sigma_i}, A'_\tau f\rangle \right. \\
&\quad \left. \cdot \operatorname{tr}_2(\tilde{U}_\tau^+ P_\psi \tilde{U}_\tau \tilde{U}_t^+ P_\psi \tilde{U}_t)\, dt \right| \\
&\leq |\sum_i m(\sigma_i)[\langle m_{\sigma_i}, A'_\tau f\rangle - (A'_\tau f)(z_i)]| \\
&\quad + |\sum_i m(\sigma_i)[(A'_\tau f)(z_i) - \operatorname{tr}_2(\tilde{U}_\tau^+ P_\psi \tilde{U}_\tau \tilde{S}(A'_\cdot f)(z_i))| \\
&\quad + \left|\sum_i \frac{D_2}{2\mathbf{T}} \int_{-\mathbf{T}}^{+\mathbf{T}} m(\sigma_i)[\langle m_{0\sigma_i}, A'_t f\rangle - (A'_t f)(z_i)] \right. \\
&\quad \left. \cdot \operatorname{tr}_2(\tilde{U}_\tau^+ P_\psi \tilde{U}_\tau P_\psi \tilde{U}_t^+)\, dt \right|.
\end{aligned}
$$

Since

$$|\langle m_{0\sigma_i}, A'_t f\rangle - (A'_t f)(z_i)| < \varepsilon_1,$$

we have

$$
\begin{aligned}
&\left|\sum_i \frac{D_2}{2\mathbf{T}} \int_{-\mathbf{T}}^{+\mathbf{T}} m(\sigma_i)[\langle m_{0\sigma_i}, A'_t f\rangle - (A'_t f)(z_i)] \operatorname{tr}_2(\tilde{U}_\tau^+ P_\psi \tilde{U}_\tau \tilde{U}_t^+ P_\psi \tilde{U}_t)\, dt\right| \\
&\quad \leq \varepsilon_1 \frac{D_2}{2\mathbf{T}} \int_{-\mathbf{T}}^{+\mathbf{T}} \operatorname{tr}_2(\tilde{U}_\tau^+ P_\psi \tilde{U}_\tau \tilde{U}_t^+ P_\psi \tilde{U}_t)\, dt = \varepsilon_1 .
\end{aligned}
$$

Hence

$$
\begin{aligned}
|\langle A_\tau m, f\rangle - \langle S[U_\tau^+ (Tm)\, U_\tau], f\rangle|
&\leq 2\varepsilon_1 + |\sum_i m(\sigma_i)[(A'_\tau f(z_i) - \operatorname{tr}_2(\tilde{U}_\tau^+ P_\psi \tilde{U}_\tau \tilde{S}(A'_\cdot f)(z_i))]| \\
&\leq 2\varepsilon_1 + (1+\pi)\, m\left(\frac{\mathbf{T}}{\pi} D_2^{-1/2}\right) \leq 2\varepsilon_1 + \varepsilon_2 . \quad \square
\end{aligned}
$$

From a physical viewpoint, T 3.3.5 seems very artificial. The formulation: "There is a continuous unitary group $\{U_\tau\}$ ... such that" seems without physical significance since the Hamiltonian in $\mathscr{PT}_{q\,\exp}$ is not at our disposal (but fixed as a law of nature). Such a "given" Hamiltonian is written down in [2] V (5.8). It is obvious that we often replace this very complicated Hamiltonian by suitable approximations. For instance for a gas we use a Hamiltonian describing bound atoms as "elementary systems" in the sense of [2] VIII. The group $\{U_\tau\}$ is therefore given and not disposeable.

Nevertheless, T 3.3.5 can tell us some properties of the spectrum of the Hamiltonian, necessary for the possibility of embedding. It is clear that these properties actually should be proved, what in practice is not possible because of the complicated Hamiltonian. Nevertheless, we have great confidence that such properties are present. But what are these properties?

The model for the Hamiltonian used in the proof of T 3.3.5 is very artificial. The eigenvalues are $\frac{\pi\nu}{\mathbf{T}}$ $(\nu=1, \ldots, D_2)$, each with degeneracy $D_1$. If we would lift the degeneracy by replacing the eigenvalue $\pi\,\nu\,\mathbf{T}^{-1}$ by $D_1$ eigenvalues $\varepsilon_{\nu i}\,(i=1, \ldots, D_1)$ spread between $\pi(\nu-1/2)\,\mathbf{T}^{-1}$ and $\pi(\nu+1/2)\,\mathbf{T}^{-1}$, then the approximation would remain good for $\tau<\mathbf{T}$. For times $\tau>\mathbf{T}$, the two Hamiltonians with the spectra $\pi\,\nu\,\mathbf{T}^{-1}$ resp. $\varepsilon_{\nu i}$ would give different time evolutions. For instance, the recurrence time for the spectrum $\{\varepsilon_{\nu i}\}$ is of the magnitude $(D_1\,\mathbf{T})$ whereas that of $\pi\,\nu\,\mathbf{T}^{-1}$ is of the magnitude $\mathbf{T}$. We must expect that a good embedding of the time evolution is possible only up to times $\mathbf{T}$ small compared with the recurrence time $D_1\,\mathbf{T}$.

Another time scale is given by the $\tau=\mathbf{T}\,D_2^{-1/2}$ in (2) of T 3.3.5. The significance of $m(\tau)$ is that a macroscopic change is not noticeable (in the sense of $\varepsilon_2$) during a time $\tau_1=T D_2^{-1/2}$. With $\Delta E$ as the imprecision of the macroscopic energy relating to the microscopic energy (see §3.2), we can define the time $\tau_2\sim\frac{1}{\Delta E}$. Here we have $\Delta E=D_2\pi\,\mathbf{T}^{-1}$ i.e. $\tau_2\sim T D_2^{-1}=\tau_1 D_2^{-1/2}$. This $\tau_2$ is the shortest time for evolutions of microscopic (not measurable) quantities which are coexistent with the macroscopic energy. Every physicist would say, that the observation of faster changes is impossible. We claim that already changes in times shorter than $\tau_1$ cannot be observed.

In this way we can interpret T 3.3.5 in the following way: There are two characteristic times $\mathbf{T}$ and $\tau_1$, where $\tau_1$ is the shortest time of macroscopic changes and $\mathbf{T}$ the time until which the macrosystem is not essentially disturbed by the surroundings (i.e. until which the embedding of the time evolution is possible). If the spectrum of the Hamiltonian has the property that the number $D_1$ of eigenvalues in an interval $1/T$ satisfies the condition (1) of T 3.3.5, and if $\Delta E\gtrsim\mathbf{T}/\tau_1^2$, then an embedding with the precision $2\varepsilon_1+\varepsilon_2$ in the sense of T 3.3.5 is possible.

What do we know of the two numbers $D_1$ and $D_2=(\mathbf{T}/\tau_1)^2$ for real systems? The time $\tau_1$ will be greater than the time which an atom needs to fly $10^{-8}$ cm, i.e. $\tau_1>10^{-8}\text{ cm}/10^5\text{ cm sec}^{-1}=10^{-13}$ sec. As $\mathbf{T}$ we may take $T\approx 10^{17}$ sec ($\sim 10^{10}$ years). Then we get $D_2=10^{30}$ or $\log D_2\approx 70$. On the other hand, $\log(D_1\,D_2)=\log D_1+\log D_2$ has the magnitude of the entropy, for 1 mm$^3$ of Helium under atmospheric conditions $\log(D_1\,D_2)\sim 4\cdot 10^{17}$ (see the example in §3.2). Clearly $\log D_2\ll\log(D_1\,D_2)$. Therefore the problem of the time $\mathbf{T}$ is physically irrelevant for the existence of approximate embeddings, which is very reassuring.

Modifications of the results of § 3.2 can therefore only be expected from the estimates of $D_1$, i.e. from the metric $d_\theta$ which enters the condition (1) of T 3.3.5. This $d_\theta$ describes the physical distinguishability of states, when not only measurements at $t=0$ are taken into account. For systems which approach an equilibrium monotonically, the difference between initial states deteriorates step by step such that $d_\theta$ can be expected to be practically equal to $d$.

If on the contrary $d_\theta$ is "very much" larger than $d$, i.e. if $D_1$ according to (1) of T 3.3.5 is so large that an embedding is not possible, then one can follow two ways.

First one can look for a better half group $A_\tau$ to describe the macroscopic dynamics. For instance, it could be that the initially chosen $A_\tau$ describe a deterministic dynamics (see also [44] V. 3.6) and that this dynamics is in some regions of $Z$ very unstable (see § 2.6). Then one must look for stochastic $\{A_\tau\}$ which allow an approximate embedding.

If this is not possible, then the conjecture is justified that the dynamics in $Z$ is not "without memory". We can try to take a finer state space $Z_f$ such that $\Sigma_{mk}$ can be identified with $\Sigma_{Z_f}$ (in the sense of $\Sigma_{mk}=r' \Sigma_{Z_f}$, see § 2.5). If also this fails, then possibly $\Sigma_{mk}$ is not a Boolean ring.

## § 3.4 A Heavy Masspoint

Through all previous sections of this chapter, as a red thread there passed the problem to find the actually possible macroscopic observables and macroscopic ensembles. Especially we could not find the actual possibilities for the observable $\chi$ in (2.5.9) ($\chi$ determines the mixture-morphism $S$ used in § 3.1 through § 3.3). Only after choosing $\chi$, we had given an axiom for $K^*$ in the form (2.5.15) ($K^*$ determines the mixture-morphism $T$ used in § 3.1 through § 3.3). Although the general solution of the problem to find the $\chi$ for real macroscopic measurements is not yet solved, we have some arguments for special cases to choose some $\chi$ as "real". We shall now treat such a special case, a heavy masspoint, i.e. the motion of the center of mass of a macroscopic body.

What are the description of the center of mass in $\mathscr{P}\mathscr{T}_{q\,\mathrm{exp}}$ and in $\mathscr{P}\mathscr{T}_m$? These two descriptions are very similar.

In $\mathscr{P}\mathscr{T}_{q\,\mathrm{exp}}$ the description is given by a dual pair $\mathscr{B}(\mathscr{H})$, $\mathscr{B}'(\mathscr{H})$ and an irreducible representation of the Galilei group (see IX § 1 and 2; or [2] VII § 2 and [2] VIII § 1), which also determines the mass $m$ of the body. The physical interpretation of the Galilei group is described in IX § 1.

In $\mathscr{P}\mathscr{T}_m$ the center of mass is described as a classical system in the sense of VII § 5.3, i.e. by a dual pair $\mathscr{B}(\Sigma)$, $\mathscr{B}'(\Sigma)$. The representation of the Galilei group is defined in $\mathscr{B}'(\Sigma)$ (similarly as in $\mathscr{B}'(\mathscr{H})$) with the same physical interpretation as in $\mathscr{P}\mathscr{T}_{q\,\mathrm{exp}}$. For the center of mass we also presume that this representation is irreducible. It is possible to prove (see [32], [33], [54]) that this leads to the following representation of $\Sigma$ and of the Galilei group.

$\Sigma$ can be identified with $\mathscr{B}(Z)/\mathscr{J}(Z)$, where $\mathscr{B}(Z)$ is the Borel field of $Z=\mathbf{R}^6$, and $\mathscr{J}(Z)$ the set of all $\sigma\in\mathscr{B}(Z)$ with Lebesque measure zero. $\mathscr{B}'(\Sigma)$ can be identified with the measurable, essentially bounded functions on $\mathbf{R}^6$. The representation of the Galilei group is generated by the well known point transformations in the $\mu$-space $\mathbf{R}^6=\{\boldsymbol{r},\boldsymbol{p}\}$, where $\boldsymbol{r}$ is transformed like the position and $\boldsymbol{p}$ like the momentum.

The trajectories in $Z$ ($=\mu$-space) are given by

$$\begin{aligned} \boldsymbol{r}(t)&=\boldsymbol{r}(0)+\frac{\boldsymbol{p}}{\tilde{m}}\,t,\\ \boldsymbol{p}(t)&=\boldsymbol{p}. \end{aligned} \tag{3.4.1}$$

The "mass" $\tilde{m}$ is a parameter characterizing the relation between momentum and velocity. This $\tilde{m}$ as a property of classical mass-points is not defined by the Galilei transformations of one mass-point alone. One has to add a method to compare the momenta of various masspoints, e.g. by the conservation law of the sum of the momenta of interacting masspoints. In this manner the classical mass values are defined only up to a common factor. For the embedding described below, it will be very convenient to choose this factor such that corresponding masses $m$ for $\mathcal{PT}_{q\,\mathrm{exp}}$ and $\tilde{m}$ from $\mathcal{PT}_m$ can be identified.

This classical description of the motion of the center of mass is so familiar from classical mechanics that we need not show the mentioned proof.

In $\mathcal{PT}_{q\,\mathrm{exp}}$ the Hilbert space $\mathcal{H}_b$ (see [2] VII §1) is irreducible relative to the position and momentum operators $\boldsymbol{Q}$, resp. $\boldsymbol{P}$, while the Hamiltonian becomes $H=\dfrac{1}{2m}\boldsymbol{P}^2$. In a well known manner, $\boldsymbol{Q}$, $\boldsymbol{P}$, $H$ determine the representations of the Galilei group (see IX §1 or [2] VII).

We have sketched the representations of the Galilei group in these two cases because we shall strengthen the condition (2.2.7b), taking for $R$ and $\delta$ "all" transformations of the Galilei group and not only the time translations as in (2.2.7b).

The dynamics (3.4.1) is a special, simple example for a deterministic dynamics. It suffices therefore to contemplate the observable (2.5.9):

$$\mathcal{B}(Z)\xrightarrow{\chi} L(\mathcal{H}_b). \tag{3.4.2}$$

From the generalized condition (2.2.7b) then follow (in the same way as for the time translation from (2.2.7b)) as conditions for $\chi$:

$$\mu(w, U(g)\,\chi(\sigma)\,U(g)^+)\approx\mu(w,\chi(g\sigma)) \tag{3.4.3}$$

for all elements $g$ of the Galilei group and for all $w\in K_m^*$. Here $U(g)$ is the representation of the Galilei group in $\mathcal{H}_b$, and $g\sigma$ is the subset generated from $\sigma$ by the point transformations in the $\mu$-space, i.e. by the representation of the Galilei group in $\mathcal{B}'(\Sigma)$ (the $\sigma\in\Sigma$ are the elements of $\partial_e L(\Sigma)$). The sign $\approx$ indicates the approximation.

We try to fulfill (3.4.3) by the constraints that

$$U(g)\,\chi(\sigma)\,U(g)^+=\chi(g\sigma) \tag{3.4.4}$$

for all elements of the "Galilei group without time translations" (a subgroup of the whole Galilei group) and

$$\mu(w, U_\tau\,\chi(\sigma)\,U_\tau^+)\approx\mu\left(w,\chi\left(\sigma-\frac{\boldsymbol{p}}{\tilde{m}}\tau\right)\right) \tag{3.4.5}$$

for all $w\in K_m^*$. Here $\sigma-\dfrac{\boldsymbol{p}}{\tilde{m}}\tau$ is an abbreviation for

$$\left\{(\boldsymbol{r},\boldsymbol{p})\,|\,\boldsymbol{r}=\boldsymbol{r}'-\frac{\boldsymbol{p}'}{\tilde{m}}\tau;\boldsymbol{p}=\boldsymbol{p}',(\boldsymbol{r}',\boldsymbol{p}')\in\sigma\right\}.$$

We have chosen the strong condition (3.4.4), since one has suceeded to find "all" solutions of (3.4.4) (see [43] and [55]). Now it is very convenient to identify the two masses $m$ and $\tilde{m}$, since then no factors $m/\tilde{m}$ enter the following formulas. (That this identification is possible for all masspoints in the same way, is implied by the fact that the embedding of heavy interacting masspoints exactly in this way conserves the sum of the momenta.)

The solutions of (3.4.4) have the form

$$\chi(\sigma)=\int_{\sigma} F(\boldsymbol{r},\boldsymbol{p})\, d^3\boldsymbol{r}\, d^3\boldsymbol{p}, \tag{3.4.6}$$

where

$$F(\boldsymbol{r},\boldsymbol{p})=\frac{1}{(2\pi)^3}\, e^{i(\boldsymbol{p}\cdot\boldsymbol{Q}-\boldsymbol{r}\cdot\boldsymbol{P})}\, W e^{-i(\boldsymbol{p}\cdot\boldsymbol{Q}-\boldsymbol{r}\cdot\boldsymbol{P})}. \tag{3.4.7}$$

Here $W$ is a rotation-invariant element of $K(\mathscr{H}_b)$.

The condition (3.4.5) takes the form

$$\mu\left(w, U_\tau \chi\left(\sigma+\frac{\boldsymbol{P}}{m}\tau\right) U_\tau^+\right)\approx\mu(w,\chi(\sigma)) \tag{3.4.8}$$

for all $w\in K_m^*$. This says, that the left side of (3.4.8) is (for $0\le\tau\le\mathbf{T}$) approximately independent of $\tau$.

With

$$U_\tau=e^{-i\frac{1}{2m}\boldsymbol{P}^2\tau},$$

(3.4.8) holds if

$$\frac{1}{(2\pi)^3}\operatorname{tr}\left(w e^{i\frac{1}{2m}\boldsymbol{P}^2\tau} e^{i\left[\boldsymbol{p}\cdot\boldsymbol{Q}-\left(\boldsymbol{r}+\frac{\boldsymbol{p}}{m}\tau\right)\cdot\boldsymbol{P}\right]} W\cdot e^{-i\left[\boldsymbol{p}\cdot\boldsymbol{Q}-\left(\boldsymbol{r}+\frac{\boldsymbol{p}}{m}\tau\right)\cdot\boldsymbol{P}\right]} e^{-i\frac{1}{2m}\boldsymbol{P}^2\tau}\right). \tag{3.4.9}$$

is approximately independent of $\tau$. With

$$e^{i\frac{1}{2m}\boldsymbol{P}^2\tau} e^{i\left[\boldsymbol{p}\cdot\boldsymbol{Q}-\left(\boldsymbol{r}+\frac{\boldsymbol{p}}{m}\tau\right)\cdot\boldsymbol{P}\right]} e^{-i\frac{1}{2m}\boldsymbol{P}^2\tau}=e^{i\left[\boldsymbol{p}\cdot\left(\boldsymbol{Q}+\frac{1}{m}\boldsymbol{P}\tau\right)-\left(\boldsymbol{r}+\frac{\boldsymbol{p}}{m}\tau\right)\cdot\boldsymbol{P}\right]},$$

(3.4.9) takes the form

$$\frac{1}{(2\pi)^3}\operatorname{tr}\left(e^{-i(\boldsymbol{p}\cdot\boldsymbol{Q}-\boldsymbol{r}\cdot\boldsymbol{P})} w e^{i(\boldsymbol{p}\cdot\boldsymbol{Q}-\boldsymbol{r}\cdot\boldsymbol{P})} W_\tau\right) \tag{3.4.10}$$

with

$$W_\tau=e^{i\frac{1}{2m}\boldsymbol{P}^2\tau} W e^{-i\frac{1}{2m}\boldsymbol{P}^2\tau}. \tag{3.4.11}$$

The embedding is possible, if (3.4.10) is approximately independent of $\tau$ $(0\le\tau\le\mathbf{T})$ for all $w\in K_m^*$.

From (3.4.6) follows the mixture-morphism $S$ given by

$$(Sw)(\sigma)=\int_{\sigma}\operatorname{tr}(wF(\boldsymbol{r},\boldsymbol{p}))\, d^3\boldsymbol{r}\, d^3\boldsymbol{p}. \tag{3.4.12}$$

Now we must choose $K_m^*$ in order to evaluate further the stated conditions and to find the mixture-morphism $T$. We take (2.5.15) for $K_m^*$, i.e. $K_m^*$ is the convex set generated by the elements

$$\mathscr{w}(\boldsymbol{r},\boldsymbol{p})=e^{i(\boldsymbol{p}\cdot\boldsymbol{Q}-\boldsymbol{r}\cdot\boldsymbol{P})} W e^{i(\boldsymbol{p}\cdot\boldsymbol{Q}-\boldsymbol{r}\cdot\boldsymbol{P})}. \tag{3.4.13}$$

This $\omega(\boldsymbol{r},\boldsymbol{p})$ coincides with $\omega(z)$ from (2.5.24), while (3.4.12) takes the form

$$(Sw)(\sigma)=\int_\sigma \operatorname{tr}(w\,\omega(\boldsymbol{r},\boldsymbol{p}))\,d\bar{m}(\boldsymbol{r},\boldsymbol{p}) \tag{3.4.14}$$

with $d\bar{m}(\boldsymbol{r},\boldsymbol{p})=\frac{1}{(2\pi)^3}d^3\boldsymbol{r}\,d^3\boldsymbol{p}$. In this notation, (3.4.14) is identical with (2.5.23b) and $\bar{m}(\sigma)$ with (2.5.19). According to (2.5.31) with

$$m(\sigma)=\int_\sigma \gamma(\boldsymbol{r},\boldsymbol{p})\,d\bar{m}(\boldsymbol{r},\boldsymbol{p}),$$

$T$ is defined by

$$Tm=\int \gamma(\boldsymbol{r},\boldsymbol{p})\,\omega(\boldsymbol{r},\boldsymbol{p})\,d\bar{m}(\boldsymbol{r},\boldsymbol{p}),$$

i.e.

$$Tm=\int \gamma(\boldsymbol{r},\boldsymbol{p})\,F(\boldsymbol{r},\boldsymbol{p})\,d^3\boldsymbol{r}\,d^3\boldsymbol{p}.$$

The smearing operator $Q=ST$ takes the form

$$(Qm)(\sigma)=\int_\sigma d\bar{m}(\boldsymbol{r},\boldsymbol{p})\int \gamma(\boldsymbol{r}',\boldsymbol{p}')\operatorname{tr}(\omega(\boldsymbol{r}',\boldsymbol{p}')\,\omega(\boldsymbol{r},\boldsymbol{p}))\,d\bar{m}(\boldsymbol{r},\boldsymbol{p})$$

with the kernel $\operatorname{tr}(\omega(\boldsymbol{r}',\boldsymbol{p}')\,\omega(\boldsymbol{r},\boldsymbol{p}))$. This kernel is a function $k(\boldsymbol{r}-\boldsymbol{r}',\boldsymbol{p}-\boldsymbol{p}')$ because of (3.4.4). We find

$$\begin{aligned} k(\boldsymbol{r},\boldsymbol{p})&=\operatorname{tr}(\omega(0,0)\,\omega(\boldsymbol{r},\boldsymbol{p}))\\ &=\operatorname{tr}(e^{-i(\boldsymbol{p}\cdot\boldsymbol{Q}-\boldsymbol{r}\cdot\boldsymbol{P})}W e^{i(\boldsymbol{p}\cdot\boldsymbol{Q}-\boldsymbol{r}\cdot\boldsymbol{P})}W), \end{aligned} \tag{3.4.15}$$

while (3.4.10) is approximately independent of $\tau$ if

$$\operatorname{tr}(e^{-i(\boldsymbol{p}\cdot\boldsymbol{Q}-\boldsymbol{r}\cdot\boldsymbol{P})}W e^{i(\boldsymbol{p}\cdot\boldsymbol{Q}-\boldsymbol{r}\cdot\boldsymbol{P})}W_\tau) \tag{3.4.16}$$

is so. Our task is to choose the rotation invariant $W$ such that the smearing is small, i.e. that $k(\boldsymbol{r},\boldsymbol{p})$ decreases rapidly with $\boldsymbol{r}$ and $\boldsymbol{p}$, and that (3.4.16) is approximately independent of $\tau$ if this is not too large. The choice of $W$ is the choice of the measuring device for the macroscopic position and velocity of the masspoint. It is very doubtful, whether all rotation-invariant $W$ can be realized by suitable devices.

The simplest choice for a rotation-invariant $W$ seems to be

$$W=P_\psi$$

with

$$\langle\boldsymbol{r}|\psi\rangle=\frac{\sigma^{-3/2}}{(2\pi)^{3/4}}e^{-\frac{r^2}{4\sigma^2}}$$

in the position representation and

$$\langle\boldsymbol{k}|\psi\rangle=\left(\frac{2}{\pi}\right)^{3/4}\sigma^{3/2}e^{-\sigma^2\boldsymbol{k}^2}$$

in the momentum representation. For this $P_\psi$, Heisenberg's uncertainty relation holds with an equal sign. To measure position and momentum (by the corresponding effect $F(\boldsymbol{r},\boldsymbol{p})$) together as precisely as Heisenberg's relation allows, is very unrealistic for the center of mass of a macroscopic body, e.g. of a tennis ball. Therefore we would prefer a $W$ with *two* parameters to represent the precision of the position and momentum independently. Such a more realistic choice seems to be that in

[61] (2.17), which we write

$$W=(4\sigma\mu)^3 e^{-\sigma^2 P^2} e^{-2\mu^2 Q^2} e^{-\sigma^2 P^2}. \tag{3.4.17}$$

Then we have

$$\langle k|\, W\, |k\rangle = \left(\frac{2}{\pi}\right)^{3/2} \sigma^3 e^{-2\sigma^2 k^2} \tag{3.4.18}$$

as the probability density for momentum, and

$$\langle r|\, W\, |r\rangle = \frac{1}{(2\pi)^{3/2} a^3}\, e^{-\frac{r^2}{2a^2}} \quad \text{with} \quad a=\sqrt{\sigma^2+\frac{1}{4\mu^2}} \tag{3.4.19}$$

as the probability density for position, i.e. as if momentum and position in the sense of $\mathscr{P}\mathscr{T}_{q\,\mathrm{exp}}$ could be measured exactly (what we do not believe). For the realistic probability distribution of measurements of the "states" $(\boldsymbol{r},\boldsymbol{p})$, we have to calculate (3.4.10) at $\tau=0$, i.e. $\frac{1}{(2\pi)^3}\, k(\boldsymbol{r},\boldsymbol{p})$ with $k$ in (3.4.15).

We find

$$k(\boldsymbol{r},\boldsymbol{p}) = \frac{(2\mu\sigma)^3}{(1+4\sigma^2\mu^2)^{3/2}}\, e^{-\sigma^2 p^2}\, e^{-\frac{r^2}{\frac{1}{\mu^2}+4\sigma^2}}. \tag{3.4.20}$$

The expression (3.4.16) takes the form

$$\frac{(2\mu\sigma)^3}{\left(1+4\sigma^2\mu^2+\dfrac{\mu^2\tau^2}{4\sigma^2 m^2}\right)^{3/2}}\, e^{-\sigma^2 p^2}\, e^{-\frac{\left(r+\frac{p}{2m}\tau\right)^2}{\frac{1}{\mu^2}+4\sigma^2+\frac{\tau^2}{4\sigma^2 m^2}}}. \tag{3.4.21}$$

(I thank the last two authors of [61] very much for the calculation of (3.4.20) and (3.4.21).) In (3.4.20) we can interpret

$$\Delta p = \frac{1}{\sigma} \tag{3.4.22}$$

and

$$\Delta r = \left(\frac{1}{\mu^2}+4\sigma^2\right)^{1/2} \tag{3.4.23}$$

as the smallest imprecisions of preparation and measurement of momentum resp. position. This contrasts to the usual "purely classical" idealization that measurements can be made more and more and thus "arbitrarily" precise. In this sense we get a theory more comprehensive than $\mathscr{P}\mathscr{T}_m$ but more realistic than $\mathscr{P}\mathscr{T}_{q\,\mathrm{exp}}$.

To demonstrate this more clearly we will contemplate the general case. In $\mathscr{P}\mathscr{T}_m$ we describe the ensemble (corresponding to a preparation method) by a density $\rho(\boldsymbol{r},\boldsymbol{p})$ and the effect (corresponding to a registration procedure) by a function $f(\boldsymbol{r},\boldsymbol{p})$. Then the probablity for this effect $f$ (at a time $\tau$ in the ensemble $\rho$) according to $\mathscr{P}\mathscr{T}_m$ equals

$$\int \rho(\boldsymbol{r},\boldsymbol{p})\, f\!\left(\boldsymbol{r}-\frac{\boldsymbol{p}}{m}\tau,\boldsymbol{p}\right) d^3 r\, d^3 p,$$

in the more comprehensive theory to be replaced by

$$\int \rho(\boldsymbol{r},\boldsymbol{p}) \frac{1}{(2\pi)^3} k(\boldsymbol{r}-\boldsymbol{r}',\boldsymbol{p}-\boldsymbol{p}')\, f\left(\boldsymbol{r}'-\frac{\boldsymbol{p}'}{m}\tau,\boldsymbol{p}'\right) d^3\boldsymbol{r}\, d^3\boldsymbol{p}\, d^3\boldsymbol{r}'\, d^3\boldsymbol{p}'.$$

This cannot be distinguished from the "embedded" expression

$$\operatorname{tr}(\tilde{w} U_\tau \tilde{F} U_\tau^+)$$

with

$$\tilde{w} = T\rho = \int \rho(\boldsymbol{r},\boldsymbol{p})\, \omega(\boldsymbol{r},\boldsymbol{p})\, d^3\boldsymbol{r}\, d^3\boldsymbol{p}$$

and

$$\tilde{F} = S' f = \int f(\boldsymbol{r},\boldsymbol{p})\, F(\boldsymbol{r},\boldsymbol{p})\, d^3\boldsymbol{r}\, d^3\boldsymbol{p},$$

if the expression (3.4.21) is in good approximation independent of $\tau$, i.e. as long as

$$\frac{\Delta p}{m} < \frac{\Delta r}{\tau}. \tag{3.4.24}$$

The expression $\Delta v = m^{-1} \Delta p$ is the imprecision in velocity; (3.4.24) says that this smallest imprecision $\Delta v$ must be smaller than $\Delta r/\tau$. This is very illustrative, since otherwise we could reduce the imprecision in velocity by measuring $\boldsymbol{r}/\tau$. Thus (3.4.24) would not be a strong postulate if we could make $\Delta p$ arbitrarily small. But this is not possible since due to (3.4.22) and (3.4.23) we have

$$\Delta p\, \Delta r > 1,$$

corresponding to Heisenberg's uncertainty relation. For realistic measurements on macroscopic bodies, we conjecture $\mu^{-1} \gg \sigma$ such that $\Delta p\, \Delta r$ is much greater than 1. This means that macroscopic measurements cannot come near the bounds from Heisenberg's relation.

We therefore find

$$\frac{1}{m \Delta r} \ll \frac{\Delta r}{\tau} \tag{3.4.25}$$

as the essential condition for the classical description of a masspoint. The time $\mathbf{T} = m(\Delta r)^2$ is the upper bound on the magnitude of time intervals during which the description of a free masspoint by trajectories of the form (3.4.1) is possible. We conjecture that this time $\mathbf{T}$ is also the characteristic time for the description of interacting masspoints if the forces do not vary much in intervals $\Delta r$.

The imprecision $\Delta r$ cannot be chosen arbitrarily. It is defined by the "best" real measuring methods (see above). Take for instance $\Delta r = 10^{-10}$ cm and $m = N \cdot 10^{12}\,\text{cm}^{-1}$ with $N$ of the magnitude $10^{24}$ of Avogadro's number and $10^{12}\,\text{cm}^{-1}$ as the magnitude of the mass of an atom. Then we find $\mathbf{T} = 10^{16}$ cm, i.e. $\mathbf{T} \sim 3 \cdot 10^5$ sec $\sim 90$ hours. Taking a body of magnitude 1 meter, we would get $\mathbf{T} \sim 10^4$ years and for a body of the magnitude of the earth $\mathbf{T} \sim 10^{24}$ years, all this for the high accuracy of $\Delta r \sim 10^{-10}$ cm. Thus we see why it is impossible to find any deviation from classical mechanics for macroscopic bodies. For instance, we can play tennis without taking into account "quantum mechanics", i.e. without the problem of embedding classical mechanics into $\mathcal{PT}_{q\,\mathrm{exp}}$.

This example of a heavy masspoint can serve to calculate all the terms introduced in §2. The trajectories (3.4.1) in the $\mu$-space $\mathbf{R}^6$ can be characterized by the "parameters" $\boldsymbol{r}(0)$ and $\boldsymbol{p}$, i.e. $S_m$ can be identified with the $\mu$-space $\mathbf{R}^6$: $\{\boldsymbol{r}(0), \boldsymbol{p}\}$. Then we have $\Sigma_m = \mathscr{B}(\mathbf{R}^6)/\mathscr{J}(\mathbf{R}^6)$ with $\mathscr{J}(\mathbf{R}^6)$ as the subset of those elements of $\mathscr{B}(\mathbf{R}^6)$ which have Lebesgue measure zero. As $F_{\tilde{b}_0}$ in (2.3.13) we can choose

$$F_{\tilde{b}_0}(\sigma) = \chi(\sigma) \tag{3.4.26}$$

with $\chi(\sigma)$ in (3.4.6), (3.4.7) and $W$ in (3.4.17). It may be left to the reader to see, that (2.3.20) holds approximately if (3.4.21) is approximately independent of $\tau$. Let us emphasize that $\sigma$ on the left side of (3.4.26) is an element of $\Sigma_m$ and on the right side the corresponding element of $\Sigma_Z = \mathscr{B}(Z)/\mathscr{J}(Z)$. Bearing this in mind, it is not difficult for the reader to calculate all the terms in §2.3 through §2.6.

In §2.6 it is illustrative to consider the "contraction" of the $\mu$-space $\mathbf{R}^6$ to $\mathbf{R}^3$ given by

$$(\boldsymbol{r}, \boldsymbol{p}) \to \boldsymbol{r}.$$

In $\mathbf{R}^3$ the dynamics is not "without memory". Moreover, if one takes as state space $\mathbf{R}^3 = \{\boldsymbol{r}\}$, the embedding conditions are fulfilled since we have to take the *same* "trajectory" observable $F_{\tilde{b}_0}$ as in (3.4.26)! In this manner we see, that it is advantageous to take the $\mu$-space as state space since there the dynamics is without memory and even deterministic.

Obviously all considerations of the entropy in §2.6 are trivial since the entropy does not change with time (the trajectories are reversible).

## §3.5 The Boltzmann Distribution Function

We had already discussed in §3.2 the general embedding possibilities in the state space $K(X)$, where $X$ is a compactification of the $\mu$-space. The physical interpretation of an $m \in K(X)$ was that $Nm(\rho)$, with $\rho$ an open subset of $X$, and with $N$ the number of particles in the macrosystem, is the "approximate number of particles" in the set $\rho$. But we had not used this interpretation as an additional condition for embedding. Let us now use this interpretation to give a special form of the macro-observable $\chi(\sigma)$ with $\sigma \in \Sigma_Z$.

In $\mathscr{P}\mathscr{T}_{q\,\mathrm{exp}}$ we take (as an approximation for an atomic gas) the $N$ identical atoms of a gas as "elementary" systems in the sense of [2] VII D2.1. Thus the atoms shall be in the ground state, described only by their positions $\boldsymbol{Q}_\nu$, momenta $\boldsymbol{P}_\nu$, and by potential interaction energies $V(|\boldsymbol{Q}_\nu - \boldsymbol{Q}_\mu|)$.

In $\mathscr{P}\mathscr{T}_m$ we take as state space $Z$ the set $NK(X)$, i.e. the measures on $X$ normed to the particle number $N$. (It is possible to use also an imprecise description of this particle number $N$; for simplicity we will not do so here.) The elements $z$ of $Z$ are called Boltzmann distribution functions since with $g \in C(X)$ one often writes

$$\langle z, g\rangle = \int f(\boldsymbol{r}, \boldsymbol{p})\, g(\boldsymbol{r}, \boldsymbol{p})\, d^3\boldsymbol{r}\, d^3\boldsymbol{p}, \tag{3.5.1}$$

where $f(\boldsymbol{p}, \boldsymbol{r})$ is the "density" belonging to $z$.

By (3.4.7) together with (3.4.17), effects to measure one atom in a region of the space $X$ are defined. If we have $N$ atoms, we introduce the $\Gamma$-space by $\Gamma = X^N$

and can define effects to measure the $N$ atoms in a region $\rho$ of the $\Gamma$-space:

$$\gamma(\rho)=\int_{\rho} F_1(x_1)\times F_2(x_2)\times \ldots \times F_N(x_N)\, dx_1\, dx_2 \ldots dx_N. \tag{3.5.2}$$

Here $F_\nu(x_\nu)$ is given by (3.4.7), (3.4.17), if we replace $x=(\boldsymbol{r},\boldsymbol{p})$ by $x_\nu=(\boldsymbol{r}_\nu,\boldsymbol{p}_\nu)$ and $\boldsymbol{Q}$, $\boldsymbol{P}$ by $\boldsymbol{Q}_\nu$, $\boldsymbol{P}_\nu$, while $dx_\nu$ stands for $d^3\boldsymbol{r}_\nu\, d^3\boldsymbol{p}_\nu$. Since the atoms are all equal, we consider only such $\rho$ which are symmetric, i.e. which contain with a point $(x_1, \ldots, x_N)$ all the points with permuted $x_\nu$. Then $\gamma(\rho)$ is a symmetric operator which leaves invariant the symmetric subspace $\{\mathscr{H}^N\}_+$ (for Bose atoms) or the antisymmetric subspace $\{\mathscr{H}^N\}_-$ (for Fermi atoms) of $\mathscr{H}^N$, with $\mathscr{H}$ as the Hilbert space of one atom.

The integrand in (3.5.2) is the effect that the $N$ atoms are at the points $x_1, \ldots, x_N$ in the $\mu$-space $X$, i.e. the effect for the special distribution function

$$f(\boldsymbol{r},\boldsymbol{p})=f(x)=\sum_{\nu=1}^{N} \delta(x-x_\nu) \tag{3.5.3}$$

in (3.5.1). For the $z$ corresponding to (3.5.3), let us briefly write $z(x_\nu)$; then (3.5.1) takes the form

$$\langle z(x_\nu), g\rangle=\sum_{\nu=1}^{N} g(x_\nu). \tag{3.5.4}$$

It is obvious that $z(x_\nu)$ is symmetric in $x_1, \ldots, x_N$.

Until now we have not used the intuitive notion that $f(x)$ is in general an "approximation" to the special distributions $z(x_\nu)$. This approximation is described by a metric $d$ in $Z$. We have seen in §3.2 that this metric in $Z$ is determined by the metric of physical imprecision in $X$, which determines the Lipschitz interval $\Lambda$. Because of the norming of $z\in Z$ by the particle number ($Z=NK(X)$), as metric in $Z$ we choose

$$d(z,z')=\frac{1}{N}\sup_{g\in\Lambda}|\langle z,g\rangle-\langle z',g\rangle|. \tag{3.5.5}$$

We choose a measure $m\in K(Z)$ with $m(\sigma)\neq 0$ for all open sets $\sigma\in\mathscr{B}(Z)$. With $\mathscr{J}(Z)=\{\sigma \mid \sigma\in\mathscr{B}(Z), m(\sigma)=0\}$ this $m$ determines $\Sigma_Z=\mathscr{B}(Z)/\mathscr{J}(Z)$.

We take a finite $\varepsilon>0$ and as macroscopic observable choose $\Sigma(Z)\xrightarrow{\chi} L$ by means of

$$\chi(\sigma)=\int_{\sigma} \hat{F}(z)\, dm(z), \tag{3.5.6}$$

where

$$\hat{F}(z)=\int_{d(z,z(x_\nu))\le\varepsilon} \frac{F_1(x_1)\times\ldots\times F_N(x_n)}{v_\varepsilon(x_1,\ldots,x_N)}\, dx_1 \ldots dx_N \tag{3.5.7}$$

with

$$v_\varepsilon(x_1,\ldots,x_N)=\int_{d(z,z(x_\nu))\le\varepsilon} dm(z). \tag{3.5.8}$$

Hence we have

$$\chi(Z)=\int_Z \hat{F}(z)\, dm(z)=\int_\Gamma F_1(x_1)\times\ldots\times F_N(x_N)\, dx_1 \ldots dx_N$$
$$=\mathbf{1}\times\mathbf{1}\times\ldots\times\mathbf{1}=\mathbf{1}.$$

This $\chi$ is characterized by the three macroscopic imprecision constants $\varepsilon$, $\mu$, $\sigma$. We must choose the imprecision in $X$ (determined by $\mu, \sigma$) smaller than that given by the metric $d$ in $X$ in the form $d(x, x') \leq \varepsilon$. This $d$ in $X$ is determined by the same set $\Lambda$ as the $d$ in $Z$. Therefore $d \leq \varepsilon$ characterizes *equivalent* imprecisions in $X$ *and* $Z$. Here $\varepsilon$ must be so big, that the entropy (defined below) is not changing essentially with $\varepsilon$, e.g. if we replace $\varepsilon$ by $2\varepsilon$. On the other hand, let us choose $\varepsilon$ as small as possible under these constraints. (We must emphasize again that the $\chi$ in (3.5.6) is not the only possible observable; there are others which cannot be distinguished macroscopically from that in (3.5.6), e.g. if we choose another effective $m \in K(\Sigma_Z)$.)

With (3.5.6), the mixture-morphism $S$ is determined by

$$(Sw)(\sigma) = \int_\sigma \operatorname{tr}(w\hat{F}(z))\, dm(z).$$

If we take $K^*$ in (2.5.15), we have $K^* = \overline{\operatorname{co}} \bigcup_{z \in Z} \omega(z)$ with

$$\omega(z) = [\operatorname{tr}(\hat{F}(z))]^{-1} \hat{F}(z) \tag{3.5.9}$$

(see (2.5.24)). We find $\bar{m}$ in (2.5.19) by

$$\bar{m}(\sigma) = \int_\sigma \operatorname{tr}(\hat{F}(z))\, dm(z). \tag{3.5.10}$$

The mixture-morphism $T$ is given by (2.5.31):

$$Tm = \int \tilde{\gamma}(z)\, \omega(z)\, dm(z) \tag{3.5.11}$$

with $\tilde{\gamma}(z)$ defined by

$$m(\sigma) = \int_\sigma \tilde{\gamma}(z)\, dm(z).$$

The kernel of the macroscopic smearing operator $Q = ST$ (see (2.5.26)) is given by

$$k(z, z') = \operatorname{tr}(\omega(z)\, \omega(z')) = \frac{\operatorname{tr}(\hat{F}(z)\, \hat{F}(z'))}{[\operatorname{tr} \hat{F}(z)][\operatorname{tr} \hat{F}(z')]}. \tag{3.5.12}$$

Since the imprecision $\varepsilon$ exceeds that determined by $\mu$, $\sigma$, the size of $k(z, z')$ will be of the order $\varepsilon$.

By (2.6.16) (resp. (2.6.17) together with (3.5.9)) the entropy $s_1(z)$ (resp. $s_2(z)$) is defined.

For calculating the traces in these formulas, we must observe that the Hilbert space is $\{\mathscr{H}^N\}_+$ resp. $\{\mathscr{H}^N\}_-$ (and not $\mathscr{H}^N$!).

Although all these terms are well defined, the necessary calculations are not so simple as in §3.4. We can only give approximate values for great numbers $N$ and not too small $\varepsilon$. Therefore let us not try to show here these extensive calculations (see the remarks to follow below after the considerations about the dynamics).

Much more difficult than these calculations is it to deduce the dynamics. Until now it is not possible to do so rigorously. But if we make the *assumption* that the dynamics is *without memory and deterministic* (proved for a heavy mass point in §3.4), then we can derive the famous Boltzmann collision equation as dynamical law in $Z$.

This derivation starts from the equation (2.6.7), i.e. from

$$\int \operatorname{tr}(\omega(z)\,\hat{F}(z'))\,y(z')\,dm(z')=\operatorname{tr}(\omega(z)\,U_\tau\,\hat{F}(d_\tau z')\,U_\tau^+)\,y(z')\,dm(d_\tau z'). \quad (3.5.13)$$

In particular choosing $y(z')=\langle d_\tau z', g\rangle$ with $g\in\Lambda$, we get

$$\begin{aligned}\int \langle d_\tau z', g\rangle \operatorname{tr}(\omega(z)\,\hat{F}(z'))\,dm(z') &= \int \langle d_\tau z', g\rangle \operatorname{tr}(\omega(z)\,U_\tau\,\hat{F}(d_\tau z')\,U_\tau^+)\,dm(d_\tau z')\\ &= \int \langle z'', g\rangle \operatorname{tr}(\omega(z)\,U_\tau\,\hat{F}(z'')\,U_\tau^+)\,dm(z'').\end{aligned}$$

Since the "smearing" by $\operatorname{tr}(w(z)\,\hat{F}(z'))$ is of the size $d(z,z')<\varepsilon$, the left side takes the approximate form $\langle d_\tau z, g\rangle$ if $d_\tau z'$ does not differ much from $d_\tau z$ for $d(z,z')<\varepsilon$, i.e. if the dynamics is stable.

Thus we have

$$\langle d_\tau z, g\rangle = \int \langle z', g\rangle \operatorname{tr}(\omega(z)\,U_\tau\,\hat{F}(z')\,U_\tau^+)\,dm(z'). \quad (3.5.14)$$

Using on the right side (3.5.7), we get

$$\langle d_\tau z, g\rangle = \int \langle z', g\rangle \operatorname{tr}\left(\omega(z)\,U_\tau \int_{d(z',z(x_\nu))\le\varepsilon} \frac{F_1(x_1)\times\ldots\times F_N(x_N)}{v_\varepsilon(x_1,\ldots,x_N)}\,U_\tau^+\right)dm(z').$$

For $g\in\Lambda$ we can replace $\langle z', g\rangle$ by $\langle z(x_\nu), g\rangle$ because of $d(z', z(x_\nu))\le\varepsilon$. With $\langle z(x_\nu), g\rangle = \sum_{\nu=1} g(x_\nu)$ and (3.5.8) we get

$$\langle d_\tau z, g\rangle = \int \operatorname{tr}(\omega(z)\,U_\tau\,F_1(x_1)\times\ldots\times F_N(x_N)\,U_\tau^+)\sum_\nu g(x)_\nu)\,dx_1,\ldots dx_N.$$

Because of $\int F(x)\,dx=\mathbf{1}$ follows

$$\begin{aligned}\langle d_\tau z, g\rangle = \int g(x)\operatorname{tr}(\omega(z)\,U_\tau[&F_1(x)\times\mathbf{1}\times\ldots\times\mathbf{1}\\ &+\mathbf{1}\times F_2(x)\times\mathbf{1}\times\ldots\times\mathbf{1}+\ldots+\mathbf{1}\times\ldots\times\mathbf{1}\times F_N(x)]\,U_\tau^+)\,dx. \quad (3.5.15)\end{aligned}$$

With $d_\tau z = f(\boldsymbol{r},\boldsymbol{p};\tau)$, the latter is equivalent to

$$\begin{aligned}f(\boldsymbol{r},\boldsymbol{p};\tau)=\operatorname{tr}(\omega\,(f(\boldsymbol{r},\boldsymbol{p};0))\,U_\tau[&F_1(\boldsymbol{r},\boldsymbol{p})\times\mathbf{1}\times\ldots\times\mathbf{1}\\ &+\mathbf{1}\times F_2(\boldsymbol{r},\boldsymbol{p})\times\mathbf{1}\times\ldots\times\mathbf{1}+\ldots+\mathbf{1}\times\ldots\times\mathbf{1}\times F_N(\boldsymbol{r},\boldsymbol{p})]\,U_\tau^+). \quad (3.5.16)\end{aligned}$$

For this formula see also [61].

It is impossible to evaluate (3.5.16) for all times $\tau$. But we can calculate the operator $A$ in the following expansion of (3.5.16):

$$f(\boldsymbol{r},\boldsymbol{p};0)+\tau\left[\frac{\partial}{\partial\tau}f(\boldsymbol{r},\boldsymbol{p};\tau)\right]_{\tau=0}+\ldots=f(\boldsymbol{r},\boldsymbol{p};0)+\tau A f(\boldsymbol{r},\boldsymbol{p};0)+\ldots. \quad (3.5.17)$$

In this expansion the time $\tau$ must be greater than $r_0/v$ where $v$ is the mean velocity of the atoms and $r_0$ the range of the potential $V(r)$. But $\tau$ must be smaller than $\lambda/v$ where $\lambda$ is the mean free path.

Since we had assumed a dynamics without memory, from (3.5.17) we get the Boltzmann equation

$$\frac{\partial}{\partial\tau}f(\boldsymbol{r},\boldsymbol{p};\tau)=Af(\boldsymbol{r},\boldsymbol{p};\tau). \quad (3.5.18)$$

The principal methods of such an evaluation of (3.5.17) are already given in [56] and [57] (the present section is only an improved version of [56]!). Essential for this evaluation is the special form (3.5.9) with (3.5.7) of $\omega(z)$. See also the simplified evaluation in [61]. A more general evaluation of $A$ is in preparation by the authors of [61] and others.

What should we think of our assumptions about the dynamics?

The assumption that the dynamics is without memory seems to be fulfilled (on the basis of experience). We cannot preclude that perhaps good mathematicians could suceed in proving this assumption.

The assumption that the dynamics is deterministic is certainly not fulfilled in the total space Z. Since the Boltzmann equation in a contracted state space provides the Navier-Stokes equations (see §2.6), and since we know that the solutions of these Navier-Stokes equations have unstable regions, the deterministic dynamics of the Boltzmann equation must be improved by a master equation of a form described in §2.5. Such an improvement has been attempted in [57].

## §4 Intermediate Systems

To the presentation developed by us in §2, where $\mathscr{P}\mathscr{T}_{q\,\mathrm{exp}}$ is not the most comprehensive theory, it is often objected that it does not appear meaningful to view $\mathscr{P}\mathscr{T}_q$ as a realistic theory for microsystems, but to declare large parts of $\mathscr{P}\mathscr{T}_{q\,\mathrm{exp}}$ as fictitious for macrosystems. One asks: "How many elementary systems must compose a system for the deviation from quantum mechanics to begin?"

Of course we must answer: "For *no definite* number of elementary systems which compose a system." Rather, everything indicates that the quantum theory of composite systems contains larger and larger domains of purely fictive preparation and registration procedures, when we raise the number of the elementary systems in the system. Therefore, $\mathscr{P}\mathscr{T}_q$ can be viewed as an appoximately g.G.-closed theory (see XIII §3.4) of microsystems only as long as the number of subsystems is "small".

Conversely, the theory presented in II with its embedding in $\mathscr{P}\mathscr{T}_{q\,\mathrm{exp}}$ described in this section can only hold if the number of subsystems is "very large". Expressed otherwise: As a theory of microsystems, $\mathscr{P}\mathscr{T}_q$ asymptotically approximates an unknown theory of systems with an arbitrary number of elementary subsystems. In fact, it provides an asymptotic approximation for small numbers of subsystems. As described in II, $\mathscr{P}\mathscr{T}_m$ is an asymptotic approximation to the unknown theory for very large numbers of subsystems. Systems for which these two asymptotic approximations no longer suffice, will briefly be called intermediate systems.

As just mentioned, of course we cannot give a theory for intermediate systems. We can only describe some viewpoints which pertain to such a theory.

First, the domain of macrosystems for which a theory of the form $\mathscr{P}\mathscr{T}_m$ holds is rather large. For instance, small crystals which are only visible under a microscope are very well describeable by a $\mathscr{P}\mathscr{T}_m$. But also large molecules, important for biophysics, are very well conceivable for theories of the same form as $\mathscr{P}\mathscr{T}_m$; this has repeatedly been demonstrated by theoretical biophysicists. Therefore, the intermediate systems are indeed not "so large" as one could at first assume. The transition from a g.G.-closed $\mathscr{P}\mathscr{T}_q$ to a $\mathscr{P}\mathscr{T}_m$ thus does not concern really large numbers of elementary subsystems. Hence indications of such a transition should appear already for small numbers of subsystems; and this is actually true.

For instance, let us reflect how to construct devices for the preparation procedures $a$ in $\mathscr{P}\mathscr{T}_q$ for various $w \in K$ (i.e. with $\varphi(a)=w$). One soon sees that there exist such $w \in K$ for which it is "easy" to construct preparation devices. For other $w$, it makes enormous expenses to construct a device for an $a$ with $\varphi(a)=w$. Of course, the "expense" of the necessary supply of material needed to construct the device could be so large that it would be "physically impossible" to realize the desired $a$ with $\varphi(a)=w$. If one has found an $a \in \mathscr{Q}'$ which is not realizable, an indication is thus obtained on the basis of experience that $\mathscr{P}\mathscr{T}_q$ cannot be g.G.-closed. Asking the question this way, one quickly realizes that one must replace $\mathscr{P}\mathscr{T}_q$ by a more comprehensive theory that can be embedded in $\mathscr{P}\mathscr{T}_q$. Such an embedding would tell us that $\mathscr{P}\mathscr{T}_q$ is the better g.G.-closed the smaller is the number of elementary components of the systems.

The physics of atomic nuclei is a beautiful experimental example for intermediate systems, since nature in the clearest way furnishes these nuclei with an increasing number of subsystems. If gravity is included, one can still increase the number of particles of the nuclei and arrive at systems (neutron stars), which can be described entirely by a theory of the form $\mathscr{P}\mathscr{T}_m$, a theory $\mathscr{P}\mathscr{T}_m$ which one tries to derive more exactly from statistical mechanics, i.e. by the methods developed in §§2 and 3.

Purely theoretically, one can indeed think of a way how one could in principle attain a theory for intermediate systems. One will also use preparation and registration procedures as the basic structures of such a theory for intermediate systems. But in order to find the desired realistic sets of preparation and registration procedures, one could refer to the embedding to be described in the next chapter XI. For the "somewhat larger" action carriers, the critical remarks to follow in XI §4 about the sets $\alpha(\mathscr{Q}'_1 \times L(\hat{S}_m))$ and $\beta(\mathscr{Q}'_2 \times L(\hat{S}_m))$ will be of decisive importance. These sets would be determinable "in principle" with the aid of the embedding methods from XI §6, if in $\mathscr{P}\mathscr{T}_{q\,\mathrm{exp}}$ one would work not with imagined interactions (i.e. imagined Hamiltonians), but rather with the realistic Hamilton operator given in [2] VIII (5.8).

In this way we could "in principle" find a g.G.-closed theory for intermediate systems from the theories $\mathscr{P}\mathscr{T}_m$ *and* their embeddings in $\mathscr{P}\mathscr{T}_{q\,\mathrm{exp}}$ (i.e. with statistical mechanics). Of course, this "in principle" approach is not practical because of its enormous difficulties. Therefore one will further advance towards a theory for intermediate systems only when guided by experience as in nuclear physics. Of course we cannot write a book on theoretical nuclear physics so that the last remarks only indicate to the reader that the many nuclear models should be examined from the viewpoint of seeking an "intermediate theory" that is compatible with $\mathscr{P}\mathscr{T}_q$.

# XI Compatibility of $\mathscr{PT}_q$ with $\mathscr{PT}_{q\,\mathrm{exp}}$

In this section we apply the basic structures presented in §2 to the special case of those macrosystems from which we proceeded in Chapter III (the fundamental domain of a theory of microsystems). But we extend the theory of experiments with microsystems insofar as we also demonstrate how one can theoretically think of more complicated device structures, e.g. a device composed of more than two macrosystems (see I §3 and III §1 and XII). For that reason, we first deal with a partial problem of such a complex device structure.

## §1 Scattering of Microsystems on Macrosystems

Here we shall find, independently of the embedding problem, a description for the scattering of microsystems on macrosystems. Let us describe the microsystems by the ensemble set $K$ and the effect set $L$, and the macrosystems in the form set forth in II. Therefore, the theory to be presented shows an analogy to the scattering theory of microsystems on microsystems, briefly sketched in IX §2 and presented in detail in many books (see the references in IX §2).

In order to describe the scattering of microsystems on macrosystems, we proceed from the situation of Fig. 1 (in I §3). There the macrosystem (0) prepares the microsystems which are "scattered" by the macrosystem (1). The microsystems "arising" from (1) (which may comprise other system types than those which "fall" from (0) onto (1)) are then registered by the macrosystem (2). One readily extends the theory from III to three coupled systems. Hence let us not discuss this extension, but give only brief clarifications.

The base sets are the three index sets $M_0$, $M_1$, $M_2$. Let $M \subset M_1 \times M_2 \times M_2$ be the set of coupled systems. In analogy to $AZ1$ from III §1 we require: If two elements $(x_0, x_1, x_2) \in M$ and $(x'_0, x'_1, x'_2) \in M$ coincide in one component, then the remaining components also coincide.

It is not difficult to carry out all further considerations analogously to III, in particular to introduce a probability function $\lambda_{012}$ in analogy to $\lambda_{12}$. Then, with the aid of $\lambda_{012}$, one can easily introduce a directedness of the interactions, namely that the probability distributions of the registrations on (0) do not depend on the procedures $a_1 \in \mathscr{Q}_1$, $a_2 \in \mathscr{Q}_2$, $b_{10} \in \mathscr{R}_{10}$, $b_{20} \in \mathscr{R}_{20}$, and that the registrations on (1) do not depend on the procedures $a_2 \in \mathscr{Q}_2$, $b_{20} \in \mathscr{R}_{20}$.

With $c_0 \in \mathscr{S}_0$, $a_1 \in \mathscr{Q}_1$, $a_2 \in \mathscr{Q}_2$, $b_{10} \in \mathscr{R}_{10}$, $b_{20} \in \mathscr{R}_{20}$, $b_1 \in \mathscr{R}_1$, $b_2 \in \mathscr{R}_2$, $b_{10} \supset b_1$, $b_{20} \supset b_2$, then

$$\lambda_{012}(c_0 \times (a_1 \cap b_{10}) \times (a_2 \cap b_{20}) \cap M, c_0 \times (a_1 \cap b_1) \times (a_2 \cap b_2) \cap M) \tag{1.1}$$

is the probability for the "response of $b_1$ and $b_2$" when the systems (0) were selected according to $c_0$ and the systems (1) and (2) were prepared according to $a_1$ and $a_2$, respectively.

Let us eliminate the systems (0), which prepare the microsystems falling on (1). To this end, we first think of all systems (0) as preparation systems and (1)+(2) as registration systems. Then, in analogy with III (4.1), by

$$\mathscr{Q}^{(i)} = \{c_0 \times M_1 \times M_2 \cap M \mid c_0 \in \mathscr{S}_0\} \tag{1.2}$$

we can introduce a set of "initial" preparation procedures for the microsystems prepared by (0) and falling on (1) (hence we have chosen the index $i$). Then, with the $\varphi$ from III D5.1.1, to each $a^{(i)} \in \mathscr{Q}^{(i)}$ there is assigned a $\varphi(a^{(i)}) \in K$ with $K$ as the ensemble set of the microsystem.

Corresponding to III (4.2), III (4.3),

$$\begin{aligned} b_0^{(i)} &= M_0 \times (a_1 \cap b_{10}) \times (a_2 \cap b_{20}), \\ b^{(i)} &= M_0 \times (a_1 \cap b_1) \times (a_2 \cap b_2) \end{aligned} \tag{1.3}$$

define a registration method $b_0^{(i)}$ and a registration procedure $b^{(i)}$ for microsystems. Then, with $\psi$ from III D5.1.2, the probability in (1.1) with $a^{(i)} = c_0 \times M_1 \times M_2 \cap M$ and (1.3) becomes

$$\begin{aligned} &\lambda_{012}(c_0 \times (a_1 \cap b_{10}) \times (a_2 \cap b_{20}) \cap M, c_0 \times (a_1 \cap b_1) \times (a_2 \cap b_2) \cap M) \\ &= \mu(\varphi(a^{(i)}), \psi(b_0^{(i)}, b^{(i)})). \end{aligned} \tag{1.4}$$

To each $a^{(i)} \in \mathscr{Q}^{(i)}$ one can bijectively assign a subset $\bar{a}^{(i)} \in M_1$ by

$$\bar{a}^{(i)} = \{x_1 \mid (x_0, x_1, x_2) \in a^{(i)}\}. \tag{1.5}$$

The set $\bar{a}^{(i)}$ then forms a structure $\bar{\mathscr{Q}}^{(i)}$ of selection procedures (over $M_1$) that is isomorphic to $\mathscr{Q}^{(i)}$. Then a mapping $\bar{\mathscr{Q}}^{(i)\prime} \xrightarrow{\bar{\varphi}} K$ is defined by $\bar{\varphi}(\bar{a}^{(i)}) = \varphi(a^{(i)})$.

If we denote by $\bar{\mathscr{Q}}_1$ the structure of selection procedures generated by $\{\bar{a}^{(i)} \cap a_1 \mid \bar{a}^{(i)} \in \bar{\mathscr{Q}}^{(i)}, a_1 \in \mathscr{Q}_1 (a^{(i)}, a_1)$ may be combined$\}$, one can interpret $\bar{\mathscr{Q}}_1$ as a sort of "extended" structure of preparation procedures of the systems (1). Thereby (1) itself is prepared according to $a_1$ *and* likewise the microsystems falling on (1) are prepared according to $a^{(i)}$. With the aid of $\bar{\mathscr{Q}}_1$, one can in practice eliminate the systems (0) by now thinking of (0)+(1) as a preparation system and (2) as a registration system. But one can easily describe the coupled system (0)+(1) *solely* by the indices of the systems (1), since to $x_1 \in M_1$ there is assigned an $x_0 \in M_0$ (bijectively by $(x_0, x_1, x_2) \in M$). For this reason, with

$$\bar{M} = \{(x_1, x_2) \mid \text{there is an } x_0 \text{ with } (x_0, x_1, x_2) \in M\}, \tag{1.6}$$

we can introduce a probability function $\bar{\lambda}_{12}$ by

$$\begin{aligned} &\bar{\lambda}_{12}((\bar{a}^{(i)} \cap a_1 \cap b_{10}) \times (a_2 \cap b_{20}) \cap \bar{M}, (\bar{a}^{(i)} \cap a_1 \cap b_1) \times (a_2 \cap b_2) \cap \bar{M}) \\ &= \lambda_{012}(c_0 \times (a_1 \cap b_{10}) \times (a_2 \cap b_{20}) \cap M, c_0 \times (a_1 \cap b_1) \times (a_2 \cap b_2) \cap M). \end{aligned} \tag{1.7}$$

This stands in complete analogy with the function $\lambda_{12}$ introduced in III, with the single difference that $\mathscr{Q}_1$ is replaced by $\bar{\mathscr{Q}}_1$. We have eliminated the indices of the macrosystems (0) by means of the left side of (1.7), so that the coupled system (0)+(1) appears as "one" preparation system for microsystems. Thus we can take over everything from III word for word if we only replace $\mathscr{Q}_1$ by $\bar{\mathscr{Q}}_1$.

For the microsystems prepared by (0)+(1), we can thus introduce the sets $\mathscr{Q}$, $\mathscr{R}_0$, $\mathscr{R}$ according to III, (4.1) through (4.3). Therefore, in particular,

$$\begin{aligned}\tilde{a}&=(\bar{a}^{(i)}\cap a_1\cap b_{10})\times M_2\cap\bar{M}\in\mathscr{Q},\\ a&=(\bar{a}^{(i)}\cap b_1)\times M_2\cap\bar{M}\in\mathscr{Q},\\ b_0&=M_1\times(a_2\cap b_{20})\cap\bar{M}\in\mathscr{R}_0,\\ b&=M_1\times(a_2\cap b_2)\cap\bar{M}\in\mathscr{R}.\end{aligned}\tag{1.8}$$

From this follows $\varphi(\tilde{a})\in K$, $\varphi(a)\in K$, $\psi(b_0,b)\in L$. Here, $\varphi(\tilde{a})$ is the ensemble of the microsystems that are prepared by (0)+(1), i.e. are coming out of (1) when microsystems prepared according to $a^{(i)}$ fall on (1). Therefore, $\varphi(\tilde{a})$ in this sense is the ensemble of the microsystems "scattered" on (1). One obtains $\varphi(a)$ as a mixture-component of $\varphi(\tilde{a})$ if one sorts out according to the indicator $b_1$. The effect $\psi(b_0,b)$ can be triggered by the scattered microsystems.

In carrying over the considerations of III §6, one must observe that the sets $K_{12m}(\hat{Y})$ and $K_m(\hat{Y}_1)$ are larger than the corresponding sets from III §6.3, since $\bar{\mathscr{Q}}_1$ contains "more preparation possibilities" for the systems (1). For this reason, in particular the support $\hat{S}_1$ is larger than if "no" microsystems fall on (1). Here, "no" says that only the "vacuum" impinges on (1).

Then the formula III (6.4.15) is significant for the description that we are striving for (of the scattering of microsystems on the systems (1)). By III (6.3.11) that formula now reads

$$\begin{aligned}&\bar{\lambda}_{12}((\bar{a}^{(i)}\cap a_1\cap b_{10})\times(a_2\cap b_{20})\cap\bar{M},(\bar{a}^{(i)}\cap a_1\cap b_1)\times(a_2\cap b_2)\cap\bar{M})\\ &\quad=\langle\varphi_{12}((\bar{a}^{(i)}\cap a_1)\times a_2\cap\bar{M}),k_1\,k_2\rangle\\ &\quad=\langle\varphi_1(\bar{a}^{(i)}\cap a_1),k_1\rangle\,\mu(\alpha(\bar{a}^{(i)}\cap a_1,k_1),\beta(a_2,k_2)).\end{aligned}\tag{1.9}$$

Here we used $k_1=\psi_{1s}(b_{10},b_1)$, $k_2=\psi_{2s}(b_{20},b_2)$ and, according to III (6.4.14), $\alpha(a_1,k_1)=\varphi\pi(\bar{a}^{(i)}\cap a_1\cap b_1)=\varphi(a)$ with $a$ according to (1.8), and $\beta(a_2,k_2)=\psi(b_0,b)$ with $b_0,b$ according to (1.8). Also $\varphi_{12}$ and $\varphi_1$ are defined as in III §6.4, except that one everywhere must replace $\mathscr{Q}_1$ by $\bar{\mathscr{Q}}_1$. Then a mapping

$$\bar{\mathscr{Q}}_1\times\psi_{1s}(\mathscr{F}_1)\xrightarrow{\tilde{\alpha}}\breve{K}$$

is defined by

$$\tilde{\alpha}(\bar{a}^{(i)}\cap a_1,k_1)=\langle\varphi_1(\bar{a}^{(i)}\cap a_1),k_1\rangle\alpha(\bar{a}^{(i)}\cap a_1,k_1).\tag{1.10}$$

This, on the basis of (1.9), is linear and norm-continuous in $k_1$.

If one combines (1.4), (1.7) and (1.9), it follows that $\tilde{\alpha}(\bar{a}^{(i)}\cap a_1,k_1)$ does not depend on the $\bar{a}^{(i)}$ explicitly, but only on the $\varphi(a^{(i)})=\tilde{\varphi}(\bar{a}^{(i)})\in K$. Therefore, by

$$\Delta(\tilde{\varphi}(\bar{a}^{(i)}),a_1,k_1)=\tilde{\alpha}(\bar{a}^{(i)}\cap a_1,k_1)\tag{1.11}$$

(and its extension from $\tilde{\varphi}(\tilde{\mathcal{Q}}^{(i)})$ to all of $K$) there is defined a mapping

$$K \times \mathcal{Q}_1 \times L(\hat{S}_1) \xrightarrow{\Delta} \check{K}. \tag{1.12}$$

This $\Delta$ is affine on $K$ and linear and norm-continuous on $L(\hat{S}_1)$. Moreover, $\Delta$ is an additive measure on a subset $\mathcal{Q}_1(a_1)$ of $\mathcal{Q}_1$.

In order to prevent misunderstandings, let us point out that the mapping $\Delta$ can describe changes of the system type by the scattering on (1). For instance, when $w^{(i)} \in K$ is an ensemble in which only a single system type (e.g. electrons) occurs, then $\Delta(w^{(i)}, a_1, k_1)$ can in general contain several system types which might differ from the original type. Of course, we can also describe the situation that the impinging systems will be stuck in (1): For example, for a definite $k_1$ it could follow that $\Delta(w^{(i)}, a_1, k_1)$ is a multiple of the "vacuum ensemble" (cf. IX §1).

We have thus shown that the application of the theory from III to the case of Fig. 1 (in I §3) implies that the scattering of microsystems on a macrosystem can be described by a mapping $\Delta$. It has the properties given by (1.12) and the interpretation that by (1.9) the measure

$$\mu(\Delta(w^{(i)}, a_1, k_1), g) \tag{1.13}$$

is the probability for "the registration of the effect $g$ on the scattered microsystems and the registration of the trajectory effect $k_1$ on the macrosystem".

Now let us show that also conversely a mapping $\Delta$, according to (1.12) with the interpretation (1.13), implies the description of an experiment composed of preparation and registration (postulated as basic in III). In XII §1 we shall see that with $\Delta$ even more complicated experiments can be described.

Therefore, let us choose $\Delta$ according to (1.12) with (1.13). Naturally this assumed $\Delta$ is not yet determined by a theory. One could perhaps obtain a theory for $\Delta$ in a way similar to X §2, where one obtains the unknown dynamics of macrosystems as a consequence of embedding (see §6).

When (1.13) is used with $g = 1$, we obtain

$$\mu(\Delta(w^{(i)}, a_1, k_1), \mathbf{1}) = \langle u_1, k_1 \rangle_1 \tag{1.14}$$

with a trajectory measure $u_1$. By (1.14) with $u_1 = \tilde{\varphi}_1(w^{(i)}, a_1)$, i.e. by $\Delta$, there is defined a mapping

$$K \times \mathcal{Q}'_1 \xrightarrow{\tilde{\varphi}_1} K_{1m}(\hat{Y}) \tag{1.15}$$

which is connected with the $\varphi_1$ in (1.9) by

$$\tilde{\varphi}_1(\tilde{\varphi}(\bar{a}^{(i)}), a_1) = \varphi_1(\bar{a}^{(i)} \cap a_1).$$

In this way we have retrieved the mapping $\varphi_1$ from the mapping $\Delta$.

Since (1.13) is a bilinear form in $(k_1, g)$, one can rewrite it in yet another way. To this end we introduce, with $\mathcal{D}$ as in IV D3.2, the Banach space $C(\hat{S}_1, \mathcal{D})$ of all continuous functions $\hat{S}_1 \to \mathcal{D}$ with the norm topology in $\mathcal{D}$. Then the products $k_1 g$, with $k_1 \in C(\hat{S}_1)$ and $g \in \mathcal{D} \cap L$, are elements of $C(\hat{S}_1, \mathcal{D})$. The probability (1.13) is norm-continuous in $k_1$ and $g$; hence it can be extended uniquely as a linear form on the whole norm-closed subspace of $C(\hat{S}_1, \mathcal{D})$ generated by the $k_1 g$, i.e.

onto all of $C(\hat{S}_1, \mathscr{D})$. Thus $\Delta$ and

$$\mu(\Delta(w^{(i)}, a_1, k_1), g) = \langle u, g k_1 \rangle_{1\mathscr{D}} \tag{1.16}$$

determine a measure $u \in K(\hat{S}_1, \mathscr{D}) \subset C'(\hat{S}_1, \mathscr{D})$. In this context, $\langle \ldots, \ldots \rangle_{1\mathscr{D}}$ is the canonical bilinear form of $C'(\hat{S}_1, \mathscr{D})$, $C(\hat{S}_1, \mathscr{D})$, with $C'(\hat{S}_1, \mathscr{D})$ as the Banach space dual to $C(\hat{S}_1, \mathscr{D})$. Let $K(\hat{S}_1, \mathscr{D})$ be the set of all positive elements $u$ in $C(\hat{S}_1, \mathscr{D})$ with $\langle u, 1 \rangle_{1\mathscr{D}} = 1$.

Therefore, (1.16) with $u = \Gamma(w^{(i)}, a_1)$ determines a mapping

$$K \times \mathscr{D}'_1 \xrightarrow{\Gamma} K(\hat{S}_1, \mathscr{D}). \tag{1.17}$$

With $g = \mathbf{1}$ and (1.14), (1.16) follows

$$\langle \Gamma(w^{(i)}, a_1), k_1 \mathbf{1} \rangle_{1\mathscr{D}} = \langle \tilde{\varphi}_1(w^{(i)}, a_1), k_1 \rangle_1 . \tag{1.18}$$

Thus the mapping

$$\Gamma(w^{(i)}, a) \to \tilde{\varphi}_1(w^{(i)}, a)$$

is given precisely by the reduction operator $C'(\hat{S}_1, \mathscr{D}) \xrightarrow{R_1} C'(\hat{S}_1)$:

$$\tilde{\varphi}_1(w^{(i)}, a) = R_1 \, \Gamma(w^{(i)}, a_1).$$

As is well known, $R_1$ is in general defined by $\langle v, k_1 \mathbf{1} \rangle_{1\mathscr{D}} = \langle R_1 v, k_1 \rangle_1$.

Therefore, $\Delta$ determines $\Gamma$ uniquely, and also conversely. The right side of (1.16), now being

$$\langle \Gamma(w^{(i)}, a_1), g k_1 \rangle_{1\mathscr{D}}, \tag{1.19}$$

is the probability that for impinging $w^{(i)}$ there is registered the trajectory effect $k_1$ on the macrosystem *and* the effect $g$ on the microsystem after the scattering. For fixed $k$, (1.19) is a norm-continuous linear form over $g$. Now let us assume that (1.19) even be $\sigma(\mathscr{B}', \mathscr{B})$-continuous relative to $g$ (necessary on physical grounds because the physical imprecision in $\mathscr{L}$ is given by the $\sigma(\mathscr{L}, \mathscr{B})$-topology; see IV §§2 and 3). Then there is a uniquely determined element $w \in K$ with

$$\langle \Gamma(w^{(i)}, a_1), g k_1 \rangle_{1\mathscr{D}} = \mu(w, g). \tag{1.20}$$

The mapping $\Delta$ is determined by $w = \Delta(w^{(i)}, a_1, k_1)$. An "operator" $k'_1$ is defined by

$$\langle u, g k_1 \bar{k}_1 \rangle_{1\mathscr{D}} = \langle k'_1 u, g \bar{k}_1 \rangle_{1\mathscr{D}} .$$

The "reduction operator" $R$ on the microsystem is defined by $\langle u, g \mathbf{1} \rangle_{1\mathscr{D}} = \mu \langle Ru, g \rangle$; therefore follows

$$\Delta(w^{(i)}, a_1, k_1) = R k'_1 \, \Gamma(w^{(i)}, a_1). \tag{1.21}$$

Also for the scattering of microsystems, one can study the case that the macrosystems are described *completely* by trajectories (similarly to III §6.5). In this case, the mapping $\Delta$ does not depend on the individual $a_1 \in \mathscr{D}'_1$ but only on $\varphi_1^{(0)}(a_1) \in K_{1m}(\hat{Y}_1)$, where $\varphi_1^{(0)}(a_1)$ is defined by

$$\varphi_1^{(0)}(a_1) = \tilde{\varphi}_1(w_0, a_1) \tag{1.22}$$

with $w_0$ as the "vacuum ensemble" (cf. IX §1). Then, with $K_{1m}^{(0)}=\overline{\mathrm{co}}\,\varphi_1^{(0)}\,\mathcal{Q}_1'$ (closure in the norm), from $\Delta$ and $\Gamma$ one can obtain the mappings

$$K\times K_{1m}^{(0)}\times L(\hat{S}_1)\xrightarrow{\tilde{\Delta}}\check{K},$$

$$K\times K_{1m}^{(0)}\xrightarrow{\tilde{\Gamma}}K(\hat{S}_1,\mathcal{D}),$$

with

$$\Delta(w^{(i)},a_1,k_1)=\tilde{\Delta}(w^{(i)},\varphi_1^{(0)}(a_1),k_1) \tag{1.23}$$

and

$$\Gamma(w^{(i)},a_1)=\tilde{\Gamma}(w^{(i)},\varphi_1^{(0)}(a_1)). \tag{1.24}$$

One may call $\tilde{\Gamma}$ a scattering operator, because one can identify $K\times K_{1m}^{(0)}$ with a subset of $K(\hat{S}_1,\mathcal{D})$. Then $\tilde{\Gamma}$ transforms the ensemble $w^{(i)}\,\varphi_1^{(0)}(a_1)$ "before" the scattering into the ensemble $\tilde{\Gamma}(w^{(i)},\varphi_1^{(0)}(a_1))$ "after" the scattering.

In the following §§2 through 5 we shall show that the mapping $\Delta$ comprises the theory of preparation and registration presented in III.

## §2 Preparation

We can obtain the preparation from $\Delta$ if in particular we set $w^{(i)}=w_0$, where $w_0$ is the vacuum ensemble (cf. IX §1). We define $\tilde{\alpha}$ as the mapping $\mathcal{Q}_1'\times L(\hat{S}_1)\xrightarrow{\tilde{\alpha}}K$ by

$$\tilde{\alpha}(a_1,k_1)=\Delta(w_0,a_1,k_1) \tag{2.1}$$

and

$$\varphi_1^{(0)}(a_1)=\tilde{\varphi}_1(w_0,a_1), \tag{2.2}$$

with $\tilde{\varphi}_1$ from (1.15).

For

$$\langle\varphi_1(a_1),k_1\rangle_1\neq 0,$$

$$\alpha(a_1,k_1)=\frac{\tilde{\alpha}(a_1,k_1)}{\langle\varphi_1(a_1),k_1\rangle_1}\in K \tag{2.3}$$

is defined. We have thus constructed the mapping $\alpha$ in III (6.4.14) with the aid of $\Delta$. With $k_1=\psi_{1s}(b_{10},b_1)$, by III (6.4.14) there also follows the mapping $\varphi$, describing the ensembles corresponding to the preparation procedures from $\mathcal{Q}$. Therefore, knowledge of $\Delta$ enables us to calculate the prepared ensembles.

The mapping $\tilde{\alpha}(a_1\,k_1)=\Delta(w_0,a_1,k_1)$ yields (for fixed $a_1$) a mapping

$$L(\hat{S}_1)\xrightarrow{k}K,$$

as introduced in V §8. From this mapping follows (see V (8.4)) the ideal preparator

$$\Sigma_{a_1}\xrightarrow{k}\check{K} \tag{2.4}$$

of the ensemble $w=\tilde{\alpha}(a_1,1)$. This preparator is fixed by $a_1$ and belongs to the preparing macrosystems.

Our goal is to construct the right side of III (6.4.15). Naturally, one cannot do this solely with the set $M_1$ of the macrosystems used for preparation. Therefore, we must see whether we can also obtain $\beta$ from III (6.4.14) by means of $\Delta$.

## §3 Registration

Let us now label the macrosystems with the index 2 instead of 1! Since we will use the systems (2) only for registration, in this section we consider the mapping $\tilde{\varphi}_1$ that follows from $\Delta$ according to (1.15) in the notation

$$K \times \mathscr{Q}'_2 \xrightarrow{\tilde{\varphi}_2} K(\hat{Y}_2). \tag{3.1}$$

Thus $\langle \tilde{\varphi}_2(w^{(i)}, a_2), k_2 \rangle_2$ is the probability for the trajectory effect $k_2$, when microsystems of the ensemble $w^{(i)}$ impinge on the system (2). This $\langle \tilde{\varphi}_2(w^{(i)}, a_2), k_2 \rangle_2$ is a positive affine functional on $K$, since $\tilde{\varphi}_2$ is affine on $K$. Hence there exists (see IV T4.7) a $g \in L$ such that

$$\langle \tilde{\varphi}_2(w^{(i)}, a_2), k_2 \rangle_2 = \mu(w^{(i)}, g). \tag{3.2}$$

By (3.2) with $g = \beta(a_2, k_2)$, a mapping

$$\mathscr{Q}'_2 \times L(\hat{S}_2) \xrightarrow{\beta} L$$

is defined, which corresponds to the $\beta$ in III (6.4.14). Therefore, the mapping $\beta$ is also determined by $\Delta$. In turn, from $\beta$ the mapping $\psi$ follows by III (6.4.14). Hence $\Delta$ also allows the calculation of "what" was measured by the macrosystem (2) (see §6).

This $\Delta$ especially allows to calculate the "ideal observable" specified by $a_2$ (see V §5). To do this we must only write the probability $\tilde{\mu}(w, \beta(a_2, k_2))$ in the form

$$\tilde{\mu}(w, \beta(a_2, k_2)) = \langle u, k_2 \rangle_2 ;$$

then we get the mapping

$$K \xrightarrow{\tilde{S}_{a_2}} K(\hat{S}_2)$$

in V (5.2) by $w \to u$. From $\tilde{S}_{a_2}$ follows the ideal observable

$$\Sigma_{a_2} \xrightarrow{F_{a_2}} L, \tag{3.3}$$

as explained in V §5. Therefore, from the knowledge of $\Delta$ follows that of the ideal observable determined by the macrosystem.

## §4 Coupling of Preparation and Registration

With $w^{(i)}$ as $\alpha(a_1, k_1)$ according to §2, from (3.2) we obtain

$$\mu(\alpha(a_1, k_1), \beta(a_2, k_2)) \tag{4.1}$$

as the probability for the trajectory effect $k_2$. Let us multiply this by the $\langle\varphi_1(a_1), k_1\rangle_1$ from §2, the probability that corresponds to the trajectory effect $k_1$ on the system (1). Then finally, with $\tilde{\alpha}$ from §2,

$$\mu(\tilde{\alpha}(a_1, k_1), \beta(a_2, k_2)) \tag{4.2}$$

is the probability that (in the coupled experiment) the trajectory effect $k_1$ appears on the system (1) and $k_2$ appears on (2).

Therefore, we should be able to obtain the right side of III (6.4.15) from (4.2).

We see immediately that (4.2) is bilinear and norm-continuous in $k_1, k_2$ since $\Delta$ is norm-continuous in $k_1$ and the left side of (3.2) is norm-continuous in $k_2$. Therefore, there is a $u \in C'(\hat{S}_1 \times \hat{S}_2)$ such that

$$\mu(\tilde{\alpha}(a_1, k_1), \beta(a_2, k_2)) = \langle u, k_1 k_2\rangle. \tag{4.3}$$

Thus $\varphi_{12}$ in III (6.4.15) is uniquely determined by (4.3), with $\varphi_{12}(a_1, a_2) = u$. Therefore we can also write the probability (4.2) in the form

$$\mu(\tilde{\alpha}(a_1, k_1), \beta(a_2, k_2)) = \langle\varphi_{12}(a_1 \times a_2 \cap M), k_1 k_2\rangle. \tag{4.4}$$

We have thus eliminated the microsystems as action carriers by the right side of (4.4).

The expression $\varphi_{12}(a_1 \times a_2 \cap M)$ defined by (4.4) also guarantees the directedness of the interaction, since (4.4) for $k_2 = \mathbf{1}$ implies

$$\langle\varphi_{12}(a_1 \times a_2 \cap M), k_1 \mathbf{1}\rangle = \mu(\tilde{\alpha}(a_1, k_1), \beta(a_2, \mathbf{1})),$$

and since $g = \beta(a_2, \mathbf{1})$ in (3.2) is determined by

$$\langle\tilde{\varphi}_2(w^{(i)}, a_2), \mathbf{1}\rangle_2 = \mu(w^{(i)}, \beta(a_2, \mathbf{1})).$$

According to (1.4) we get $\tilde{\varphi}_2(w^{(i)}, a_2) \in K(\hat{Y}_2)$ and hence $\langle\tilde{\varphi}_2(w^{(i)}, a_2), \mathbf{1}\rangle_2 = 1$, whence $\beta(a_2, \mathbf{1}) = 1$ follows. This gives

$$\langle\varphi_{12}(a_1 \times a_2 \cap M), k_1 \mathbf{1}\rangle = \mu(\tilde{\alpha}(\alpha_1, k_1), \mathbf{1}),$$

which is independent of $a_2$. According to (2.2) we explicitly find

$$\mu(\tilde{\alpha}(a_1, k_1), \mathbf{1}) = \langle\varphi(a_1), k_1\rangle_1.$$

With III (6.3.11), from (4.4) with $k_1 = \psi_{1s}(b_{10}, b_1)$, $k_2 = \psi_{2s}(b_{20}, b_2)$ we obtain the values

$$\lambda_{12}(a_1 \times a_2 \cap M \cap b_{10} \times b_{20}, a_1 \times b_{20}, a_1 \times a_2 \cap M \cap b_1 \times b_2).$$

This due to III T2.8 determines the function $\lambda_{12}$ completely.

Thus the description of the scattering of microsystems on macrosystems by the mapping $\Delta$ from §1 encompasses the theory of preparation and registration presented in III.

From these considerations, one could get the impression that it would *suffice* to calculate the mapping $\Delta$ in order to prove backwards the "correctness" of the main laws introduced in VI. This is a widespread error! The wrong view causing it can be formulated as follows:

If one establishes the usual form of quantum mechanics with Hilbert spaces and superselection rules, then one can (by the methods of statistical mechanics) construct the mapping $\Delta$ from §1 for the scattering of microsystems (also see §6). With the aid of $\Delta$, one then succeeds in describing all experiments with microsystems by means of quantum mechanics (as we have just constructed $\lambda_{12}$). Experiments do not *contradict* these theoretical considerations. One believes that by this compatibility with experiments also such remarkable things as complementary observables (see V §4) and complementary de-mixtures (see V §7) would be established.

But the *existence* of the mapping $\Delta$ from §1 *alone* still does not guarantee the main laws in VI, as we shall presently see.

According to §2, $\Delta$ determines the mapping $\mathscr{Q}'_1 \times L(\hat{S}_1) \xrightarrow{\alpha} K$ (to be precise, on the subset of all $(a_1, k_1)$ for which $\langle \varphi_1(a_1), k_1 \rangle_1 \neq 0$). But there is no guarantee that the set $\alpha(\mathscr{Q}'_1 \times L(\hat{S}_1))$ is a subset that in "physical approximation" is norm-dense in $K$.

Similarly, the $\beta$ that follows by (3.3) from $\Delta$ does not guarantee that $\beta(\mathscr{Q}'_2 \times L(\hat{S}_2))$ is "physically" dense in $L = [\mathbf{0}, \mathbf{1}]$ in the $\sigma(\mathscr{B}', \mathscr{B})$-topology. Or, to make it still clearer: With (4.1), we still are far from showing that

$$\mu(w, g_1) = \mu(w, g_2) \qquad \text{for all } w \in \alpha(\mathscr{Q}'_1 \times L(\hat{S}_1))$$

already implies $g_1 = g_2$ and that

$$\mu(w_1, g) = \mu(w_2, g) \qquad \text{for all } g \in \beta(\mathscr{Q}'_2 \times L(\hat{S}_2))$$

already implies $w_1 = w_2$.

Therefore, the *existence* of $\Delta$ in itself does not yet guarantee that the partition into classes of $\mathscr{Q}'$ and $\mathscr{F}$ in III §5.1 is again reproduced. On the contrary, we must expect that already for large molecules the partition into classes will *not* be reproduced. We pointed this out already in X §4.

In § 6 the calculation of $\Delta$ will seem possible for given macrosystems. Nevertheless it appears remote to state something about the sets $\alpha(\mathscr{Q}'_1 \times L(\hat{S}_1))$ and $\beta(\mathscr{Q}'_2 \times L(\hat{S}_2))$ on the basis of the embedding in $\mathscr{PT}_{q\,\mathrm{exp}}$ (to be described in §6). In fact till now we know no theoretical methods to survey all physically possible macrosystems and their trajectories. If one thinks about "all" macroscopic devices that are constructible then it appears quite hopeless that we humans could ever obtain a systematic overview of "all possible" devices (a difficult problem of experimental physics just is continually to invent new devices). Rather, we must be content with a theoretical description of *special* macrosystems by means of statistical mechanics.

Conversely, one will perhaps succeed in proving, with the aid of the considerations in §6, that for large molecules the partition into classes of $\mathscr{Q}'$ and $\mathscr{F}$ must be revised, i.e. that for large molecules quantum mechanics can be improved by a more comprehensive theory (see X §4).

## §5 Macrosystems as Transpreparators

In IX §3, we saw how scattering processes of microsystems on microsystems can be used as transpreparators. Here we shall show that $\Delta$ from §1 determines a transpreparator.

One sees immediately that for fixed $a_1$ and $k_1$ the mapping

$$K \xrightarrow{\Delta(\ldots, a_1, k_1)} \check{K} \tag{5.1}$$

is an operation in the sense of V D11.1. This operation is a function of $k_1$ (for fixed $a_1$). Similarly as in V §5 and in V §8, this function can be extended to the elements $\eta$ of the characteristic functions of Borel sets.

In §1 we denoted the support of $\tilde{\varphi}_1(K \times \mathcal{Q}'_1)$ by $\hat{S}_1$. Let $\mathcal{B}(\hat{S}_1)$ be the Borel field corresponding to $\hat{S}_1$, so that we can define $\Delta(w, a_1, \eta_1)$ also for the elements $\eta_1 \in \mathcal{B}(\hat{S}_1)$. Then (for fixed $a_1$),

$$K \xrightarrow{\Delta(\ldots, a_1, \eta_1)} \breve{K} \tag{5.2}$$

to each $\eta_1$ assigns an operation $\mathcal{O}_{a_1}(\eta_1)$ with $\mathcal{O}_{a_1}(\eta_1) w = \Delta(w, a_1, \eta_1)$. Thus $\mathcal{O}_{a_1}(\eta)$ is a $\sigma$-additive measure over $\mathcal{B}(\hat{S}_1)$, such that (see V D11.3)

$$\mathcal{B}(\hat{S}_1) \xrightarrow{\mathcal{O}_{a_1}} \Pi \tag{5.3}$$

represents a transpreparator. We call it the ideal transpreparator generated by the macrosystem prepared according to $a_1$.

Transpreparators played a large role in the discussion of the measurement process, especially after the rise of quantum mechanics and in the efforts to find its interpretation.

J. v. Neumann in his famous book "Mathematical Foundations of Quantum Mechanics" (see [25] IV.3) introduced an unnecessarily narrow definition of the concept of measurement. This measurement process of J. v. Neumann is today called measurement of the first kind (also see [2] XVII §5). In this context, requirements are placed on a transpreparator which are not discussed in more detail here (see [2] XVII §5). But here arises a question similar to that discussed in V §5 and V §8, namely about the "realization" of transpreparators. We regard an approximate "realization" of a transpreparator $\Sigma \to \Pi$ as given if there is a homomorphic mapping $h$ of a finite subring $\Sigma'$ of $\Sigma$ into a $\mathcal{B}(\hat{S}_1)$, such that (4.5.2) restricted to $h\Sigma'$ represents the prescribed transpreparator approximately. Here we shall not discuss how one can conceive "approximately" in analogy to V §§5 and 8.

We introduced the axiom AOb in V §5 and the axiom Apr in V §8 to emphasize that one can realize approximately "all" observables resp. "all" preparators. Here we consciously will impose *no* requirements on the realizability of transpreparators. Therefore in particular we do not require that transpreparators corresponding to measurements of the first kind are approximately realizable.

Rather, we leave the question of the physically possible transpreparators for microsystems (physically possible in the sense of [3] §10.4 and XIII §4.6) to a theory of the mapping $\Delta$ from §1. Such a theory cannot be established solely by an extrapolated quantum mechanics, but only by a theory more comprehensive than $\mathcal{PT}_{q\,\mathrm{exp}}$, for which we shall set down some basic theorems in §6. Therefore, concerning the transpreparators we suspect that $\mathcal{PT}_q$ is also *not* a g.G.-closed theory for microsystems (cf. [3] § 10.3 and XIII §4.3).

In concluding this subsection, let it again be emphasized, as discussed in detail in [2] XVII §4.3, that the "jump" from $w^{(i)}$ to $w = \Delta(w^{(i)}, a, k)$ is only a mathematical jump from the impinging ensemble $w^{(i)}$ to the scattered ensemble $w$. "In nature" a complicated interaction process with the macroscopic system lies between $w^{(i)}$ and $w$.

Unfortunately, the expositions of J. v. Neumann have been misunderstood (see [25] IV.3), since one conceived of the *idealizations* of a measurement process (that he consciously introduced) as *real* postulates. In addition one misinterpreted the jump from $w^{(i)}$ to $w$ as a "jump" of each individual microsystem. Thus there arose the apparent problem of a "collapse of the wave packet of a microsystem" (also see [2] XVII §4.3). When one looks at the mountains of literature that have been written about these collapse processes that do not even exist in nature, one cannot blame the author of this book for wishing that this apparent problem might vanish quite soon into oblivion. And if the present book together with [2] would only dissolve this apparent problem, that would already be an achievement!

## §6 The Problem of Embedding the Scattering Theory of Microsystems on Macrosystems in $\mathscr{PT}_{q\,\mathrm{exp}}$

We can proceed relatively quickly in the formulation of the embedding problem of the theory presented in §1 (of the scattering of microsystems on macrosystems in an extrapolated quantum mechanics $\mathscr{PT}_{q\,\mathrm{exp}}$), since we have discussed in detail the simpler case of an embedding in X §§2.2 and 2.3.

Concerning scattering theory, we proceed from the formula (1.16) which together with the definition of $\Gamma$ (see also (1.19)) reads

$$\mu(\Delta(w^{(i)}, a_1, k_1), g) = \langle \Gamma(w^{(i)}, a_1), g k_1 \rangle_{1\mathscr{D}} \tag{6.1}$$

with $k_1 \in \psi_{1s}(\mathscr{F}_1) \subset L(\hat{S}_1)$. This $\mu$ gives the probability for the simultaneous response of the trajectory effect $k_1$ (for the macrosystem) and of the effect $g$ (for the scattered microsystem). This probability (6.1) depends on the impinging microsystem (i.e. on $w^{(i)}$) and on the preparation $a_1$ of the macrosystem.

It was also shown in §1 that $\Gamma$ and $\Delta$ determine each other uniquely. For our embedding problem, it turns out that the right side of (6.1) is more suitable.

In (6.1) one can not yet recognize how a time displacement of the registration should be described; but we can easily make up for this. Due to the physical meaning of such a time displacement (as described for macrosystems in II §§4.1 and 4.2 and for microsystems in IX §1), it is easy to find this description of the time displacement for the right side of (6.1). With $V_\tau^{(s)}$ as in II (4.2.14), and $U_\tau^{(0)}$ as the $U(\tau)$ given in IX (2.5) for microsystems alone (on that account the index (0)),

$$\langle \Gamma(w^{(i)}, a_1), (U_\tau^{(0)}\, g\, U_\tau^{(0)+}) V_\tau^{(s)} k_1 \rangle_{1\mathscr{D}} \tag{6.2}$$

is the probability for the registration displaced by a time $\tau$ relative to the case in (6.1).

According to §1, in (6.2) we have $k_1 = \psi_{1s}(b_{10}, b_1)$ and $g = \psi(b_0, b)$, with $b_0, b$ as in (1.8). Thus (6.2) takes the form

$$\langle \Gamma(w^{(i)}, a_1),\, U_\tau^{(0)}\, \psi(b_0, b)\, U_\tau^{(0)+}\, \psi_{1s}(R_\tau b_{10}, R_\tau b_1) \rangle_{1\mathscr{D}}. \tag{6.3}$$

According to (1.9), now (6.1) equals

$$\bar{\lambda}_{12}((\bar{a}^{(i)} \cap a_1 \cap b_{10}) \times (a_2 \cap b_{20}) \cap \bar{M},\ (\bar{a}^{(i)} \cap a_1 \cap b_1) \times (a_2 \cap b_2) \cap \bar{M}) \tag{6.4}$$

with $w^{(i)}=\tilde{\varphi}(\bar{a}^{(i)})=\varphi(a^{(i)})$. In order to introduce the time displacement in (6.4), we must note that the *entire* registration device (2) must be displaced by $\tau$. Having been described in III §7, this displacement was used to introduce the time displacement for microsystems in IX §1.

In III §7, the transformations of the system (2) relative to the system (1) were denoted by $\delta$. In this sense let $\delta_\tau$ be such a displacement by the time $\tau$. Then (6.3) equals

$$\bar{\lambda}_{12}((\bar{a}^{(i)}\cap a_1\cap R_\tau b_{10})\times(\delta_\tau a_2\cap\delta_\tau b_{20})\cap\bar{M},\\ (\bar{a}^{(i)}\cap a_1\cap R_\tau b_1)\times(\delta_\tau a_2\cap\delta_\tau b_2)\cap\bar{M}). \tag{6.5}$$

This expression is appropriate for formulating the embedding of the scattering theory of microsystems on macrosystems in a $\mathscr{P}\mathscr{T}_{q\,\mathrm{exp}}$.

In this connection, one must observe that in an extrapolated quantum mechanics we want to embed only the macrosystems (1) with the impinging and the scattered microsystems, and *not* the macrosystems (2). Furthermore, the systems (2) serve only as auxiliary systems also in $\mathscr{P}\mathscr{T}_{q\,\mathrm{exp}}$, in order to define the effect $g=\psi(b_0,b)$. For this reason, analogous to the transition from $a^{(i)}$ to $\bar{a}^{(i)}$, let us think of $a_2$ and $b_2$ as selection procedures over $M_1$ (the set of systems (1)).

The mapping $(x_1,x_2)\to x_1$ is a bijective mapping $\bar{M}\to M_1$ which leads in a canonical way to a mapping $M_1\times b_{20}\to\bar{b}_{20}$, $M_1\times b_2\to\bar{b}_2$, $M_1\times a_2\to\bar{a}_2$, where the $\bar{b}_{20}$, $\bar{b}_2$ and $\bar{a}_2$ generate selection procedures over $M_1$, isomorphic to $\mathscr{R}_{20}$, $\mathscr{R}_2$ and $\mathscr{Q}_2$ respectively. (The mapping $\bar{M}\to M_1$ just introduced was already used in III §6.6. There denoted by $\pi^{-1}$, it also served to interpret the systems (2) as registration devices for the systems (1)!)

As in III §6.6, we now interpret the $b_{10}\cap\bar{a}_2\cap\bar{b}_{10}$ as registration methods and the $b_1\cap\bar{a}_2\cap\bar{b}_2$ as registration procedures. (Strictly speaking, we should have considered the structures generated by the $b_{10}\cap\bar{a}_2\cap\bar{b}_{10}$ resp. by the $b_1\cap\bar{a}_2\cap\bar{b}_2$ as selection procedures over $M_1$. This shall not be done here, since such extensions were treated several times in detail in III.) It is entirely in this sense that we denote the images of $\delta_\tau a_2$, $\delta_\tau b_{20}$, $\delta_\tau b_2$ by $R_\tau\bar{a}_2$, $R_\tau\bar{b}_{20}$, $R_\tau\bar{b}_2$, since we have interpreted $R_\tau$ as time displacement of the registrations of the systems (1).

Thus (6.4) defines a probability function

$$\lambda_1(\bar{a}^{(i)}\cap a_1\cap b_{10}\cap\bar{a}_2\cap\bar{b}_{20},\bar{a}^{(i)}\cap a_1\cap b_1\cap\bar{a}_2\cap\bar{b}_2)\\ =\bar{\lambda}_{12}((\bar{a}^{(i)}\cap a_1\cap b_{10})\times(a_2\cap b_{20})\cap\bar{M},(\bar{a}^{(i)}\cap a_1\cap b_1)\times(a_2\cap b_2)\cap\bar{M}) \tag{6.6}$$

for the systems (1), for which (6.5) and (6.3) imply

$$\lambda_1(\bar{a}^{(i)}\cap a_1\cap R_\tau b_{10}\cap R_\tau\bar{a}_2\cap R_\tau\bar{b}_{20},\bar{a}^{(i)}\cap a_1\cap R_\tau b_1\cap R_\tau\bar{a}_2\cap R_\tau\bar{b}_2)\\ =\langle\Gamma(\tilde{\varphi}(\bar{a}^{(i)}),a_1),U_\tau^{(0)}\,\varphi(b_0,b)\,U_\tau^{(0)+}\,\psi_{1s}(R_\tau b_{10},R_\tau b_1)\rangle_{1\mathscr{D}}. \tag{6.7}$$

The left side of (6.7) is completely analogous to that of X (2.2.8), if we there replace the preparation procedure $a$ by the extended preparation procedure $\bar{a}^{(i)}\cap a_1$, the registration method $b_0$ by the extended registration method $b_{10}\cap\bar{a}_2\cap\bar{b}_{20}$, and the registration procedure $b$ by the extended registration procedure $b_1\cap\bar{a}_2\cap\bar{b}_2$. Hence we can immediately carry over the considerations from X §2.2.

The ensemble sets and effect sets from $\mathscr{P}\mathscr{T}_{q\,\mathrm{exp}}$ (i.e. of the systems composed of microsystems and many-particle systems) must be distinguished from the corresponding sets $K$ and $L$ of the microsystems alone. For this distinction, let us denote the sets from $\mathscr{PT}_{q\,\mathrm{exp}}$ by $K_c$ and $L_c$ and the corresponding Banach spaces by $\mathscr{B}_c$ and $\mathscr{B}'_c$. In analogy to X §2.2, for an embedding $i$ we require

$$\begin{aligned}&\lambda_1(\bar{a}^{(i)}\cap a_1\cap R_\tau b_{10}\cap R_\tau \bar{a}_2\cap R_\tau \bar{b}_{20}, \bar{a}^{(i)}\cap a_1\cap R_\tau b_1\cap R_\tau \bar{a}_2\cap R_\tau \bar{b}_2)\\ &\quad=\mu(\varphi_c(i(\bar{a}^{(i)}\cap a_1)), U_\tau \psi_c(i(b_{01}\cap \bar{a}_2\cap \bar{b}_{02}), i(b_1\cap \bar{b}_1\cap \bar{a}_2\cap \bar{b}_2))\, U_\tau^+). \qquad (6.8)\end{aligned}$$

Here, $\varphi_c$ is the mapping of the preparation procedure from $\mathscr{PT}_{q\,\mathrm{exp}}$ into $K_c$, and $\psi_c$ is the mapping of the effect procedure from $\mathscr{PT}_{q\,\mathrm{exp}}$ into $L_c$. The time displacement operator defined in $\mathscr{PT}_{q\,\mathrm{exp}}$ takes $g$ into $U_\tau g U_\tau^+$. Therefore, $U_\tau$ contains the interaction of the microsystem with the macrosystem on which the microsystems are scattered.

Naturally, there must exist a relation between $K$ and $K_c$, resp. between $L$ and $L_c$ which declares that microsystems described by $K, L$ form a "subsystem" of the system described by $K_c$, $L_c$. This problem is well known within the framework of normal quantum mechanics (see IX §2). Therefore it is well defined within $\mathscr{PT}_{q\,\mathrm{exp}}$ how the connection between $K, L$ and $K_c$, $L_c$ must be formulated. We shall return to this connection below.

From (6.8) together with (6.7) we get the embedding condition

$$\begin{aligned}&\langle \Gamma(\tilde{\varphi}(\bar{a}^{(i)}), a_1), U_\tau^{(0)}\psi(b_0, b)\, U_\tau^{(0)+}\, V_\tau^{(s)}\psi_{1s}(b_{10}, b_1)\rangle_{1\mathscr{D}}\\ &\quad=\mu(\varphi_c(i(\bar{a}^{(i)}\cap a_1)), U_\tau \psi_c(i(b_{10}\cap \bar{a}\cap \bar{b}_{20}), i(b_1\cap \bar{a}_2\cap \bar{b}_2)\, U_\tau^+), \qquad (6.9)\end{aligned}$$

where $b_0, b$ are given by (1.8). With the mapping $\tilde{\psi}$ defined by

$$\tilde{\psi}(\bar{a}_2\cap \bar{b}_{20}, \bar{a}_2\cap \bar{b}_2)=\psi(b_0, b),$$

one can finally rewrite (6.9) as

$$\begin{aligned}&\langle \Gamma(\tilde{\varphi}(\bar{a}^{(i)}, a_1), U_\tau^{(0)}\tilde{\psi}(\bar{a}_2\cap \bar{b}_{20}, \bar{a}_2\cap \bar{b}_2)\, U_\tau^{(0)+}\, V_\tau^{(s)}\psi_{1s}(b_{10}, b_1)\rangle_{1\mathscr{D}}\\ &\quad=\mu(\varphi_c(i(\bar{a}^{(i)}\cap a_1)), U_\tau \psi_c(i(b_{10}\cap \bar{a}_2\cap \bar{b}_{20}), i(b_1\cap \bar{a}_2\cap \bar{b}_2))\, U_\tau^+), \qquad (6.10)\end{aligned}$$

analogously to X (2.3.4a).

From the *right* side of (6.10) it still cannot be recognized that $\bar{a}^{(i)}$ refers to the preparation of the impinging microsystems and that $\bar{a}_2\cap \bar{b}_{20}$, $\bar{a}_2\cap \bar{b}_2$ refer to the registration of the scattered microsystems. But it can be recognized on the left side from the ensemble $w^{(i)}=\tilde{\varphi}(\bar{a}^{(i)})$ of the impinging microsystems, and from the effect $g=\tilde{\psi}(\bar{a}_2\cap \bar{b}_{20}, \bar{a}_2\cap \bar{b}_2)$ triggered by the scattered microsystems. Therefore, we must yet introduce the mentioned relation between $K$ and $K_c$ (and afterwards that between $L$ and $L_c$). To this end, we consider the system composed of microsystems and macrosystems in $\mathscr{PT}_{q\,\mathrm{exp}}$. Still more specifically, we think of the total system as composed (see IX §2) of an *impinging* microsystem and a many-particle system as the "scatterer". Hence, for $\varphi_c(i(\bar{a}^{(i)}\cap a_1))$ "at time $t=0$" (i.e. "before" the scattering, see IX §2), we require

$$\varphi_c(i(\bar{a}^{(i)}\cap a_1))\approx\{w^{(i)}\times \varphi_{1c}(ia_1)\}_s.$$

The interaction of the impinging systems with the scatterer determines a "wave operator" $\Omega$ (as in IX (2.10)) so that the ensemble prepared according to $i(\bar{a}^{(i)}\cap a_1)$ is given by

$$\varphi_c(i(\bar{a}^{(i)}\cap a_1))=\Omega\{w^{(i)}\times \varphi_{1c}(ia_1)\}_s. \qquad (6.11)$$

where $w^{(i)}=\tilde{\varphi}(\bar{a}^{(i)})$. Therefore (6.11) is a supplementary condition to (6.10) which describes the connection between $w^{(i)}\in K$ and $\varphi_c(i(\bar{a}^{(i)}\cap a_1))\in K_c$.

(Attention should be paid to the fact that $\Omega$ works as the unit operator if $w^{(i)}$ is the vacuum, i.e. for the case of a preparation device. We may replace $\varphi_c(i(\bar{a}^{(i)} \cap a_1))$ by $\varphi_{1c}(ia_1)$; and this has the character of a decaying state ([2] XVII §6.5). This is essential if we want to get $\tilde{\alpha}$ and $k$ in §2.)

We assumed that $w^{(i)}$ and $\varphi_{1c}(ia_1)$ as ensembles so "fit one another" that the time $t=0$ falls "before" the collision. This assumption is not valid for all $w^{(i)}$ and all $\varphi_{1c}(ia_1)$, i.e. not for all $a_1$. In the general scattering theory for microsystems (as described briefly in IX §2), one can always find an appropriate time $\tau^{(i)}$ "before" the scattering (see [2] XVI). But for the scattering of microsystems on macrosystems it is essential that $t=0$ is such a time before scattering since for macrosystems a limit for $t \to -\infty$ in general makes no sense (see II §1). Hence such a limit is *no* realistic part of $\mathscr{PT}_{q\,\exp}$. This is justified by an embedding of $\mathscr{PT}_m$! Indeed, the fact that not all $w^{(i)}$ and $\varphi_{1c}(ia_1)$ "fit one another" reflects the combination problem between the preparation device (0) (see §1) and the device (1) used for registration. Therefore, the combination problem (III §5.1) is and remains an essential aspect of a theory of microsystems.

Finally, together with (6.11) we obtain from (6.10) the embdding condition

$$\begin{aligned}&\langle \Gamma(w^{(i)}, a_1), U_\tau^{(0)} \tilde{\psi}(\bar{a}_2 \cap \bar{b}_{20}, \bar{a}_2 \cap \bar{b}_2)\, U_\tau^{(0)+}\, V_\tau^{(s)}\, \psi_{1s}(b_{10}, b_1)\rangle_{1\mathscr{D}}\\ &\quad = \mu(\Omega\{w^{(i)} \times \varphi_{1c}(ia_1)\}_s, U_\tau\, \psi_c(i(b_{10} \cap \bar{a}_2 \cap \bar{b}_{20}), i(b_1 \cap \bar{a}_2 \cap \bar{b}_2))\, U_\tau^+). \qquad (6.12)\end{aligned}$$

For abbreviation, we define the operators

$$\mathscr{U}_\tau^{(0)} g = U_\tau^{(0)} g\, U_\tau^{(0)+} \qquad \text{for } g \in L$$

and

$$\mathscr{U}_\tau g = U_\tau g\, U_\tau^+ \qquad \text{for } g \in L_c.$$

With $g \in L$ and $k \in L(\hat{S}_1)$,

$$(\mathscr{U}_\tau^{(0)} \times V_\tau^{(s)})\, g k = (\mathscr{U}_\tau^{(0)} g)(V_\tau^{(s)} k)$$

defines an operator $\mathscr{U}_\tau^{(0)} \times V_\tau^{(s)}$ in the space $C(\hat{S}_1, \mathscr{D})$. Then we can rewrite (6.12) in the form

$$\begin{aligned}&\langle (\mathscr{U}_\tau^{(0)\prime} \times V_\tau^{(s)\prime})\, \Gamma(w^{(i)}, a_1), \tilde{\psi}(\bar{a}_2 \cap \bar{b}_{20}, \bar{a}_2 \cap \bar{b}_2)\, \psi_{1s}(b_{10}, b_1)\rangle_{1\mathscr{D}}\\ &\quad = \mu(\mathscr{U}_\tau'\, \Omega\{w^{(i)} \times \varphi_{1c}(ia_1)\}_s, \psi_c(i(b_{10} \cap \bar{a}_2 \cap \bar{b}_{20}), i(b_1 \cap \bar{a}_2 \cap \bar{b}_2)). \qquad (6.13)\end{aligned}$$

Let $\mathscr{R}(b_{10} \cap \bar{a}_2 \cap \bar{b}_{20})$ denote the Boolean ring generated by all the $b_1 \cap \bar{a}_2 \cap \bar{b}_2$ (with $b_1 \subset b_{10}$, $\bar{b}_2 \subset \bar{b}_{20}$). An additive measure

$$\mathscr{R}(b_{10} \cap \bar{a}_2 \cap \bar{b}_{20}) \to L(\hat{Y}_1, \mathscr{D}) \qquad (6.14)$$

is uniquely determined by

$$b_1 \cap \bar{a}_2 \cap \bar{b}_2 \to \tilde{\psi}(\bar{a}_2 \cap \bar{b}_2)\, \psi_{1s}(b_{10}, b_1),$$

which we abbreviate by

$$\tilde{\psi}_{\bar{a}_2 \cap \bar{b}_{20}} \times \psi_{1sb_{10}}.$$

Similarly, an additive measure

$$\mathscr{R}(b_{10} \cap \bar{a}_2 \cap \bar{b}_{20}) \to L_c \qquad (6.15)$$

is determined by

$$b_1 \cap \bar{a}_2 \cap \bar{b}_2 \to \psi(i(b_{10} \cap \bar{a}_2 \cap \bar{b}_{20}), i(b_1 \cap \bar{a}_2 \cap \bar{b}_2)),$$

which we write

$$\psi_{i(b_{10} \cap \bar{a}_2 \cap \bar{b}_{20})}\, i.$$

In order to complete $\mathscr{R}(b_{10} \cap \bar{a}_2 \cap \bar{b}_{20})$ in a way similar to $\mathscr{R}(b_0)$ in X §2.3 and to extend the mappings, we must extend $L(\hat{Y}_1, \mathscr{D})$. To this end, we consider the same set $\Sigma_m$ as in X §2.3, only now denote it by $\Sigma_1$. The set $K(\Sigma_1, \breve{K})$ of all (in the norm topology of $\mathscr{B}$) $\sigma$-additive measures $\Sigma_1 \xrightarrow{W} \breve{K}$ (with $W(\varepsilon) \in K$) forms a basis of a base-normed Banach space $\mathscr{B}(\Sigma_1, \mathscr{B})$. With the Banach space $\mathscr{B}'(\Sigma_1, \mathscr{B})$ dual to $\mathscr{B}(\Sigma_1, \mathscr{B})$ and with $L(\Sigma_1, \mathscr{B})$ as the order interval $[\mathbf{0}, \mathbf{1}]$ of $\mathscr{B}'(\Sigma_1, \mathscr{B})$, one can identify $L(\hat{Y}_1, \mathscr{D})$ with a subset of $L(\Sigma_1, \mathscr{B})$ that is $\sigma(\mathscr{B}'(\Sigma_1, \mathscr{B}), \mathscr{B}(\Sigma_1, \mathscr{B}))$-dense in $L(\Sigma_1, \mathscr{B})$.

If one now completes $\mathscr{R}(b_{10} \cap \bar{a}_2 \cap \bar{b}_{20})$ to a ring $\Sigma_{b_{10} \cap \bar{a} \cap \bar{b}_{20}}$ in the same way as in X §2.3, then one can extend the mappings (6.14) and (6.15) to

$$\Sigma_{b_{10} \cap \bar{a}_2 \cap \bar{b}_{20}} \xrightarrow{\tilde{\psi}_{\bar{a}_2 \cap \bar{b}_{20}} \times \psi_{1 s b_{10}}} L(\Sigma_1, \mathscr{B}), \tag{6.16}$$

$$\Sigma_{b_{10} \cap \bar{a}_2 \cap \bar{b}_{20}} \xrightarrow{\psi_{i(b_{10} \cap \bar{a}_2 \cap \bar{b}_{20})}\, i} L_c. \tag{6.17}$$

The set of all elements $b_1 \cap \bar{a}_2 \cap \bar{b}_{20}$ from $\mathscr{R}(b_{10} \cap \bar{a}_2 \cap \bar{b}_{20})$ with $b_1 \subset b_{10}$ forms a Boolean subring of $\mathscr{R}(b_{10} \cap \bar{a}_2 \cap \bar{b}_{20})$ and hence also of $\Sigma_{b_{10} \cap \bar{a}_2 \cap \bar{b}_{20}}$. Let the closure of this subring be denoted by $\Sigma_{b_{10}}$. In exactly the same way, the set of all $b_{10} \cap \bar{a}_2 \cap \bar{b}_2$ forms a Boolean subring whose closure will be denoted by $\Sigma_{\bar{a}_2 \cap \bar{b}_{20}}$. Let the subset of all elements from $\Sigma_{b_{10} \cap \bar{a}_2 \cap \bar{b}_{20}}$ of the form $\sigma_2 \wedge \sigma_1$ with $\sigma_1 \in \Sigma_{b_{10}}$ and $\sigma_2 \in \Sigma_{\bar{a}_2 \cap \bar{b}_{20}}$ be denoted by $\{\Sigma_{\bar{a}_2 \cap \bar{b}_{20}} \times \Sigma_{b_{10}}\}$. The measures (6.16) and (6.17) are already determined by their values on this subset. Because of

$$(\tilde{\psi}_{\bar{a}_2 \cap \bar{b}_{20}} \times \psi_{1 s b_{10}})(\sigma_2 \wedge \sigma_1) = \tilde{\psi}_{\bar{a}_2 \cap \bar{b}_{20}}(\sigma_2)\, \psi_{1 s b_{10}}(\sigma_1),$$

the measure (6.16) is already determined by its value on $\Sigma_{\bar{a}_2 \cap \bar{b}_{20}}$ and $\Sigma_{b_{10}}$. There follows

$$(\tilde{\psi}_{\bar{a}_2 \cap \bar{b}_{20}} \times \psi_{1 s b_{10}})\, \Sigma_{b_{10}} \subset \mathbf{1}\, L(\Sigma_1)$$

(where $\mathbf{1}$ is the unit operator in $L$) and

$$(\tilde{\psi}_{\bar{a}_2 \cap \bar{b}_{20}} \times \psi_{1 s b_{10}})\, \Sigma_{\bar{a}_2 \cap \bar{b}_{20}} \subset L\, \mathbf{1}$$

(where $\mathbf{1}$ is the "unit function" on $\hat{S}_1$).

We will require the existence of soft trajectory measurements, i.e. of a $\bar{b}_{10}$ such that

$$\overline{\mathrm{co}}^{\sigma}(\tilde{\psi}_{\bar{a}_2 \cap \bar{b}_{20}} \times \psi_{1 s \bar{b}_{10}})\, \Sigma_{\bar{b}_{10}} = \mathbf{1}\, L(\Sigma_1) \tag{6.18}$$

is satisfied (see (2.3.12)). As in §2.3, we can identify a subring of $\Sigma_{\bar{b}_{10}}$ with $\Sigma_1$. In this sense, we introduce a macro-observable $F_{\bar{b}_0}$ by

$$F_{\bar{b}_0}(\sigma_1) \overset{\text{def}}{=} \psi_{i(\bar{b}_{10} \cap \bar{a}_2 \cap \bar{b}_{20})}\, i \sigma_1 \tag{6.19}$$

for $\sigma_1 \in \Sigma_1 \subset \Sigma_{\bar{b}_{10} \cap \bar{a}_2 \cap \bar{b}_{20}}$.

Moreover, as measurements on the microsystem after scattering we can consider such measurements for which

$$\Sigma_{\bar{a}_2 \cap \bar{b}_{20}} \xrightarrow{\tilde{\psi}_{\bar{a}_2 \cap \bar{b}_{20}} \times \psi_{1 s \bar{b}_{10}}} L\,1$$

(considered as a mapping in $L$) has a decision observable as kernel (cf. V §3.3 and VII §3). Then one can identify a subring $\Sigma_2$ of $\Sigma_{\bar{a} \cap \bar{b}_{20}}$ with the range of values of this kernel observable (i.e. with a complete Boolean sublattice of $G$, see VII T2.2). In ths sense we find

$$(\tilde{\psi}_{\bar{a}_2 \cap \bar{b}_{20}} \times \psi_{1 s \bar{b}_{10}})\, \sigma_2 = e\,\mathbf{1} \tag{6.20}$$

with $e \in G$ for $\sigma_2 \in \Sigma_2$.

Let the complete Boolean subring generated by $\{\sigma_1 \wedge \sigma_2 | \sigma_1 \in \Sigma_1,\ \sigma_2 \in \Sigma_2\}$ in $\Sigma_{\bar{b}_{10} \cap \bar{a}_2 \cap \bar{b}_{20}}$ be called $\Sigma_{12}$. Then the mapping

$$\Sigma_{12} \xrightarrow{\psi_{i(\bar{b}_{10} \cap \bar{a}_2 \cap \bar{b}_{20})}\, i} L_c \tag{6.21}$$

is determined by the special values

$$\varphi_{i(\bar{b}_{10} \cap \bar{a}_2 \cap \bar{b}_{20})}\, i(\sigma_1 \wedge \sigma_2) \quad \text{for } \sigma_1 \in \Sigma_1, \sigma_2 \in \Sigma_2 . \tag{6.22}$$

Instead of considering the whole mapping (6.17), it suffices to consider the observable (6.21). Since in $\mathscr{P}\mathscr{T}_{q\,\exp}$ we describe the interaction of the impinging microsystems with the many-particle system (1) as scattering, we can view $\Omega$ as well defined.

Similarly, we can regard as well defined what it means to measure effects $g \in L$ on the scattered microsystems *after* the collision under the following conditions: (a) The microsystems no longer interact with the remaining macrosystem. (b) One performs no measurements on the macrosystem, i.e. measures only the scattered microsystems. Hence, we assume that an injective affine (and $(\sigma(\mathscr{B}', \mathscr{B}) - \sigma(\mathscr{B}'_c, \mathscr{B}_c))$-continuous) mapping $L \xrightarrow{\kappa} L_c$ is well defined. It means that to a measurement of $g \in L$ on the microsystems (in the description of $\mathscr{P}\mathscr{T}_q$ as a theory of microsystems) there corresponds a measurement $\kappa g \in L_c$ in the description of $\mathscr{P}\mathscr{T}_{q\,\exp}$ (one often assumes $\kappa(g) = g \times 1$ in the product representation of a Hilbert space in which $w^{(i)} \times \varphi_1(i a_1)$ is defined).

Condition (6.18) means that a measurement of the paths has no influence on the statistics of the paths. Here let us in addition require that a registration of paths of the macrosystem also has no influence on (i.e. is also soft relative to) the measurement of microsystems *after* (!) scattering. We express this by the requirement

$$\psi_{i(\bar{b}_{10} \cap \bar{a}_2 \cap \bar{b}_{20})}(i(\bar{b}_{10} \cap \bar{a}_2 \cap \bar{b}_2)) = \kappa \tilde{\psi}(\bar{a}_2 \cap \bar{b}_{20}, \bar{a}_2 \cap \bar{b}_2).$$

With $e$ from (6.20), this in particular implies

$$\psi_{i(\bar{b}_{10} \cap \bar{a}_2 \cap \bar{b}_{20})}\, i \sigma_2 = \kappa e \quad \text{for } \sigma_2 \in \Sigma_2 . \tag{6.23}$$

Therefore the mapping (6.21) satisfies the special conditions (6.19) and (6.23), whereby $e$ is determined in (6.23) by (6.20).

In $\mathscr{P}\mathscr{T}_{q\,\exp}$ let the mapping $\kappa$ satisfy the further condition $\kappa G \subset G_c$, where the $G_c$ are the decision effects of $L_c$ (i.e. $G_c = \partial_e L_c$).

If $\kappa(e)$ is a decision effect, we obtain (cf. VIII T4.3.1)

$$\psi_{i(\breve{b}_{10}\cap\breve{a}_2\cap\breve{b}_{20})}\, i(\sigma_1\wedge\sigma_2)=F_{\breve{b}_0}(\sigma_1)\,\kappa(e)=\kappa(e)\,F_{\breve{b}_0}(\sigma_1),$$

with $e$ from (6.20).

Thus (6.13) goes over into

$$\langle(\mathscr{U}_\tau^{(0)'}\times V_\tau^{(s)'})\,\hat{\Gamma}(w^{(i)},w_1),e\sigma_1\rangle_{1\mathscr{D}}=\mu(\mathscr{U}_\tau'\,\Omega\{w^{(i)}\times w_1\}_s,\kappa(e)F_{\breve{b}_0}(\sigma_1)) \qquad (6.24)$$

for all $w_1$ from $K_{1m}$. While $K_{1m}$ is the norm-closure of co $\varphi_{1c}(i\mathscr{Q}_1')$, we used

$$\hat{\Gamma}(w^{(i)},\varphi_{1c}(ia_1))=\Gamma(w^{(i)},a_1)$$

and identified $\Sigma_1$ with $\partial_e L(\Sigma_1)$. Since the measurement on the microsystems after the scattering should be arbitrary, we require (6.24) for all $e\in G$. For all $\sigma_1\in\Sigma_1$, the $F_{\breve{b}_0}(\sigma_1)$ must commute with all elements of $\kappa G$, hence with all elements of $\kappa L$, and thus also with all elements of $\kappa\mathscr{B}$. By

$$\mu(w,\kappa(e)\,F_{\breve{b}_0}(\sigma_1))=\langle u,e\sigma_1\rangle_{1\mathscr{D}}$$

with $w\in K_c$ and $u\in K(\Sigma_1,\check{K})$, there is defined a mapping $(\kappa'\times S_{\breve{b}_0})\,w=u$. With $\kappa\times S_{\breve{b}_0}'$ as the mapping dual to $\kappa'\times S_{\breve{b}_0}$, in particular we have $(\kappa'\times S_{\breve{b}_0}')(e\sigma_1)=\kappa(e)F_{\breve{b}_0}(\sigma_1)$. Finally, (6.24) goes over into

$$(\mathscr{U}_\tau^{(0)'}\times V_\tau^{(s)'})\,\hat{\Gamma}(w^{(i)},w_1)=(\kappa'\times S_{\breve{b}_0})\,\mathscr{U}_\tau'\,\Omega\{w^{(i)}\times w_1\}_s \qquad (6.25)$$

for all $w_1\in K_{1m}$; this for $\tau=0$ in particular implies

$$\hat{\Gamma}(w^{(i)},w_1)=(\kappa'\times S_{\breve{b}_0}')\,\Omega\{w^{(i)}\times w_1\}_s. \qquad (6.26)$$

The "embedding condition"

$$(\mathscr{U}_\tau^{(0)'}\times V_\tau^{(s)'})(\kappa\times S_{\breve{b}_0})\,\Omega\{w^{(i)}\times w_1\}_s=(\kappa'\times S_{\breve{b}_0}')\,\mathscr{U}_\tau'\,\Omega\{w^{(i)}\times w_1\} \quad \text{for all } w_1\in K_{1m} \qquad (6.27)$$

together with (6.26) is equivalent to (6.25).

The notation $(\kappa'\times S_{\breve{b}_0})$ is to say that for $gk\in L(\Sigma_1,\mathscr{B})$ (with $g\in L$, $k\in L(\Sigma_1)$), the relation

$$(\kappa\times S_{\breve{b}_0}')\,gk=\kappa(g)\,S_{\breve{b}_0}'\,k$$

holds, where the $S_{\breve{b}_0}'\,k$ commute with $\kappa\mathscr{B}'$.

Therefore, the problem of the compatibility of the scattering of microsystems on macrosystems with $\mathscr{PT}_{q\,\exp}$ consists in giving a macro-observable and thus an $S_{\breve{b}_0}$ so that the embedding condition (6.27) is satisfied (at least in a very good approximation). According to §2 and §3, this procedure also solves the problem of the compatibility of preparation and registration with $\mathscr{PT}_{q\,\exp}$. If $S_{\breve{b}_0}$ is known, the operator $\hat{\Gamma}(w^{(i)},w_1)$ follows from (6.26).

If one has $\hat{\Gamma}$, the mappings $\varphi$ and $\psi$ for the microsystems are obtained as given in §2 and §3. The complicated analysis of experiments in fact proceeds as follows. Aided by a statistical mechanics of preparation and registration devices, one tries to determine the mappings $\hat{\Gamma}$ for these devices (if only as coarse approximations) and thus to derive approximate values of $\varphi(a)$ and $\psi(b_0,b)$ for the constructed preparation and registration procedures, respectively. One seeks to improve the approximations for $\varphi(a)$, $\psi(b_0,b)$ found this way by test experiments: For example, one tries to improve the approximation for $\varphi(a)$ by testing with various $\psi(b_0,b)$ that are well known. It is just the art of the experimental physicist, in this way to obtain the best possible $\varphi(a)$ and $\psi(b_0,b)$.

It cannot be the task of the author to demonstrate by examples the applications of (6.26), (6.27). Since in X §§2 and 3 we have discussed the structure of the simpler (relative to (6.27)) embedding condition X (2.3.22), here we shall omit an analogous discussion. By experiments and theory, physicists are convinced that the compatibility conditions are fulfilled.

The theory of this section (as that of X) is incomplete since we have no *general* theory, neither for state spaces of macrosystems nor for the embedding maps. Only in many special examples, the state spaces and sometimes more or less good embedding mappings (see e.g. X §3) are known. Nevertheless, to simplify the representation we shall in the next section do as if the theory of the scattering of microsystems on macrosystems were complete.

At the end of §4 we had already mentioned that the possibility to calculate the interaction operator $\Delta$ (resp. $\Gamma$) by the embedding procedure does not guarantee that $\alpha(\mathscr{Q}'_1 \times L(\hat{S}_1))$ is norm dense in $K$ and that $\beta(\mathscr{Q}'_2 \times L(\hat{S}_2))$ is $\sigma(\mathscr{B}', \mathscr{B})$-dense in $[\mathbf{0}, \mathbf{1}]$. In X §4 we presented arguments which made us expect greater differences between $K$ and $\alpha(\mathscr{Q}'_1 \times L(\hat{S}_1))$ resp. $[\mathbf{0}, \mathbf{1}]$ and $\beta(\mathscr{Q}'_2 \times L(\hat{S}_2))$ for large molecules. But a general constraint for preparation and registration is already imposed by the conservation laws of energy and momentum. Let us briefly discuss the conservation of energy because only the energy has in X §3 been treated as "macroscopic observable".

The trajectories of the preparing device also determine the change of the energy of the macrosystem during the preparation. Hence we can demix the generated ensemble according to the values $E$ of this energy change. Thus the prepared ensemble can only be a *mixture* of ensembles $w_E$ with various $E$.

These mixture components $w_E$ must have an "imprecise" energy $E$ since the macro-observable energy resembles an imprecise measurement of the non-measurable microscopic energy (given by the Hamiltonian) and since this microscopic energy is conserved. To each value $E$ there must be an effect $F_E$ which corresponds to an imprecise measurement of the energy of the prepared microsystems such that $\operatorname{tr}(w_E F_E) = 1$. The effect $F_E$ of the prepared microsystems is determined by the macro-observable energy of the preparing device (and the conservation law).

Thus we conclude: An ensemble which is a superposition (not a mixture) of "very" different energies, cannot be prepared. Here "very" different means that energy differences greater than the width of $F_E$ enter the superposition. The width of $F_E$ will be the greater, the greater the preparing macrosystems are. Therefore, for the nonrelativistic quantum mechanics of microsystems (where the energies are limited by normative axioms) we cannot expect that energy conservation severely constrains the preparation possibilities. But this can be otherwise for relativistic quantum theory.

In fact the law of energy conservation gives also constraints for the possible registrations: Every registration comprises a registration of the change in the macroscopic energy of the registering device. We may also measure the energy of the microsystems which come out of the device. Therefore all effects $F$ registered by the device must coexist with a imprecise measurement of the energy of the incoming microsystems. Hence they must coexist with effects $\tilde{F}_E$, where $E$ is the sum of the energy change of the registration device and of the energy of the outcoming microsystems. The "width" of $\tilde{F}_E$ is determined by the difference between the macroscopic and the microscopic energy of the device.

Thus all effects coexist with those of an imprecise measurement of the energy. Because of the "great width" of $\tilde{F}_E$, however, also this coexistence will not severely constrain nonrelativistic microsystems.

The influence of conservation laws on measurements was already investigated in connection with earlier descriptions of the measuring process [59]. A more comprehensive investigation would be a new task connected with the embedding theory presented here.

Let us still discuss a problem that is sometimes used as an objection to our claim that macrosystems can be described objectively. This problem was already described in III §6.5.

Let us write III (6.5.5) with III (6.5.8) as

$$\langle u_1, k_1\rangle\, \mu(\alpha_1(u_1, k_1), g) = \langle u_1, l_1\rangle, \tag{6.28}$$

where $g$ is an arbitrary effect produced by the prepared microsystems (we dropped the $\sim$ from $\mu$). Various $g$ can be non-coexistent, although all $l_1$ are coexistent. This is no objection to the right side of (6.28)!

Since $\tilde{\alpha}_1(u_1, k_1) = \langle u_1, k_1\rangle\, \alpha_1(u_1, k_1)$ is affine in $u_1$ and in $k_1$, it has the form

$$\tilde{\alpha}_1(u_1, k_1) = \int_{\hat{S}_1} k_1(y)\, w(y)\, d u_1(y), \tag{6.29}$$

where $w(y)$ are elements of $\bar{K}^\sigma$ (see IV D3.7). Thus the left side of (6.28) for $g \in L \cap \mathscr{D}$ takes the form

$$\int_{\hat{S}_1} k_1(y)\, \mu(w(y), g)\, d u_1(y) = \langle u_1, k_1\, p(g)\rangle \tag{6.30}$$

with

$$p(g)(y) = \mu(w(y), g). \tag{6.31}$$

Thus in (6.28) we have

$$l_1 = k_1\, p(g), \tag{6.32}$$

where $p$ defines a mapping $L \cap \mathscr{D} \xrightarrow{p} L(\hat{S}_1)$. It also maps non-coexistent effects into coexistent trajectory effects. This is no contradiction to the embedding.

$S'_{\hat{b}_0}\, p(g)$ must not be equal to $\kappa g$ although

$$\begin{aligned}\langle u_1, k_1\rangle\, \mu(\alpha_1(u_1, k_1), g) &= \mu(\alpha_1(u_1, 1), \kappa g\, S'_{\hat{b}_0}\, k_1)\\ &= \mu(\alpha_1(u_1, 1), S'_{\hat{b}_0}[k_1\, p(g)])\\ &= \langle u_1, k_1\, p(g)\rangle. \end{aligned}\tag{6.33}$$

This equation demonstrates that a registration of the emitted microsystems (i.e. a registration of $g$) is "equivalent" to a registration of $p(g)$ on the preparation device (see also III §6.6).

A mathematically similar problem arises if a macrosystem contains a microsystem that can be *distinguished* from all other microsystems composing the macrosystem. Such a possibility is often taken as an objection against the objective description of macrosystems, since according to quantum mechanics one can measure this individual micro-subsystem.

In §2 we have only considered trajectory registration methods for which we postulated the embedding condition X (2.3.7). We must extend these considerations to a more comprehensive set of registration methods (see III §6.6), also containing methods which register trajectories together with registrations of the "distinguished" micro-subsystem.

For such a problem we can adopt the whole theory of §1 to §6, admitting the simplification that we have "no impinging microsystems" as for preparation procedures (see §2), and the complication that the distinguishable micro-subsystem is not emitted. But this complication is not essential since we presumed that these microsystems can be "distinguished". This makes it possible to give the Hilbert space for the whole system in $\mathcal{PT}_{q\,\mathrm{exp}}$ the form $\mathscr{H} = \mathscr{H}_1 \times \mathscr{H}_2$ where $\mathscr{H}_2$ is the space for the "distinguished" micro-subsystem. The form $\mathscr{H} = \mathscr{H} \times \mathscr{H}_2$ can be taken invariant under time translations if we use the "interaction picture" (see [2] X §3). Thus also the embedding can be taken over without essential changes We have only to replace the "prepared" microsystems by the "distinguished" ones.

Thus the left side of (6.28) can be interpreted as the probability for the registration of the trajectory effect $k_1$ together with an effect $g$ produced by the distinguished micro-subsystem. If the state space is so comprehensive that (6.28) holds, we find again that a registration of $g$ is "equivalent" to the registration of the trajectory effect $p(g)$ (see also the partition into equivalence classes by the mapping $l'$ in X §2.4).

In both cases ("prepared" or "distinguished" microsystems) we have a mapping $L \cap \mathscr{D} \xrightarrow{p} L(\hat{S}_1)$, which can be extended to a mapping $L \xrightarrow{p} L(\Sigma_m)$. This mapping is of course not surjective (it can be injective). For two non-coexistent effects $g_1, g_2$, the trajectory effect $p(g_1)\,p(g_2)$ (the product of the functions $p(g_1)$ and $p(g_2)$) cannot be in the range of $p$! It would be an error to interpret the probabilities for $p(g_1)\,p(g_2)$ as the probabilities that $g_1$ and $g_2$ "occur together"; just this was presumed impossible.

The description of a distinguishable microscopic part of a macrosystem by quantum mechanics not only presents no contradiction to the objective description of the total macrosystem in a state space $Z$. Rather this description in a state space can also comprise (by means of the mapping $p$) the quantum mechanical description of the distinguishable microscopic part (if the state space $Z$ is comprehensive enough). Only the quantum mechanical description in $\mathcal{PT}_{q\,\mathrm{exp}}$ of the "total" macrosystem (with many unrealistic preparation and registration procedures) cannot be given objectively (by trajectories in a state space $Z$).

## §7 The Problem of the Desired Observables and Preparators

After we have seen in §6 how one can in principle find the mapping $\psi$ and the ideal observable (3.3) for a given registration device, it may seem that the measurement problem is totally solved. But it was only half the problem.

The other half of the problem is: Where do the devices come from? They are constructed by physicists and technicians. But what devices are constructed and why are these and not others constructed?

Some say that all devices are constructed which are possible. This is false. Firstly there are so many possiblities that all human beings throughout their whole existence cannot realize what is possible. Secondly, everyone a little acquainted with experimental physics knows that no macrosystem is constructed which does not (at least approximately) fulfill the intentions of a physicist.

Others will say, all right, a device is only then a measurement device if it fulfills the intention to detect the objective properties and structures of the microsystems. The registration devices are only aids to "see" what we cannot see with our own eyes.

Also this is not true, as we have seen in the previous chapters. There are no objective properties (others than the superselection rules). The microsystems cannot be completely separated from the preparation and registration procedures. The microsystems are rather "made" by the preparation procedures and "act" on the registration devices, an action which cannot be explained solely by objective structures of the microsystems. Quantum mechanics shows drastically that physics is not a voyage of discovery through a given world. Physics is rather an action of human beings confronted with the world and in this sense a craft, a very ingenious craft. A desire to separate technology from physics (as an application of the physical knowledge about the real structure of the world) misrepresents the essence of physics. Physics and technology cannot be separated from one another: Without physics no technology, but also without technology no physics.

The background of the intentions of the physicists to construct certain devices and no others is a widespread field. Motives of importance can be to make money, to get reputation, to seek something extraordinary. Clearly we cannot discuss such motives here although these and similar motives must be queried by the conscience of every physicist. I hope that physicists will not use as registration devices e.g. an atomic bomb or "Schrödingers cat" (see [2] XVIII §3), or human beings if these could take injury.

But there are other motives to influence the choice of devices, motives which we can discuss in connection with the theory represented in the previous chapters and in [2]. Theoretical physicists have special desires for registration procedures.

A theory of the microsystems in which it would be necessary to treat every registration device by the methods of §6 would be very cumbersome. Therefore it is understandable that theoretical physicists seek for definitions of observables without the necessity to go into the technical details of the devices. If possible, the "defined" observables should not be complicated in order that the calculations of probabilities are not too difficult. They should be very appropriate to test the theory. Such observables, defined solely in the framework of the theory from [2] (i.e. without the more comprehensive theory from §§1 through 6 and of X) may be called "desired" observables.

In this way the problem of the measured observables is divided into two parts: (1) to find "desired" observables; (2) to find a real device that delivers as ideal observable (3.3) an approximation to the desired observable and in this sense (V §5) a realization of the desired observable.

The step (2) can be treated only by the theory of §6. But this theory is not sufficient to *find* a useful realization for the desired observable. It is left to the knack of the experimental physicists to find useful devices. After constructing the devices, however, the physicists must discuss (with the help of the theories in §6 and IX §3) how well the ideal observable of this real device approximates the desired observable. The differences between the ideal observable of the device and the desired observable are called "measurement errors". We know the extensive "error evaluations" (Fehlerabschätzungen) done by experimental physicists.

The step (1) to find desired observables has been treated in [2]. But there we had not named the problem in this manner. We had described it in another way, e.g. at the beginning of [2] XI. There we declared it necessary to introduce "as new axioms" effect operators for special registration devices. This introduction of "axioms" was necessary since we had not at hand the more comprehensive theory of §6. But after the development of this more comprehensive theory we can interpret the procedure from [2] in a new way:

The effects and observables introduced in [2] are to be called "desired observables" and no longer "axioms". Since they are no axioms we must supplement this step (1) by the step (2) described above.

Typical desired observables are the position and momentum observables, defined in [2] VII §4, and the angular momentum observables defined in [2] VII §5. The observable of the impact of a microsystem on a surface in a time interval has been defined by the effect measure $\bar{F}$ in [2] XVI §6.1.

The position and momentum observables could be defined by some properties of the corresponding registration procedures. It would in principle be possible to test these properties for a given device in order to see, whether this device measures position resp. momentum. Therefore, already in [2] it was unnecessary to introduce these observables as axioms. A severe disadvantage of this method is that it does not at all indicate how to construct devices which measure these observables. One can use the theory of §6 and [2] XVII to see whether a device fulfills the postulated properties, or to prove directly that the ideal observable of the device is an approximation to the position resp. momentum observable.

The definition of the angular momentum observable in [2] VII §5 by infinitesimal transformations already in [2] allowed to introduce these desired observables without axioms. But this definition is so indirect, that the theory of §6 and [2] XVII is needed to prove that a constructed device is appropriate to measure the angular momentum. In this way one has to make a theory e.g. of the Stern-Gerlach experiment to see that this device measures the spin. This device is composed of a measurement transformation ([2] XVII §3) by an inhomogeneous magnetic field and a macroscopic registration process (see e.g. [1] XII §2.2 for such an approximative theory, and [49] for an improvement).

For defining the observable of impact on a surface in [2] XVI §6.1, we started with the idea of a device with a sensitive surface $\mathscr{F}$. We hoped that the intuitively found operators $\bar{F}$ in good approximation describe devices with "very" sensitive surfaces $\mathscr{F}$. Therefore we introduced the effect operator $\bar{F}$ by [2] VI (6.1.13) as an axiom with the correspondence rule (see [2] I §1): $\bar{F}$ corresponds to devices the "principal" structure of which is just such a sensitive surface $\mathscr{F}$. With the theory of §6 we can now analyze any special device and see whether it yields a good approximation to the observable defined by [2] VI (6.1.13). Therefore we can now denote this observable as a desired one and omit its characterization as axiom.

The effect of a photon emission (introduced in [2] XI §1) has a totally different character. The formula [2] XI (1.16) is not only a "desired" one. We cannot derive this formula by the theory of §6 alone.

First one has to add (to the theory of atoms and molecules) an approximation to quantum electrodynamics. We have already sketched such an approximation for the emission of an electromagnetic "operator" wave in [2] XI (1.1) through (1.11). This must be complemented by a theory describing the influence of this emitted wave on other systems. It can be attained by replacing the "external fields" in VIII (5.8) by the emitted wave. We will not develop such a theory in detail. We solely need to make it clear that such a theory introduces additional axioms.

After this addition of an approximation to quantum electrodynamics, in a second step we can use the theory of §6 to get the influence of the emitted waves on macrosystems. We must only complement this theory for the case that $w^{(i)}$ also describes the possibility of impinging electromagnetic waves. Such a theory can also describe the function of spectral apparatuses.

In this way the "axiom" [2] XI (1.16) can be interpreted as a desired observable, derivable for suitable devices called spectrometers.

A frequently misunderstood observable is the "energy". This observable was introduced in [2] VII and VIII by an infinitesimal time translation. As such it is a desired observable. But the concrete form can only be derived for elementary systems (see [2] VII §5). For composite systems the specific form must be introduced as an axiom (see [2] VIII (5.8)). This axiom is completely misunderstood if one has the opinion that the measurement of the observable "energy" is defined by a pretheory, such that the correctness of [2] VIII (5.8) can be tested by measurements of the energy.

If one had defined a desired observable by [2] VIII (5.8), one could not say that this definition is wrong if one had a device the ideal observable of which is not [2] VIII (5.8). One rather should say that the device does not measure the desired observable. But we have *not* introduced [2] VIII (5.8) as a desired observable; we have introduced it as an infinitesimal time translation. As this it has another physical interpretation and by this interpretation it can be tested. It can be tested e.g. by the frequency spectra of atoms and molecules (see [2] XI to XV).

To measure the desired observable energy by a device is a much more complicated problem than defining it by infinitesimal time translation. The definition of this energy (similar to that of the angular momentum) does not give any indication how to construct a suitable device for its measurement. The experimental physicists have found such devices using measurement scattering morphisms ([2] XVII §2.2), measurement transformation morphisms in external fields ([2] XVII §3), and sensitive macroscopic systems.

The task of a collaboration between theoretical and experimental physicists is (for the theorists) to elaborate more precisely the desired observables and (for the experimentalists) to seek for devices which allow to approximate as well as possible the desired observables.

We have defined many desired ensembles in [2]. These definitions had various intentions. In [2] XI §2 we had discussed the problem to prepare bound states. It is not difficult to define bound states as desired ensembles; but this definition gives no indication for a suitable preparing device. Therefore we had introduced the procedure "ideal gas at a temperature $\theta$" and characterized this procedure by the axioms [2] XI (2.1) to (2.3). This form of axioms is not necessary if we use the theory of §6.

Scattering theory is characterized by many prescriptions for the ensembles, beginning with the postulates Coll 1 and Coll 2 in [2] XVI §1 up to the many additional presumptions for applying the formulas in [2] XVI §§6.2 to 6.3. All postulates and presumptions are nothing but formulations for "desired ensembles". Only in [2] XVI §1 we have tied Coll 1 and Coll 2 to a *qualitative* structure of the relative positions of two preparation devices. Having the theory of §6 at hand, we can take all postulates and presumptions (Coll 1 and Coll 2 included) as structures which should be fulfilled by suitable preparation devices. It is just this what the experimental physicists do when they construct devices for scattering experiments, devices which sometimes cost millions of dollars.

# XII Special Structures in Preparation and Registration Devices

## §1 Measurement Chains

In experimental physics it is very usual to build up measurement chains in order to perform complicated measurements. In IX §3 and in [2] XVII we have investigated measurement scatterings and pointed out how one can build up measurement scattering chains. Hence it will suffice here to discuss how three macrosystems can be composed into a measurement chain.

We consider only a case where the subsystem (1) acts directedly on the systems (2) and (3). With an easily recognized generalization to III, for the probability of registering on all three systems we write

$$\lambda_{123}(a_1 \times a_2 \times a_3 \cap M \cap b_{10} \times b_{20} \times b_{30}, a_1 \times a_2 \times a_3 \cap M \cap b_1 \times b_2 \times b_3). \quad (1.1)$$

That (1) acts directedly on (2)+(3), in analogy with III (3.1) means that

$$\lambda_{123}(a_1 \times a_2 \times a_3 \cap M \cap b_{10} \times b_{20} \times b_{30}, a_1 \times a_2 \times a_3 \cap M \cap b_1 \times b_{20} \times b_{30}) \quad (1.2)$$

does not depend on $a_2, a_3, b_{20}, b_{30}$.

Let us also assume that (1)+(2) acts directedly on (3). Then

$$\lambda_{123}(a_1 \times a_2 \times a_3 \cap M \cap b_{10} \times b_{20} \times b_{30}, a_1 \times a_2 \times a_3 \cap M \cap b_1 \times b_2 \times b_{30}) \quad (1.3)$$

does not depend on $a_3, b_{30}$.

Expressing both assumptions together by the symbolic Figure 9, we briefly call it a measurement chain (1)→(2)→(3).

Obviously, one can describe this measurement chain in two ways:

$\alpha$: (1) as a preparation device and (2)+(3) as a registration device.

$\beta$: (1)+(2) as a preparation device and (3) as a registration device.

In a way similar to the derivations in XI §2 through §4, we obtain $\lambda_{123}$ in the form

$$\begin{aligned}
&\lambda_{123}(a_1 \times a_2 \times a_3 \cap M \cap b_{10} \times b_{20} \times b_{30}, a_1 \times a_2 \times a_3 \cap M \cap b_1 \times b_2 \times b_3) \\
&\quad = \mu(\Delta(\Delta(\Delta(w_0, a_1, k_1), a_2, k_2), a_3, k_3), 1) \\
&\quad = \mu(\Delta(\Delta(w_0, a_1, k_1), a_2, k_2), \beta(a_3, k_3)) \\
&\quad = \langle \Gamma(\Delta(w_0, a_1, k_1), a_2), k_2\, \beta(a_3, k_3)\rangle_{2\mathscr{D}} \\
&\quad = \mu(\Delta(w_0, a_1, k_1), \eta(a_2, k_2\, \beta(a_3, k_3))),
\end{aligned} \quad (1.4)$$

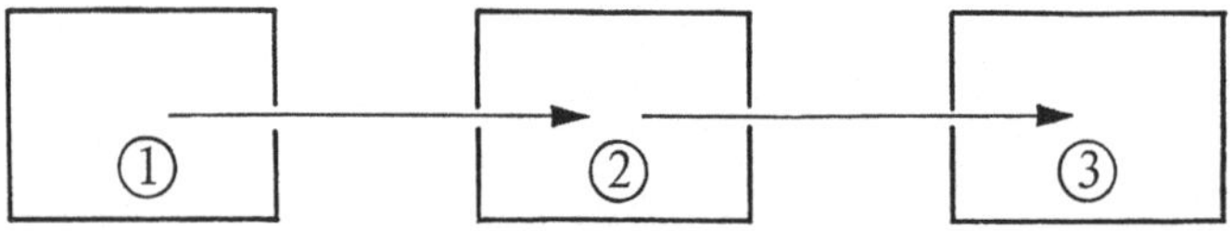


Fig. 9

where $k_i = \psi_{is}(b_{10}, b_i)$, with $\beta$ as in XI §3 and $\eta$ from

$$\langle \Gamma(w^{(i)}, a_2), k_2\, g\rangle_{2\mathcal{D}} = \mu(w^{(i)}, \eta(a_2, k_2\, g)).$$

If we describe the measurement chain in the form $\alpha$, then the ensemble $w_1$ prepared by (1) is given by

$$\lambda_1\, w_1 = \Delta(w_0, a_1, k_1). \tag{1.5}$$

where $\lambda_1$ is a normalization factor. The effect registered by (2)+(3) is

$$g_{23} = \eta(a_2, k_2\, \beta(a_3, k_2)). \tag{1.6}$$

If we describe the measurement chain in the form $\beta$, then the ensemble prepared by (1)+(2) is

$$\lambda_2\, w_2 = \Delta(w_1, a_2, k_2) \tag{1.7}$$

with $w_1$ as in (1.5), where $\lambda_2$ is a normalization factor. The effect registered by (3) is

$$g_3 = \beta(a_3, k_3). \tag{1.8}$$

The equivalence of the two descriptions is expressed, according to (1.4), by

$$\mu(w_1, \eta(a_2, k_2\, \beta(a_3, k_3))) = \mu(\Delta(w_1, a_2, k_2), \beta(a_3, k_3)). \tag{1.9}$$

One can also write the effect $g_{23}$ in the form

$$g_{23} = \eta(a_2, k_2\, \beta(a_3, k_3)) = \beta(a_2 \times a_3, k_2\, k_3), \tag{1.10}$$

whence it follows that all effects $g_{23}$ coexist for fixed $a_2, a_3$.

This sketch shows that with the scattering of microsystems on macrosystems according to XI §1 we also can describe more complicated measurement chains.

## §2 The Einstein-Podolsky-Rosen Paradox

The Einstein-Podolsky-Rosen paradox (briefly EPR paradox) plays a large role in the philosophical discussion of quantum mechanics. We assume that the reader has already had contact with examples of the EPR paradox so that here we can draw on a single example as an illustration! It can already be gleaned from [2] VIII §4.4, how the "EPR experiments" can be described consistently and without contradictions by preparation and registration procedures. Here, we shall mainly emphasize the fact that one can also describe these experiments as dealing with macroscopic systems, i.e. the microsystems as action carriers can be "forgotten".

The EPR experiments are measurement chains of special structures, indicated by Fig. 10 or briefly by (2)←(1)→(3). Here (1) is to act directedly on (2)+(3). Hence we can think of Fig. 10 as a specialization of Fig. 9, namely as the case that no

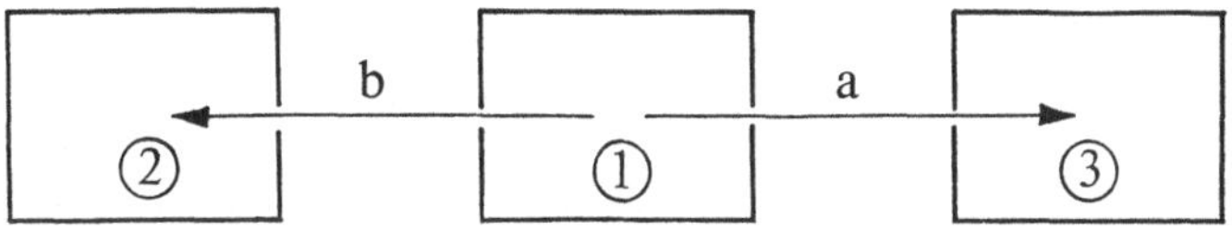


Fig. 10

microsystem really scattered by the system (2) can reach the system (3). But one can also formulate this last statement without using the concept of a microsystem:

To the requirements set forth in (1.2) and (1.3) we add that

$$\lambda_{123}(a_1 \times a_2 \times a_3 \cap M \cap b_{10} \times b_{20} \times b_{30}, a_1 \times a_2 \times a_3 \cap M \cap b_1 \times b_{20} \times b_3) \quad (2.1)$$

does not depend on $a_2, b_{20}$.

Hence, the requirements (1.2), (1.3), (2.1) characterize an EPR experiment.

Therefore, one now has three possibilities for describing an experiment with the concepts of preparation and registration; namely, besides the two possibilities $\alpha$ and $\beta$ discussed in §1 yet the possibility $\gamma$: (1)+(3) as preparation device and (2) as registration device.

Whereas it is important in Fig. 9 that (3) registers "after" the scattering on (2), this is not necessary in Fig. 10. In order to emphasize this clearly, in Fig. 10 we have drawn the systems (2) and (3) to the left and right of the system (1).

In order to illustrate our general considerations, let us use the example described *in detail* in [2] XVII §4.4. Then we can be content with some intuitive references to this example:

The preparation device (1) shall produce as a microsystem a system composed of two elementary systems $a$ and $b$. Each of the elementary systems shall have an eigenspin 1/2. Let the preparation procedure (1) be so directed that the total spin of the prepared systems composed of $a$ and $b$ equals zero. The subsystem $a$ shall leave the device (1) to the "right" (Fig. 10), the subsystem $b$ to the "left". The subsystem $b$ shall reach the device (2) to register its spin in a definite space direction $\boldsymbol{e}$. The device (3) shall register the spin of the subsystem $a$ in a space direction $\boldsymbol{d}$.

Quantum mechanics says that for $\boldsymbol{d}=\boldsymbol{e}$ the response of the two devices (2) and (3) is strongly correlated. Thus (2) shows a positive spin, when (3) shows a negative spin. This correlation is the result if one views (1) as the preparation device and (2)+(3) as the registration device. But one can also think of (1)+(2) as the preparation device for the system $a$ (the system $b$ may be absorbed in (2)). Then the digital indicators of (2) can be viewed as various "trajectories" of the preparation device (1)+(2). Using these indicators one can demix the ensemble of the systems $a$ prepared by (1)+(2) acording to their spins in the $\boldsymbol{e}$-direction.

Let us generally formulate this line of reasoning with the $\lambda_{123}$ introduced in §1:

According to (1.4),

$$\begin{aligned}&\lambda_{123}(a_1 \times a_2 \times a_3 \cap M \cap b_{10} \times b_{20} \times b_{30}, a_1 \times a_2 \times a_3 \cap M \cap b_1 \times b_2 \times b_3)\\&=\lambda_1\, \mu(\Delta(w_1, a_2, k_2), \beta(a_3, k_3))\\&=\lambda_1\, \mu(w_1, \beta(a_2 \times a_3, k_2\, k_3)) \qquad (2.2)\end{aligned}$$

with $\lambda_1\, w_1$ as in (1.5).

Because of $\beta(a_3, \mathbf{1})=\mathbf{1}$, from (2.2) conversely follows that the left side of (2.2) is independent of $a_3$ for $b_3=b_{30}$. Similarly, because of $\beta(a_2 \times a_3, \mathbf{1}\ \mathbf{1})=\mathbf{1}$, the left side of (2.2) is independent of $a_2$ and $a_3$ for $b=b_{20}$, $b_3=b_{30}$. Hence the relation (2.2) expresses what we symbolically depicted in Fig. 9. But an EPR experiment must in addition obey (2.1); i.e. in the above reasoning one can interchange (2) and (3). This implies that

$$\Delta(w_1, a_2, \mathbf{1}) \tag{2.3}$$

must be independent of $a_2$. In our example: The ensemble prepared by (1)+(2) does not depend on the direction $\boldsymbol{e}$ in which the spin is registered by (2). Naturally, the demixability of the ensemble (2.3) depends decisively on this $\boldsymbol{e}$. Let us now formulate this generally.

Analogously to III (4.1), the ensemble prepared by (1)+(2) is given by

$$\varphi(a) = \Delta(w_1, a_2, \mathbf{1}), \tag{2.4}$$

where $a = (a_1 \times a_2 \times M_3) \cap M$.

In an EPR experiment, this $\varphi(a) = \varphi((a_1 \times a_2 \times M_3) \cap M)$ is independent of $a_2$ because (2.3) is so. But $\varphi(a)$ can be demixed according to the trajectory effects $k_2$ of (2).

In order to clarify such demixings still more, let us replace (in a limit considered frequently) the $k_2$ by elements $\sigma_2 \in \mathscr{B}(\hat{S}_{2m})$, i.e. by Borel sets of trajectories (resp. by their characteristic functions, which we also denote by $\sigma_2$). Therefore, $\Delta(w_1, a_2, \sigma_2)$ is that demixture component of $\Delta(w_1, a_2, \mathbf{1})$ for which the trajectories of (2) lie in $\sigma_2$. If $\sigma_{2\nu}$ is a decomposition of $\hat{S}_{2m}$ (i.e. $\bigcup_\nu \sigma_{2\nu} = \hat{S}_{2m}$ and $\sigma_{2\nu} \cap \sigma_{2\mu} = \emptyset$ for $\nu \neq \mu$), then

$$\Delta(w_1, a_2, \mathbf{1}) = \sum_\nu \Delta(w_1, a_2, \sigma_{2\nu}) \tag{2.5}$$

is a demixture of $\Delta(w_1, a_2, \mathbf{1}) = \varphi(a_1 \times a_2 \times M_3 \cap M)$. For fixed $a_2$, all demixtures of the type (2.5) are coexistent demixtures. By changing $a_2$ one can, according to the EPR experiments, also obtain non-coexistent demixtures *of the same* ensemble (since $\Delta(w_1, a_2, \mathbf{1})$ does not depend on $a_2$).

In the example of the spin directions one can demix the ensemble $\Delta(w_1, a_2, \mathbf{1})$ by a choice of $a_2$ in any arbitrary spin direction $\boldsymbol{e}$. Two distinct spin directions $\boldsymbol{e}_1$ and $\boldsymbol{e}_2$ lead to coexistent demixtures if and only if $\boldsymbol{e}_1 = -\boldsymbol{e}_2$.

Therefore EPR experiments present no contradiction to quantum mechanics. What was felt so "paradoxical" about these experiments?

One can arrive at contradictions if one attempts to interpret the EPR experiments ontologically. An ontological interpretation is a description in which an objective structure is assigned to the macrosystems in *addition* to the structure assigned to them physically on the basis of the trajectories. Alternatively, to the *microsystems* one assigns ontological structures which *exceed* what can be expressed by preparation and registration procedures.

We first consider the usual attempt to assign ontological structures to the microsystems: If these (the pairs $(a, b)$) have "left" the device (1) (a sufficiently long time has elapsed since the preparation process has run its course), then the microsystems no longer interact with the preparation device. Hence, its ontological structure is fixed with respect to what and how each of the systems (the pairs $(a, b)$) "is". Indeed, the device (2) acts on the system (on the subsystem $b$), but it can no longer act on the systems reaching the device (3) (the subsystems $a$). Such an action would be transferred faster than light (the devices (2) and (3) may be very far apart and respond, say, at the same time – in the adopted reference system). An influence of the device (2) on the microsystem reaching the device (3) is no longer possible and indeed could not be achieved with any known interaction. Hence the action of the microsystem on the device (3) can only depend on the ontological structure of the microsystem (already determined after it left the device (1)) and on the ontological structure of the device (3).

The choice of the device (2) as well as what happens to it, cannot influence what happens on the device (3). But one obviously can demix an ensemble on the basis of the indicators of the "not influencing" device (2) so that the mixture components make one of the indicators (corresponding to decision effects $e_\nu$ with $\sum_\nu e_\nu = 1$) respond with certainty. Then this "certainty" should already be present as an ontological structure of the microsystems after these leave the preparation. It would remain impossible, on the basis of the trajectories of the preparation device (1)+(2) to determine all the ontological structures of the individual microsystems; but it should become possible (at least in thought) to sort out the microsystems according to their ontological structures. Let us express this mathematically:

In an EPR-experiment we take (1)+(2) as a preparation device. This preparation device is characterized by a set $Q(a)$ of preparation procedures for the microsystems leaving the device (1)+(2). The preparation procedure $a$ is given by $(a_1 \times a_2 \times M_3) \cap M$ and the corresponding $\varphi(a)$ by (2.4). Other procedures $a' \in Q(a)$ are e.g. $(a_1 \cap b_1 \times a_2 \cap b_2 \times M_3) \cap M$; these are characterized by registrations $b_1$, $b_2$ of the trajectories of (1) and (2) (see II (4.1)) where we replace $c_1 = a_1 \cap b_1$ by $a_1 \cap b_1 \times a_2 \cap b_2$). It is also possible to take (in a special EPR-experiment) a fixed $(a_1 \cap b_1 \times a_2 \times M_3) \cap M$ as the basic preparation device $a$ for $\mathcal{Q}(a)$. For simplicity we discuss only the case of $a = a_1 \times a_2 \times M_3 \cap M$.

Our imagination that the procedures of $\mathcal{Q}(a)$ may be decomposed according to ontological structures of the microsystems, may be expressed by additional subsets $\tilde{a}$ of $M$ with $\tilde{a} \subset a$. Here $\tilde{a}$ characterizes a finer sorting possibility, according to which even the elements $a'$ of $\mathcal{Q}(a)$ can be decomposed:

$$a' = (a' \cap \tilde{a}) \cup [a' \backslash (a' \cap \tilde{a})].$$

Therefore we imagine a set $\tilde{\mathcal{Q}}(a)$ with $\tilde{\mathcal{Q}}(a) \supset \mathcal{Q}(a)$ and require

1) $\tilde{\mathcal{Q}}(a)$ is a Boolean ring with $a$ as unit element. Therefore $\tilde{\mathcal{Q}}(a)$ is also a structure of species selection procedure. We do not require that $\tilde{\mathcal{Q}}(a)$ is a structure of species statistical selection procedure since experimentally we cannot take any statistics with decompositions possible only in thought. The following is of decisive significance for describing the meaning of the $\tilde{a} \in \tilde{\mathcal{Q}}(a)$:

Let $e_0 \in G$ be the support of $\varphi(a)$. If a registration method $b_0$ may be combined with $a$, then $a' \cap b_0 \neq \emptyset$ for all $a' \in \mathcal{Q}(a)$ with $a' \neq \emptyset$. We also presume $\tilde{a} \cap b_0 \neq \emptyset$ for all $\tilde{a} \in \tilde{\mathcal{Q}}(a)$ with $\tilde{a} \neq \emptyset$. For each $e \in G$ with $e \leq e_0$, an element $\tilde{a} \in \tilde{\mathcal{Q}}(a)$ ($\tilde{a} \neq \emptyset$) shall be called "$e$-certain" if the relation $\tilde{a} \cap b_0 = \tilde{a} \cap b$ holds for each effect procedure $(b_0, b)$ with $\psi(b_0, b) \geq e$ and $b_0$ combinable with $a$. Let the subset of the $e$-certain $\tilde{a} \in \tilde{\mathcal{Q}}(a)$ be denoted by $\tilde{\mathcal{Q}}(a; e)$. It is obvious that $e_1 \geq e_2$ implies $\tilde{\mathcal{Q}}(a; e_1) \supset \tilde{\mathcal{Q}}(a; e_2)$. To simplify the applicability of our concepts, we consider not only the registration procedures from $\mathcal{R}$ but also all the "idealized" registration procedures according to V §10. Moreover, for simplicity let us assume that for each $e \in [\mathbf{0}, e_0]$ there are a registration method $b_0$ and an idealized registration procedure $b \subset b_0$ with $\psi(b_0, b) = e$ (weaker conditions would suffice; see Bell's inequality [38]).

For each pair of $\tilde{a}_1 \in \tilde{\mathcal{Q}}(a;e)$ and $\tilde{a}_2 \in \tilde{\mathcal{Q}}(a;e^\perp)$, we find $\tilde{a}_1 \cap \tilde{a}_2 = \emptyset$. Proof: If we had $\tilde{a}_1 \cap \tilde{a}_2 \neq \emptyset$, there would exist a $b_0$ and a $b$ such that $\tilde{a}_1 \cap \tilde{a}_2 \cap b_0 \neq \emptyset$ (because $\tilde{a}_1 \cap \tilde{a}_2 \in \tilde{\mathcal{Q}}(a)$!) and $\psi(b_0, b) = e$. Then $\psi(b_0, b_0 \setminus b) = e^\perp$. From $\tilde{a}_1 \in \mathcal{Q}(a;e)$ follows $\tilde{a}_1 \cap b_0 = \tilde{a}_1 \cap b$; from $\tilde{a}_2 \in \tilde{\mathcal{Q}}(a;e^\perp)$ follows $\tilde{a}_2 \cap b_0 = \tilde{a}_2 \cap (b_0 \setminus b)$. Hence $\tilde{a}_1 \cap \tilde{a}_1 \cap b_0 = \tilde{a}_1 \cap \tilde{a}_2 \cap b \cap (b_0 \setminus b) = \emptyset$, in contradiction to $a_1 \cap a_2 \cap b_0 \neq \emptyset$.

For $\tilde{\mathcal{Q}}(a)$ it is crucial to formulate the ontological demixability in thought:

2) For each pair of $e_1, e_2 \in G$, with $e_1, e_2 \leq e_0$ and $e_1 \perp e_2$, there exist three elements $\tilde{a}_1 \in \tilde{\mathcal{Q}}(a;e_1)$, $\tilde{a}_2 \in \tilde{\mathcal{Q}}(a;e_2)$, $\tilde{a}_3 \in \tilde{\mathcal{Q}}(a;e_0 - e_1 - e_2)$ with $a = \tilde{a}_1 \cup \tilde{a}_2 \cup \tilde{a}_3$.

But just this requirement of ontological de-mixability leads to a contradicton with quantum mechanics, as will be shown now.

First let us cast the requirements 1), 2) into a more suitable form. For this purpose we introduce a mapping $[\mathbf{0}, e_0] \xrightarrow{\phi} \mathcal{P}(a)$ by

$$\phi(e) = \bigcup_{\tilde{a} \in \tilde{\mathcal{Q}}(a;e)} \tilde{a}. \tag{2.6}$$

Since $e_1 \geq e_2$ implies $\tilde{\mathcal{Q}}(a;e_1) \supset \tilde{\mathcal{Q}}(a;e_2)$, we have

$$e_1 \geq e_2 \Rightarrow \phi(e_1) \supset \phi(e_2). \tag{2.7}$$

Since $\tilde{a}_1 \in \tilde{\mathcal{Q}}(a;e)$ and $\tilde{a}_2 \in \tilde{\mathcal{Q}}(a;e^\perp)$ imply $\tilde{a}_1 \cap \tilde{a}_2 = \emptyset$, we have

$$\phi(e) \cap \phi(e^\perp) = \emptyset. \tag{2.8}$$

From 2) follows $\phi(e_1) \supset \tilde{a}_1$, $\phi(e_2) \supset \tilde{a}_2$, and $\phi(e_0 - e_1 - e_2) \supset \tilde{a}_3$, and therefore

$$\phi(e_1) \cup (e_2) \cup \phi(e_0 - e_1 - e_2) = a. \tag{2.9}$$

According to (2.7) and (2.8) we have

$$\phi(e_1) \cap \phi(e_2) = \emptyset, \quad \phi(e_1) \cap \phi(e_0 - e_1 - e_2) = \emptyset, \quad \phi(e_2) \cap \phi(e_0 - e_1 - e_2) = \emptyset;$$

therefore $\tilde{a}_2 \cup \tilde{a}_2 \cup \tilde{a}_3 = a$ implies

$$\tilde{a}_1 = \phi(e_1), \quad \tilde{a}_2 = \phi(e_2) \quad \text{and} \quad \tilde{a}_3 = \phi(e_0 - e_1 - e_2).$$

Since $e_0$ is the support of $\varphi(a)$, for $0 \neq e < e_0$ we have $\mu(\varphi(a), e) \neq 0$. For an ideal registration procedure $b$ with $\psi(b_0, b) = e$ follows $\lambda_{\mathcal{S}}(a \cap b_0, a \cap b) \neq 0$ and therefore $a \cap b \neq \emptyset$. With $e_1 = e$ and $e_2 \perp e$, (2.9) implies $a = \phi(e) \cup \phi(e_2) \cup \phi(e_0 - e_2 - e)$. Since $e_2$ and $e_0 - e_1 - e$ are orthogonal to $e$, from $\tilde{a} \in \tilde{\mathcal{Q}}(a;e_2)$ (resp. $\tilde{a} \in \tilde{\mathcal{Q}}(a;e_0 - e_2 - e)$) follows $\tilde{a} \cap b = \emptyset$ and hence $\phi(e_2) \cap b = \emptyset$ and $\phi(e_0 - e_2 - e) \cap b = \emptyset$. Therefore $a \cap b = \phi(e) \cap b \neq \emptyset$ and thus

$$\mathbf{0} \neq e \leq e_0 \Rightarrow \phi(e) \neq \emptyset. \tag{2.10}$$

We shall now see that we arrive at the same mathematical requirements if we attempt to interpret only the *macroscopic* situation ontologically. The new departure starts from the three conditions (1.2), (1.3) and (2.1) for an EPR-experiment. (1.2) defines that the systems (2) and (3) do not act on (1). (1.3) defines that the system (3) does not act on the systems (1) and (2). (2.1) defines that the system (2) does not act on the systems (1) and (3). Hence there is no interaction at all between the systems (2) and (3). The causes for changes in the systems (2) and (3) can therefore lie only in the system (1). In the EPR experiment we can "determine" the action of the system (1) on (3) by considering the action of (1) on an *appropriate* auxiliary system (2). Hence the action of the system (1) on (3) must be determined by causes lying only in (1), also if all these causes cannot be discovered on the system (1) alone or by a *fixed* auxiliary system (2). The action of (1) on various systems (2) can "reveal" once this and once that cause but never all causes. The actions of (1) on (3) can be determined by the behavior of appropriate systems (2), if the system (3) registers decision effects in the sense of quantum mechanics.

One can define the decision effects without speaking of microsystems as action carriers. This is demonstrated by the definitions III (4.2), III (4.1), the definitions of "effects" in III D 5.1.2, III D 5.2.2, and finally by the definition VI D 1.3.1. Obviously all these conditions are difficult to verify for an experimental physicist. Therefore let us add here another possible method to characterize the decision effects. $e=\psi(b_0, b)$ with $b_0, b$ according to III (4.2), III (4.1) is a decision effect (or approximately so), if there is no other effect $g \neq \mathbf{0}$ with $g \leq e = \psi(b_0, b)$ and $g \leq \mathbf{1} - e = \psi(b_0, b_0 \setminus b)$ (resp. if such a $g$ is approximately $\mathbf{0}$). We may define: The effects $g_1, g_2$ exclude each other if $\mathbf{0} \leq g \leq g_1, g_2$ imply $g = \mathbf{0}$. An effect $e$ is a decision effect if and only if $e$ and $\mathbf{1} - e$ exclude each other.*)

For a directed action of the system (1) on other systems, the action of the causes lying in (1) is determined if the other systems register decision effects. "Mentally" $(a_1 \times a_2 \times M_3) \cap M$ must therefore allow finer sorting according to the ontological causes. Sets thus sorted out would show determined actions in the registration of decision effects $e \leq e_0$. But mathematically such a sorting out of $(a_1 \times a_2 \times M_3) \cap M$ is equivalent to a sorting out of $a$ as a preparation device of the microsystems. Whether we perform the ontological foundation of the above requirements 1), 2) more macroscopically or more microscopically, is therefore without importance to the contradiction with quantum mechanics.

Let us comprise the requirements 1), 2) in an axiom using (2.7) through (2.10):

OEPR: There is a preparation procedure $a \in \mathcal{Q}$ for which the following relations are fulfilled:

(i) With $e_{0\nu}$ as the components (see VII § 5.4) of the support $e_0$ of $\varphi(a)$, there is at least one $\nu$ where $e_{0\nu} \mathcal{H}_\nu$ is not less than three-dimensional.

There is a mapping $\phi$ of $[\mathbf{0}, e_0] \subset G$ in $\mathcal{P}(a)$ with:

(ii) $e_1 \leq e_2$ implies $\phi(e_1) \subset \phi(e_2)$,

(iii) $\phi(e) \cap \phi(e^\perp) = \emptyset$,

(iv) $e_1, e_2 \in [\mathbf{0}, e_0]$ and $e_1 \perp e_2$ imply

$$a = \phi(e_1) \cup \phi(e_2) \cup \phi(e_0 - e_1 - e_2).$$

(v) $\mathbf{0} \neq e \leq e_0$ implies $\phi(e) \neq \emptyset$.

There are many EPR-experiments which obey the condition (i).

For classical systems (in sense of VII § 5.3) no additional ontological structures are necessary to "explain" EPR-experiments, since all demixtures of ensembles coex-

*) If we define: $e \in L$ separates if $e$ and $\mathbf{1} - e$ exclude each other, we find, without any axioms for $L$, that all extreme points of $L$ separate. If $G$ is defined (according to AV 1.1) then $G \subset \partial_e L$ (see VI T 1.3.6). We cannot prove more without the representation theorem in VIII, i.e. without the spectral theorem for the elements of $L$.

ist (VII T 5.3.3). Therefore it was reasonable to formulate OEPR only for the case of microsystems as action carriers.

Let us show that OEPR leads to a contradiction.

A mapping $[0, e_0] \xrightarrow{\phi} \mathscr{P}(a)$ which satisfies (ii) and (iii) of OEPR, shall briefly be called an order orthomorphism of $[\mathbf{0}, e_0]$ in $\mathscr{P}(a)$. A function $m(e)$ defined on $[\mathbf{0}, e_0]$ with $m(e)=1$ or $=0$ for which $e_1 \leq e_2 \Rightarrow m(e_1) \leq m(e_2)$ holds and also $m(e_1)=1$, $e_2 \perp e_1 \Rightarrow m(e_2)=0$, shall be called a discrete measure over $[\mathbf{0}, e]$.

If $\phi$ is an order orthomorphism, then a discrete measure is defined by $m_x(e)=\{1$ for $x \in \phi(e)$ and 0 for $x \notin \phi(e)\}$ for each $x \in a$. Conversely, if a discrete measure $m_x(e)$ is defined for each $x \in a$, then an order orthomorphism $[0, e_0] \xrightarrow{\phi} \mathscr{P}(a)$ is defined by $\phi(e)=\{x \mid m_x(e)=1\}$.

For an atom $p$ of the lattice $[\mathbf{0}, e_0]$, the components of $p$ (see VII, §5.4) are equal to zero except for a single component $p_\nu$, where $p_\nu$ is the projection operator on a vector $\varphi \in e_{0\nu} \mathscr{H}_\nu$. For this reason, to characterize the atoms of $[\mathbf{0}, e_0]$ we shall denote them by $e_{\nu\varphi}$.

We present the proof of contradiction in several steps which we shall arrange in theorems.

**T 2.1** For an order orthomorphism and the described corresponding measures, the following two relations are equivalent:

(i) There are an $e_{0\nu} \mathscr{H}_\nu$, and a $\varphi \in e_{0\nu} \mathscr{H}_\nu$, and an $x \in M$, such that $m_x(e_{\nu\varphi})=1$ and $m_x(e_{\nu\psi}+e_{\nu\psi'})=m_x(e_{\nu\psi})+m_x(e_{\nu\psi'})$ hold for all $\psi, \psi' \in e_{0\nu} \mathscr{H}_\nu$, with $\psi \perp \psi'$.

(ii) With $\Delta_\nu(\psi, \psi')=\phi(e_{\nu\psi}+e_{\nu\psi'}) \backslash [\phi(e_{\nu\psi}) \cup \phi(e_{\nu\psi'})]$, we have

$$\bigcup_{\psi \in e_{0\nu} \mathscr{H}_\nu} \phi(e_{\nu\psi}) \not\subset \bigcup_{\substack{\psi, \psi' \in e_{0\nu} \mathscr{H}_\nu \\ \psi \perp \psi'}} \Delta_\nu(\psi, \psi').$$

*Proof.* (i)$\Rightarrow$(ii): According to (i), there is an $x \in \phi(e_{\nu\varphi})$ which obeys $x \notin \phi(e_{\nu\psi})$ and $x \notin \phi(e_{\nu\psi'}) \Rightarrow x \notin \phi(e_{\nu\psi}+e_{\nu\psi'})$. Therefore, from (i) follows $x \notin \Delta_\nu(\psi, \psi')$ and hence

$$x \in \bigcup_{\psi \in e_{0\nu} \mathscr{H}_\nu} \phi(e_{\nu\psi}),$$

and

$$x \notin \bigcup_{\substack{\psi, \psi' \in \mathscr{H}_\nu \\ \psi \perp \psi'}} \Delta_\nu(\psi, \psi').$$

(ii)$\Rightarrow$(i): There is an $x \in M$ such that

$$x \in \bigcup_{\psi \in e_{0\nu} \mathscr{H}_\nu} \phi(e_{\nu\psi})$$

and

$$x \notin \bigcup_{\substack{\psi, \psi' \in e_{0\nu} \mathscr{H}_\nu \\ \psi \perp \psi'}} \Delta_\nu(\psi, \psi').$$

From $x \in \bigcup_{\psi \in e_0 \vee \mathcal{H}_\nu} \phi(e_{\nu\psi})$ follows that there is a $\varphi \in e_0 \vee \mathcal{H}_\nu$ with $x \in \phi(e_{\nu\varphi})$, i.e. $m_x(e_{\nu\varphi})=1$.
Because of

$$x \notin \bigcup_{\substack{\psi, \psi' \in e_0 \vee \mathcal{H}_\nu \\ \psi \perp \psi'}} \Delta_\nu(\psi, \psi'),$$

for all pairs $\psi$, $\psi'$ we have $x \notin \Delta_\nu(\psi, \psi')$.

If $m_x(e_{\nu\psi})=1$, i.e. $x \in \phi(e_{\nu\psi})$, we find $m_x(e_{\nu\psi'})=0$ (since $m_x$ is a discrete measure) and because $e_{\nu\psi}+e_{\nu\psi'} \geq e_{\nu\psi}$ also $m_x(e_{\nu\psi}+e_{\nu\psi'})=1$. If $m_x(e_{\nu\psi})=0$ and $m_x(e_{\nu\psi'})=0$, i.e. $x \notin \phi(e_{\nu\psi}) \cup \phi(e_{\nu\psi'})$, then $x \notin \Delta_\nu(\psi, \psi')$ also implies $x \notin \phi(e_{\nu\psi}+e_{\nu\psi'})$, i.e. $m_x(e_{\nu\psi}+e_{\nu\psi'})=0$. □

**T 2.2** If one of the Hilbert spaces $e_0 \vee \mathcal{H}_\nu$ is at least three-dimensional, then there is no discrete measure $m$ over $[\mathbf{0}, e_0]$ for which there is a $\varphi \in e_0 \vee \mathcal{H}_\nu$ with $m(e_{\nu\varphi})=1$ and $m(e_{\nu\psi}+e_{\nu\psi'})=m(e_{\nu\psi})+m(e_{\nu\psi'})$ for all pairs $\psi, \psi' \in e_0 \vee \mathcal{H}_\nu$ with $\psi \perp \psi'$.

*Proof.* Let both conditions be fulfilled. Then, for $\psi, \psi' \in e_0 \vee \mathcal{H}_\nu$, $\psi \perp \psi'$ the following two relations hold:

1. $m(e_{\nu\psi})=1 \Rightarrow m(e_{\nu\psi'})=0$ (since $m$ is a discrete measure)
2. $m(e_{\nu\psi})=0$ and $m(e_{\nu\psi'})=0$ imply $m(e_{\nu\chi})=0$ for each vector $\chi=\alpha\psi+\beta\psi'$ (we assume $\|\psi\|=\|\psi'\|=1$ and $|\alpha|^2+|\beta|^2=1$), since $0=m(e_{\nu\psi})+m(e_{\nu\psi'})=m(e_{\nu\psi}+e_{\nu\psi'}) \geq (e_{\nu\chi})$.

If one chooses a vector $\varphi'$ orthogonal to $\varphi$, then $m(e_{\nu\varphi})=1$ implies $m(e_{\nu\varphi'})=0$. Therefore, since each vector $\eta=\alpha\varphi+\beta\varphi'$ makes either $m(e_{\nu\eta})=1$ or $m(e_{\nu\eta})=0$, for each $\varepsilon>0$ there are two vectors $\eta_1, \eta_2$ with $\|\eta_1-\eta_2\|<\varepsilon$ and $m(e_{\nu\eta_1})=1$, $m(e_{\nu\eta_2})=0$.

We construct a contradiction by deducing from $m(e_{\nu\eta_1})=1$ and $m(e_{\nu\eta_2})=0$ the conclusion $m(e_{\nu\eta_1})=0$:

Without changing $e_{\nu\eta_2}$, we can change $\eta_2$ by a phase factor $e^{i\gamma}$ so that $\langle\eta_1, \eta_2\rangle$ is non-negative and real. We set $\langle\eta_1, \eta_2\rangle=\cos\vartheta$. Therefore, $\vartheta$ can be chosen arbitrarily small because $\|\eta_1-\eta_2\|<\varepsilon$. In the subspace of $e_0 \vee \mathcal{H}_\nu$ spanned by $\eta_1, \eta_2$, we choose a vector $\eta_3$ perpendicular to $\eta_1$, e.g. $\eta_3=-\eta_1\cos\vartheta+\eta_2\frac{1}{\sin\vartheta}$; then $\eta_2=\eta_1\cos\vartheta+\eta_3\sin\vartheta$. Moreover, we choose a vector $\eta_4$ that is orthogonal to $\eta_1$, $\eta_3$ ($e_0 \vee \mathcal{H}_\nu$ is at least three-dimensional!). Because of $m(e_{\nu\eta_1})=1$, we have $m(e_{\nu\eta_3})=0$ and $m(e_{\nu\eta_4})=0$. According to relation 2, for each vector $\eta$ from the space spanned by $\eta_2$, $\eta_4$ as well as from the space spanned by $\eta_3$, $\eta_4$, we have $m(e_{\nu\eta})=0$. We now choose as follows two vectors $\eta$ and $\eta'$ from these two subspaces:

With $\delta=\pm 1$ and with $z$ as a solution of $z^2-z\cos\vartheta+\sin^2\vartheta=0$ ($z$ is real provided $\sin^2\vartheta \leq \frac{1}{4}\cos^2\vartheta$, which we can assume since $\vartheta$ can be chosen arbitrarily small!), we set

$$\eta=\frac{1}{\sqrt{1+z^2}}\eta_2+\frac{z}{\sqrt{1+z^2}}\eta_4, \qquad \eta'=\frac{z}{\sqrt{z^2+\sin^2\vartheta}}-\frac{\sin\eta}{\sqrt{z^2+\sin^2\vartheta}}\eta_4.$$

There follows $\langle\eta, \eta'\rangle=0$ and

$$\frac{1}{\sqrt{2}}\frac{1}{\cos\vartheta}\sqrt{1+z^2}\,\eta-\frac{1}{\sqrt{2}}\tan\vartheta\frac{\sqrt{z^2+\sin^2\vartheta}}{z}\eta'=\frac{1}{\sqrt{2}}(\eta_1+\delta\eta_4).$$

Because of $m(e_{v\eta}) = m(e_{v\eta'}) = 0$, for $\delta = \pm 1$ then follows

$$m(e_{v\rho}) = m(e_{v\sigma}) = 0 \quad \text{for } \rho = \frac{1}{\sqrt{2}}(\eta_1 + \eta_4) \quad \text{and} \quad \sigma = \frac{1}{\sqrt{2}}(\eta_1 - \eta_4).$$

On the basis of the above relation 2, from this follows $m(e_{v\eta_1}) = 0$. □

The next theorem follows immediately from T 2.1 and T 2.2:

**T 2.3** For each of the Hilbert spaces $e_{0v}\mathscr{H}_v$ (having at least three dimensions) with $\Delta_v$ from T 2.1 we get

$$\bigcup_{\psi \in e_{0v}\mathscr{H}_v} \phi(e_{v\psi}) \subset \bigcup_{\substack{\psi, \psi' \in e_{0v}\mathscr{H}_v \\ \psi \perp \psi'}} \Delta_v(\psi, \psi'). \quad \square$$

**T 2.4** If an $e_{0v}\mathscr{H}_v$ and an $\eta \in e_{0v}\mathscr{H}_v$ obey $\phi(e_{v\eta}) \neq \emptyset$, the relation

$$\phi(e_1) \cup \phi(e_2) \cup \phi(e_0 - e_1 - e_2) = a$$

cannot hold for all orthogonal pairs $e_1, e_2 \leq e_0$.

*Proof.* We choose $e_1 = e_{v\psi}$, $e_2 = e_{v\psi'}$, with $\psi \perp \psi'$. Because of $\phi(e_{v\eta}) \neq \emptyset$ we have

$$\bigcup_{\psi \in e_{0v}\mathscr{H}_v} \phi(e_{v\psi}) \neq \emptyset.$$

Therefore, according to T 2.3 there is a pair $\psi, \psi' \in e_{0v}\mathscr{H}_v$ with $\Delta_v(\psi, \psi') \neq \emptyset$. Because of $\phi(e_0 - e_1 - e_2) \subset a \backslash \phi(e_{v\psi} + e_{v\psi'})$ we have $\Delta_v(\psi, \psi') \cap \phi(e_0 - e_1 - e_2) = \emptyset$. Since $\Delta_v(\psi, \psi) \cap \phi(e_i) = \emptyset$ holds for $i = 1, 2$, we have

$$\begin{aligned} \Delta_v(\psi, \psi') &= \Delta_v(\psi, \psi') \cap a \\ &= \Delta_v(\psi, \psi') \cap [\phi(e_1) \cup \phi(e_2) \cup \phi(e_0 - e_1 - e_2)] = \emptyset \end{aligned}$$

in contradiction to $\Delta_v(\psi, \psi') \neq \emptyset$. □

Therefore, by T 2.4 the condition (iv) of OEPR leads to a contradiction with quantum mechanics if at least one of the Hilbert spaces $e_{0v}\mathscr{H}_v$ contains an $\eta$ which makes $\phi(e_{v\eta}) \neq \emptyset$. This last condition is fulfilled by (v).

From this proof of the contradiction, one recognizes with the aid of T 2.1 through T 2.4 that one can weaken the assumptions in OEPR. We need not assume that the mapping $\phi$ is defined on the whole lattice $[\mathbf{0}, e_0]$; it suffices to replace $[\mathbf{0}, e]$ by $[\mathbf{0}, e_{0v}]$ where $e_{0v}\mathscr{H}_v$ is no less than three-dimensional.

If $e_{0v}\mathscr{H}_v$ can be represented as a product space $\mathscr{H}^{(1)} \times \mathscr{H}^{(2)}$ and $\mathscr{H}^{(2)}$ is not less than three-dimensional, it suffices to replace $[\mathbf{0}, e_{0v}]$ by the subset of all the $e \leq e_{0v}$ of the form $\mathbf{1} \times e^{(2)}$, where $e^{(2)}$ is a projection in $\mathscr{H}^{(2)}$.

We may weaken OEPR still more by replacing $[\mathbf{0}, e_{0v}]$ or the set $\{e^{(2)}\}$ by suitable subsets, for example by denumerable ortho-complemented and atomic sublattices of $[\mathbf{0}, e_{0v}]$ or of $\{e^{(2)}\}$ which are $\sigma(\mathscr{B}', \mathscr{B})$-dense in $[\mathbf{0}, e_{0v}]$ or $\{e^{(2)}\}$ respectively.

All this shows that an ontological interpretation of a single (suitable) EPR-experiment contradicts quantum mechanics.

Now what should one think of this contradiction? There are two possibilities: 1) one can deny that the ontological interpretations are meaningful. 2) one insists on this interpretation and seeks for a mistake in the conclusions leading to OEPR.

Most easily one can recognize such a mistake in the last "macroscopic" argumentation. The relation (2.1) says only that systems (2) do not change the frequency of the trajectories on (1) and (3), and we had defined by (2.1) that (2) "does not act" on (1) and (3).

This is a very physical definition: No physically testable influence of (2) on (1) and on (3) exists. But if we allow an ontological decomposition according to invisible causes lying in (1), then it is inconsequent to use (2.1) as the condition that there is no action of (2) on (1) and (3). In spite of (2.1) there may be an ontological action of (2) on (1) and (3), an ontological action which cannot be detected by the trajectories of (2) and (3). Similarly, from the relations (1.2) and (2.3) one cannot deduce that there is no ontological interaction between (2) and (3). Why did we not consider these possibilities in the course of our first hasty conclusions?

Obviously, on the basis of many experiences with directed action transfers by macrosystems (e.g. the transfer of the action of a weapon on a windowpane by a bullet), we tend to conclude that III (3.1) means that *no ontological* action of the systems (2) on (1) is possible. Then the statistical dependence of the trajectories of the system (1) and (2) rests solely on the fact that the action emanating from (1) can differ for different trajectories of (1). But how may we conclude this? Perhaps only because this notion in the case of a weapon as system (1) and a windowpane as system (2) appeared to be correct? Or perhaps because in many (but not all!) experiments the trajectories on (2) run their course later than those on (1) and no actions can reach backwards in time?

But if in discussing EPR-experiments we allow ontological structures of a system which do not show up in the trajectories of this system, then we also can no longer conclude from III (3.1) that the causes for ontological structures of (1) can lie only in (1) itself, but those of (2) can lie in (1) and (2). Rather, we must then also allow that in spite of III (3.1) there is a true ontological interaction of (1) on (2) and of (2) on (1), only with the special constraint III (3.1). Thus the cause for an ontological structure of (1) could also lie in what has happened or will happen to (2). This "reverse ontological action" of (2) on (1) (also possibly from the future into the past) is indeed "unusual" (we are not accustomed to something like this in our normal surroundings), but need not be logically excluded.

But if we allow such a reverse action, then the above conclusions from the EPR experiment are impossible. Namely, (2) can act in reverse on (1) (in the past) so that the trajectory on (3) in fact is "caused" by the trajectory on (2). In the spin example, the indicator on (3) of the spin in the direction $\boldsymbol{e}$ can be produced by means of the indicator on (2). The process on (2) characterized by the indicator can act in reverse on (1) in the past (on (1) not recognizable in the trajectories) and act further from (1) into the future on (3). Of course, also (3) can correspondingly act on (2). Hence, between (2) and (3) there can very well exist an interaction that runs its course over (1) (in the past). Of course this interaction, because of the conditions (1.2), (1.3), (2.1), cannot "transmit information" from (2) to (3) or from (3) to (2). For, by transmission of an information from (2) to (3) we understand that we *arbitrarily* make an adjustment (preparation) on (2) which can be determined by processes on (3).

But transmission of information with directed interaction from (1) to (2) is possible under the condition III (3.1). In fact, with given $a_2, b_{20}, b_2$,

$$\lambda_{12}(a_1 \times a_2 \cap M \cap b_{10} \times b_{20}, a_1 \times a_2 \cap M \cap b_{10} \times b_2)$$

depends as well on the "arbitrarily" chosen $a_1$.

After these critical remarks on the problem of ontological interactions, everything resolves very naturally. In the first conclusions we had "unknowingly" allowed our experiences with classical systems as action carriers to influence us. Hence conversely we must not be astonished that the consequences 1) through 3) deduced for $\tilde{\mathcal{Q}}(a)$ are allowed only for classical systems as action carriers but not for microsystems.

If one thinks of (1)+(2) as a preparation device, and the indicator on (2) responds later than the indicator on the registration device (2), then we have an example where probability propositions are also made backwards in time. Since often "a priori" a special connection of probability and time direction is assumed, we shall devote the next section separately to this problem.

## §3 Microsystems and Time Direction

The EPR experiments discussed in §2 can serve as examples for the situation that probabilities can also be meaningfully defined backwards in time. Namely, if one considers (1)+(2) together as a preparation device and (3) as a registration device, then one can construct the devices so that the trajectories on (2) run their course later than those on (3). But from the trajectories of (2) one can infer the behaviour of the trajectories on (3) that have already run their course earlier.

The possibility of a meaningful definition of probabilities in both time directions exists still much more generally. It can already be read off from simple situations of preparation and registration.

If $a_1, a_2$ are elements of $\mathcal{Q}'$ with $a_1 \supset a_2$, and if $(b_0, b)$ is a procedure that may be combined with $a_1$, the following probabilities are well defined:

$$\lambda(a_1 \cap b_0, a_1 \cap b) = \mu(\varphi(a_1), \psi(b_0, b)), \tag{3.1}$$

$$\lambda(a_2 \cap b_0, a_2 \cap b) = \mu(\varphi(a_2), \psi(b_0, b)), \tag{3.2}$$

$$\lambda(a_1 \cap b_0, a_2 \cap b_0) = \lambda_{\mathcal{Q}}(a_1, a_2), \tag{3.3}$$

$$\lambda(a_1 \cap b, a_2 \cap b) = \frac{\lambda_{\mathcal{Q}}(a_1, a_2)\, \mu(\varphi(a_2), \psi(b_0, b))}{\mu(\varphi(a_1), \psi(b_0, b))}. \tag{3.4}$$

If we assume that the registration procedure $b$ happens at times essentially later than the preparation procedures $a_1$ and $a_2$, then (3.1) and (3.2) represent probabilities in the direction of the future. They give the probability of the "later" indicator $b$ after preparation according to $a_1$ resp. $a_2$ has happened earlier. The probability (3.3) does not depend on the registration. The probability (3.4) represents the probability of $a_2$ relative to $a_1$ under the condition that the indication $b$ responds "later" relative to $a_1$ and $a_2$. The probability (3.4) can also be easily tested experimentally;

from the experiments $x \in a_1 \cap b$ one needs only to sort out those with $x \in a_2 \cap b$ and to note the corresponding frequency.

Therefore, it is an error to think that the concept of probability in physics requires a time scale (perhaps a generally valid one in contrast to relativity) and can be understood only as predicting events that have not yet occurred.

Despite this, from the beginning there is distinguished a *time direction* in the theory of preparation and registration as the basis of quantum mechanics (as it was described in III). But in what way?

We have already rejected the opinion that in the formula $\mu(w, g)$ for the probability the symbol $w$ means our knowledge, $g$ a possible effect, and $\mu(w, g)$ a measure for our expectation that $g$ occurs. According to this opinion it would be senseless to speak of an expectation for something that has already happened. Therefore, this concept of probability presumes the existence of at time direction.

For us, probability is the picture for reproducible frequencies; and it is the *prescription* for a *correct* experiment, that the frequencies are reproducible. In this sense, physical probability is a property of devices. A roulette has a physical property, namely the frequencies for the various numbers. If it was well manufactured by a craftsman, these frequencies are equal for all numbers. From the physical property "frequencies" we must distinguish what a gambler calls his chance. In our interpretation of quantum mechanics in III, there is no symbol of our knowledge nor our expectations. Therefore we have to look for other structures which perhaps distinguish a time direction.

If we take quantum mechanics in the form $\mathscr{P}\mathscr{T}_3$ of III §5.2 (forget the preparation and registration procedures and take only $K$ and $L$ as base sets), we are tempted to define the time direction from $w \in K$ to $g \in L$.

But this is not correct. The EPR-Experiments (XII, §2) demonstrate that there can be processes which partially determine the ensemble $w$ *and* occur later than the registration processes. The direction from $w$ to $g$ in the first place determines the direction of the action of the preparation device on the registration device (see remarks after III (6.4.9)). Nevertheless there is something in this directed action which corresponds to a time direction. To elaborate this structure is more complicated than to pretend that $g$ is later than $w$.

The key to find the right connection between directed action and directed time was already mentioned in §3: the transmission of information. In this connection, information was defined as an "arbitrary" structure, which we produce. By the directed action, information obviously can only be transmitted from the preparation to the registration device. But not all that occurs on the preparation device is information, since not all is "arbitrary". Information in this sense can only be transmitted in the time direction which we call future. How can we see what part of $w$ is information?

For this purpose we must return to the description in III. There the preparation of the microsystems by the device (1) is described by a preparation procedure $a_1$ for the device (1) *and* trajectories from the space $Y_1$. It was essential for the description in III that the trajectories are not arbitrary; they start after we have prepared the experiment. The procedure $a_1$ is exactly what describes the structure of the device (1) as far as it is arbitrarily produced by us, i.e. $a_1$ describes the information put into (1). This information, or parts of it, can be transmitted to the registration device

(2); this can be seen by the influence of $a_1$ on the trajectories of (2). We know that information can only be transmitted into the future. Therefore, for "good" experiments we had introduced in III the prescription that the trajectories should only originate at the time "zero", where "zero" was the time "before" which the preparation $a_I$ has been finished. Hence there is indeed defined a time direction by the transmission of information from the preparation device to the registration device. In this sense, the time direction to the future is inherent in the direction from $w$ to $g$, even though this inherence cannot be seen immediately.

By this inherent time direction we are again confronted with the fact (already described at the beginning of XI §7) that physics cannot be separated from human actions. The time direction in quantum mechanics is nothing but the fundamental fact that we can only influence the world in the future.

In the description of the experiments in III, there is another structure connected with the time direction: the irreversibility of the trajectories. This does not affect $w$ or $g$; but it is essential for the discussion of motion reversal since there is no such reversal for all processes on the devices.

It is well known (see IX §2) that for microsystems there is a motion reversal transformation which transforms a $w \in K$ into a motion reversed $w' = CwC^{-1}$ (see IX (2.19)). If quantum mechanics is a g.G.-closed theory (in the sense of [3] §§8 and 10 and of XIII §4.3), then there should be preparation procedures $a \in \mathcal{Q}$ and $a' \in \mathcal{Q}$ for $w$ and $w'$, respectively, such that $\varphi(a) = w$ and $\varphi(a') = w'$. But it is completely unknown how in reality to construct a device for the procedure $a'$ if one for $a$ is given. Certainly one cannot obtain the device for $a'$ as a motion reversed device for $a$. To begin with, there does not exist such a device with motion reversed trajectories. Secondly such an imagined motion reversed device would transmit information into the past; hence it would rather resemble a registration device which absorbs impinging microsystems. But motion reversal also cannot produce a preparation device corresponding to $a'$ from an absorbing registration device since there does not exist a macroscopic device which runs in motion reversed sense relative to the absorbing registration device. The macroscopic irreversibility prevents us from giving the motion reversal transformation $w' = CwC^{-1}$ an experimentally easily conceivable sense. Rather, we appear to have encountered a thoroughly *nontrivial* problem, that of constructing two devices $a$ and $a'$ such that $\varphi(a') = C\,\varphi(a)\,C^{-1}$.

The existence of the motion reversal transformation $w' = CwC^{-1}$ is frequently interpreted by the idea that the "dynamics of the microsystem" itself is motion reversal invariant. Then the irreversibility of the macroscopic processes (with whose aid one prepares and registers the microsystem), would not react on the microsystem itself. But this reasoning veils the actual problem, since one thinks as if one could speak of the microsystems and their dynamics "in themselves" (without the preparation and registration procedures). The considerations of properties and pseudo-properties (in [2] II §4, [2] IV §8 and here in V §10 and VII §5.3) show that only for physical objects can we discard the preparation and registration procedures completely from a theory of the systems. For microsystems it is not possible; this was drastically brought to our attention in §2. To interpret IX (2.5) as describing the dynamics of the systems in itself is also a (wide-spread) misunderstanding. The operator $U_\tau$ in IX (2.5) describes a displacement of the registration device (relative to the preparation device) by a time $\tau$ (see IX, §1). Therefore we cannot say that a microdynamics in itself is motion reversal invariant. Motion reversal invariant is the time translation of the registration devices relative to the preparation devices *if* it is possible, to a given preparation procedure $a$ to construct another $a'$ with $\varphi(a')=C\,\varphi(a)\,C^{-1}$.

It this really possible? Can enough elements $w\in K$ ("physically" dense in $K$) really be prepared by suitable procedures $a$? To pose the question means to entertain doubts that $\mathscr{P}\mathscr{T}_q$ is a g.G.-closed theory (in the sense of [3] §10.3 and XIII §4.3).

If one looks for preparation devices for the various ensembles $w\in K$, it is obvious that the expenditure of material, technical finesse and also money can be extremely different for various $a\in\mathscr{Q}$: There are ensembles $w$ for which it is "easy" to construct devices $a$ with $\varphi(a)=w$. For other $w\in K$ it may be necessary to drive the size and complexity of the preparation device very high; there may exist such $w$ for which the size and complexity is in principle no longer realizable.

All this seems to indicate that the structure of $K$, $L$, and of the time displacement operator in $\mathscr{P}\mathscr{T}_q$ is an idealization. Though very good for systems composed of "few" elementary systems, it becomes worse and worse for bigger systems. In particular the "real" elements of $K$ will form a subset $K_r$ for which $K_r=CK_rC^{-1}$ is no longer valid. To ask for $K_r$ is to ask for a more comprehensive theory than $\mathscr{P}\mathscr{T}_q$; we encountered this already at several places in our development of quantum mechanics.

## §4 Macrosystems and Time Direction

Of course also in $\mathscr{P}\mathscr{T}_m$ (we have described such theories in II) there is determined a time direction from the preparation to the registration. This direction was introduced in II by the prescription that the preparation should be terminated "before $t=0$" and that the registration should begin "after $t=0$". But this distinction of a time direction does not imply that the trajectories are irreversible, i.e. that not every trajectory motion is the reverse to another motion. Classical point mechanics (e.g. in the Hamiltonian form) is an example for a motion reversal invariant theory.

To formulate the question for motion reversed trajectories more thoroughly, we start with the motion reversal transformation $\tilde{C}$ introduced in X §2.6. To define a motion reversed trajectory we must introduce a time $T$. The interval 0 to $T$ shall be the time during which we are interested in the trajectories. It makes no physical sense to introduce the idealization $T=+\infty$. A motion reversal of a total trajectory $\{z(t)|0<t<+\infty\}$ would be unphysical because only for a finite time interval the considered systems can be isolated. We therefore define a motion reversed trajectory only for a finite time interval: The trajectory

$$y'=\tilde{C}y \quad \text{with} \quad y'=z'(t)=\tilde{C}z(T-t) \tag{4.1}$$

is called motion reversed to $y$. Since the elements $f$ of $C(\hat{Y};\le T)$ (defined in II §4.1) depend only on the parts of the trajectories between 0 and $T$, a linear norm-continuous map $\tilde{C}$ of $C(\hat{Y};\le T)$ onto itself is defined by $\tilde{C}f(y)=f(\tilde{C}y)$. The Banach space dual to $C(\hat{Y};\le T)$ is $C'(\hat{Y})/C(\hat{Y};\le T)^{\perp}$. The set $K_m(\hat{Y};\le T)=K_m(\hat{Y})/C(\hat{Y};\le T)^{\perp}$ (with $K_m(Y)$ defined in II §3.2) can be interpreted as formed by the physically possible ensembles for the trajectory parts between 0 and $T$; it is a subset of the base $K(\hat{Y};$

$\leq T$) of $C'(\hat{Y}; \leq T)$. Then we also have $\bar{K}_m^\sigma(\hat{Y}; \leq T) \subset K(\hat{Y}; \leq T)$ and $\tilde{C}' \bar{K}_m^\sigma(\hat{Y}; \leq T) \subset K(\hat{Y}; \leq T)$, with the dual $\tilde{C}'$ of the map $\tilde{C}$. The set $\bar{K}_m^\sigma(\hat{Y}; \leq T) \leq T) \cap \tilde{C}' \bar{K}_m^\sigma(\hat{Y}; \leq T)$ is called the motion reversal invariant part of $\bar{K}_m^\sigma(\hat{Y}; \leq T)$. In many cases (if the systems are in a box) we have as support of this part only constant trajectories $z(t) = z_0$, the equilibrium states. In these cases we call the systems irreversible. What has this irreversible behavior to do with the embedding described in X §2? (Therein X §2.6 we mentioned already a special case of irreversibility.)

First one easily recognizes that irreversibility in $\mathscr{P}\mathscr{T}_m$ can be compatible with the embedding of $\mathscr{P}\mathscr{T}_m$ in $\mathscr{P}\mathscr{T}_{q\,\exp}$. The fact that $\bar{K}_m^\sigma(\hat{Y}; \leq T) \cap \tilde{C}' \bar{K}_m^\sigma(\hat{Y}; \leq T)$ is essentially smaller than $\bar{K}_m^\sigma(\hat{Y}; \leq T)$ need not contradict the fact that $K$ from $\mathscr{P}\mathscr{T}_{q\,\exp}$ shows the motion reversal invariance $CKC^{-1} = K$ (the subset $K_m$ of $K$ need not be invariant). Already the definition of

$$\tilde{K}_m = \overline{\mathrm{co}} \bigcup_{\tau \geq 0} \mathscr{U}_\tau' K_m$$

makes us expect $C\tilde{K}_m C^{-1} \neq \tilde{K}_m$.

The set $K_m(\hat{Y})$ and the corresponding sets $K_m$, $\tilde{K}_m$ are determined by the physically possible preparation procedures for macrosystems. Therefore, irreversibility enters $\mathscr{P}\mathscr{T}_m$ via the determination of the physically possible preparation procedures. This determination cannot be deduced from $\mathscr{P}\mathscr{T}_{q\,\exp}$ (thus $\mathscr{P}\mathscr{T}_{q\,\exp}$ is not comprehensive enough!), but rather must be put into $\mathscr{P}\mathscr{T}_{q\,\exp}$, e.g. by the embedding procedure described in X §2. The discussions in X §§2.5 and 2.6 show that irreversibility and embedding are strongly correlated.

Let $K_m(\hat{Y})$ be characterized as the set of ensembles which can be prepared by physically possible preparation procedures. The question arises whether we can describe these procedures themselves by physical theories and thus "explain" the size of $K_m(\hat{Y})$. This question is again a problem of compatibility: Is a macroscopic theory of a macroscopic part of a larger system compatible with the macroscopic theory of this total system? We have no doubt that this is true; see remarks to this problem in III §6.6. But this compatibility does not allow a deduction of $K_m(\hat{Y})$ for the parts (for the reasons mentioned at the end of XI §4). Also if we could estimate the size of $K_M(\hat{Y})$ for some systems as parts of larger systems we would have to presuppose the set $K_m(\hat{Y})$ for the larger systems.

In this sense, the irreversibility of smaller systems possibly can be reduced to that of larger systems. The larger the systems the smaller the set of possible preparations. The earth, the sun are given and cannot be prepared in many specimens. Perhaps we may take stars in the universe "as if" they were prepared by procedures (in regard to this "as if" see the next section)! The cosmos does not exist in more then one specimen and can only be regarded "as if" prepared once by the Big Bang.

The uniqueness of the cosmos poses the problem whether we can derive the set $K_m(\hat{Y})$ for a part of the cosmos from the structure of the cosmos. Then we could derive the irreversibility of the parts from the uniqueness of the development of the cosmos. Because of the uniqueness of the cosmos it makes no sense to derive something like an irreversibility of the cosmos. It makes only sense to derive in a theory that e.g. the motion reverse of a burning match contradicts the existing cosmos.

## §5 The Place of Human Beings in Quantum Mechanics

It is undoubted that we make experiments and theories and that we know what has happened in the experiments. A question is whether our knowledge and (or) our actions have reflection in the theories. There are three basically different view points.

(a) Our knowledge must be represented in a theory as soon as we use the concept of probability. In quantum mechanics our knowledge is represented by the elements of $K$ (or $\partial_e K$). Often this view regards the statistical description by quantum mechanics as basic for all of physics, so that classical physics is only a specialization.

We reject this view as we have mentioned often in this book. Our first argument is that we have never seen physicists who adopt this view during their every day work. It seems to us that this viewpoint was invented to "solve" pseudo problems of the measurement process. Our second and as we hope stringent argument is that we can develop quantum mechanics without representing our knowledge in the theory, as demonstrated in III to XI.

(b) As a basis for all theories ($\mathscr{PT}_m$ and $\mathscr{PT}_q$) we introduced the selection procedures (especially preparation and registration procedures). The word "procedure" expresses very clearly that we describe here human actions. We need not review this viewpoint here, since we have described it throughout this book. In some cases (fundamental domains) one can forget the procedures and describe objectively the systems (as done in II §1). But if we want to use statistics (unavoidable for systems with indeterministic dynamics) we must introduce human actions in the form of statistical selection procedures. Therefore the axioms for such selection procedures are prescriptions for "correct" acting. They are laws of nature only in the sense that the prescriptions can be fulfilled. For $\mathscr{PT}_q$ it was not possible to eliminate totally the procedures, i.e. to make the microsystems physical objects.

(c) For the last reason, many are disappointed with quantum mechanics as a physical theory. They anticipate that a physical theory should describe the real world as it is and that all experiments (of course human actions) are only justified for detecting the structure of the world.

There are several objections to our viewpoint (b). Exponents of (a) say that frequencies are not reproducible, that they are reproducible only with probability and that we have in this sense already presumed the concept of probability to introduce probability by frequencies. This objection is not correct, since the word probability is used with different meanings (from our viewpoint). If we make many experiments, the frequencies $N_+/N$ ($N_+$ selected cases in $N$ experiments) are reproducible within a small imprecision. It is impossible to find frequencies $N_+/N$ which differ considerably from a reproducible value $\alpha$. Therefore in our viewpoint there is "no" probability for the possibility that $N_+/N$ is not approximately equal to $\alpha$. It make no sense that $N_+/N$ may be "in principle" considerably different from $\alpha$ since this never happens.

For us "certainty" of an event has only the meaning that the contrary has never been seen. It is physically senseless to say that an event is not certain but has only a high probability (e.g. $1-10^{-100}$) and that because of this high probability no other case has been observed. This makes no more physical sense than the proposition that on the tip of a needle there is an angel who can never be observed.

(To the question of the reproducibility of frequencies see also [3] § 11 and XIII § 2.4.)

Exponents of (c) object to (b): It is doubtless that we make experiments and that we may describe also the experiments by physical theories. But the aim should be to eliminate all structures where human activities enter and also all structures where the accidental size of humans plays a role. One has to elaborate an objective structure also in regions as small as the microsystems.

The opinion of this objection is not that we have made a mistake but that we have stopped the development of the theory too early, before elaborating the objective reality of the microsystems. But we have not stopped too early. We have eliminated special structures of the preparation and registration devices as much as ever possible. A concrete device in a laboratory is a real object. That we cannot separate the microsystems completely from the devices is an objective fact and has nothing to do with the size of human beings. (That this size is macroscopic and not atomic is surely no accident!)

It is possible to manufacture macrosystems in such a way that they may be described well as if they where separated (for a time) from the rest of the world. To speak of microsystems as isolated from the world makes no sense. The microsystems "exist" because they are produced by macrosystems and act on macrosystems. This fact destroys the dream to deduce the macrosystems from their smallest parts, the atoms.

It is one of the most severe errors in physics (still widespread today) that the behavior of a total system can be explained by its parts; in the extreme form, that particle physics will explain the structure of the world. As physicists we must cease to dream and take the world as it is *and* as it can be changed by our activities. To describe a macrosystem as composed by atoms has *only* the sense described in $X$ as a compatibility between $\mathscr{P}\mathscr{T}$ and $\mathscr{P}\mathscr{T}_{q\,\mathrm{exp}}$. The consequences of this compatibility for the reality of atoms as parts of a macrosystem are discussed in XIII §4.8.

But we apply quantum mechanics not only to artifacts as devices for experiments with microsystems, but also to natural things as e.g. stars or biological objects. How can we do this if the theory is based on artifacts? This is a general objection, not only to quantum mechanics.

First we observe that physics could never have developed if we only had walked through the world and philosophized over the things we came across. Physics *has* developed as a sequel of craft. It is easy to see also that little of the physics of natural things would remain if we had no registration devices as e.g. telescopes or modern radar antennas to observe stars. It is no accident that Galilei has constructed telescopes and has made experiments on a new way to physics, new versus the philosophies in his time.

One will perhaps agree that no modern physics is possible without the modern technology of registration, i.e. measurement and observation. But how can we describe given natural objects by a physics which is based on preparation. We do not prepare e.g. stars or cats. Nevertheless we make physics of stars and also biophysics of cats. How can we do so?

We only succeed if we can do so "as if" the objects under consideration had been prepared. This "as if" was the starting point for the application of physics to natural objects since the beginning of physics. Already Newton's mechanics of the planetary system argues "as if" this planetary system had been prepared a long time ago. If we want to explain this preparation of the planetary system we can try to describe it by condensing clouds of material "as if" these clouds had been prepared a long time ago. If we consider cosmology, also the "Big Bang" is nothing but an "as if" preparation of the total cosmos.

If we register processes in the brain of a cat, we take this brain "as if" it has been prepared. The biotechnology demonstrates that we can prepare much more than we formerly expected.

In this connection the viewpoint (c) seduces to the following widespread error: By the registration procedures applied to the brain we find the "real" structure of the brain and its processes. We find that there are "in reality" only chemical and electrical processes. This error is based on the imagination that there are processes in the brain itself which can be detected *completely* by the devices we couple with the brain and that the devices have only this task. The right view is that "chemical and electrical processes" are only defined by registration (and preparation) devices. It is *this* reality which we meet if we act, e.g. if we couple the brain with registration devices. This view allows realities which cannot be defined by the indication of devices. And we know very well such realities.

First there is what we call the moment "now". There is no indication on any device which can tell us what is the moment "now". Every indication of a clock is of the same type, none of these indications is distinguished from the others as the moment "now". We meet with this "now" only in our consciousness. This "now" is very essential for our life! Nevertheless there is no such "now" in any physical theory. Nobody will maintain that there is no "now" in our life because there are no indications on devices by which such a "now" could be defined.

There are many other realities which cannot be defined by indications on devices. If we speak of colors and sounds, we do not mean physical quantities. Colors and sounds are already known before physicists couple registration devices with brains. There is no possibility to define by the indications on devices (as we have defined in [2] VIII §2 and [2] VII §4 what electrons are and what the position observable of such electrons is) what a color and what a sound is. We know what it is before we make biophysics of the brain. This impossibility of definition does not deny that seeing of colors and hearing of sound may be correlated with certain indications on registration devices. But these registrations cannot explain what cannot be defined by registrations.

It is obvious that all the more there is in principle no possibility to define by indications on devices what is love or hatred and even what is the free will.

In connection with the problem of the free will there is told another fairy-tale: The microsystems behave indeterministically. But according to the law of great numbers, the statistical behaviour of systems composed of many particles goes over to a deterministic dynamics for all macrosystems, also for brains. This is in contradiction to the free will.

First it is not proved that there is a statistics of the microsystems themselves (we will return to this point below). The statistics is a structure in macroscopic experiments that we have described by statistical selection procedures. Secondly the law of great numbers has nothing to do with the composition of systems by "many" atoms. Rather the discussion of the embedding problem in X makes us expect that the case of a deterministic dynamics as described in X §2.6 is only an exception. And this is what also experience shows. Natural (i.e. not artificial) systems as we see them on earth have an indeterministic dynamics. Only such exceptions as the motion of the centers of mass of the planets are deterministic over a long time. That artifacts, i.e. system manufactured by technical procedures, have a deterministic behavior (as much as possible) is not astonishing. This deterministic behavior is often the aim of our technical efforts. Who wishes to use a gambling machine as automobil?

An indeterministic dynamics does not contradict the free will. It also does not prove the existence of a free will since the concept of the free will cannot be defined by indications on devices.

All these examples show that the viewpoint (b) realistically describes the relation between physics and technology as well as the application of physics and technology to given structures in the world such as stars or biological systems. This viewpoint cannot contradict other concepts as e.g. our impressions of colors or sound or our conscience. On the contrary, biophysics can give us new technical methods to heal disease. Who is not glad that there are special physical-technical laboratories (called operation theatres) and great experts (called surgeons with their teams) if a life saving operation is necessary?

In spite of all our objections to the viewpoint (c), its proponents will not give it up since they are enamored of it. Therefore they will try to eliminate step by step all difficulties of this viewpoint. We cannot exclude that this might be possible. Obviously one may add to our viewpoint (b) imagined structures if one makes the addition carefully so that no contradictions to the propositions of (b) arise. But how shall we know that these imaginations have something to do with reality since all our physical work can be done without them?

Finally we want to give an example for such an additional imagination. We have introduced probability as a function $\lambda(c_1, c_2)$ describing relative frequencies of the selection procedures $c_1, c_2$. It is then possible to *define* a probability for an individual case: For $x \in c_1$ we define $\lambda(c_1, c_2)$ as the probability for $x \in c_2$. This probability is a relative one: It is the probability for $x \in c_2$ if one considers $x$ as element of $c_1$. The probability for $x \in c_2$ may change if we consider $x$ as an element of a stronger selection procedure $c_3$ with $c_1 \supset c_3 \supset c_2$: $\lambda(c_3, c_2) = \lambda(c_1, c_2)\, \lambda(c_1, c_3)^{-1}$.

One often uses this definition of a probability for individual cases in quantum mechanics for $c_1 = a \cap b_0$ ($a \in \mathcal{Q}$, $b_0 \in \mathcal{R}_0$) and $c_2 = a \cap b$ ($b \in \mathcal{R}(b_0)$). Assuming $x \in a \cap b_0$, as probability for $x \in a \cap b$ we get

$$\lambda(a \cap b_0, a \cap b) = \mu(\varphi(a), \psi(b_0, b)). \tag{5.1}$$

If one does not use these mathematical formulas it is easy to make mistakes. For instance one says that for $x \in a$ there is a probability $\mu(\varphi(a), \psi(b_0, b))$ of the effect $\psi(b_0, b)$. If $x \in a$ and $x \in \tilde{b}_0$ where $\tilde{b}_0 \cap b_0 = \emptyset$ ($\tilde{b}_0$ complementary to $b_0$), then there does not exist a probability for the effect $\psi(b_0, b)$. If only $x \in a$ is realized ($x$ is prepared by the procedure $a$) we have the possibility to *decide* what registration method $b_0$ we want to apply. This is a possibility of which we can freely dispose. Therefore there is no probability for the selection procedure $a \cap b_0$ relative to the selection procedure $a$.

The experimental physicist tries to make $a$ and $b_0$ in (5.1) as small as possible. For simplicity, we do not want here to use idealized registration and preparation procedures as in [2] III §4. It suffices to presume that the set $\{a|a\in\mathcal{Q}, x\in a\}$ has a smallest element $a_s$ and likewise $\{b_0|b_0\in\mathcal{R}_0, x\in b_0\}$ has a smallest element $b_{0s}$, so that $x\in a_s\cap b_{0s}$. Then

$$\lambda(a_s\cap b_{0s}, a_s\cap b) \tag{5.2}$$

is the probability that also $x\in a_s\cap b$ (for $b\subset b_{0s}$). All this is correct in the sense of the *definition* of a probability for a single event.

In the sense of the viewpoint (c), i.e. wishing to eliminate the technical work of the experimental physicist, one *adds* to (5.2) an "objective" interpretation: (5.2) is a measure for the "propensity" that $x$ will trigger the indication $b$. Here propensity is meant as a property lying in the microsystems and is considered as the cause of reproducible frequencies to occur in experiments.

Since the imagination of such a propensity has no influence on the work of the physicists described by the viewpoint (b), it is a matter of taste whether one regards such a propensity as an "explanation". The author of this book does not believe that such a propensity describes anything real. It appears to him a psychological problem why such concepts are introduced.

To conclude, the viewpoint (b), taken throughout this book, describes physics as an enterprise of human beings to live in the world around them by *reshaping* this world. (Humans cannot live as animals in biological equilibrium with the world.) Therefore our viewpoint urges us to the question: *How* shall we reshape the world? An undisturbed environment and human life are in contradiction. Not conservation but cultivation of the environment, that is the problem. At every state of human development, this cultivation requires the highest possible technology, but is nevertheless not an automatic consequence of technology.

Here one may terminate the reading of this book. Only readers should read the last chapter who more rigorously want to use such concepts as physically real, physically possible, a more comprehensive theory, experimental tests of a theory, etc. (and are not afraid of very abstract formulations).

# XIII Relations Between Different Forms of Quantum Mechanics and the Reality Problem

To conclude our presentation of quantum mechanics, let us once more discuss whether and how we have attained the goal of an axiomatic basis and how this presentation is connected with others. In particular, we will explain precisely the connection with the form that is regarded as "standard", which proceeds from the basic concepts of "state" and "observable". For these investigations, we use information from [3], [30] and [48]. We will collect those parts of these publications which are essential for the explanation of such concepts as axiomatic basis and "more comprehensive" theories. Thus the reader may follow the next sections if not afraid of very abstract investigations.

We have sketched relations between various forms of quantum mechanics in III §4, III §5.2 and III §6.7, in order to give the reader an indication that in III through XII we have in fact presented the "known" quantum mechanics, only in a new form. In this form, connections otherwise described in everyday language are presented theoretically, i.e. in mathematical form.

In §4 we will give a foundation for physical statements that concern the *reality* of "unobserved" facts or the *possibility* for the realization of facts. We have used such statements intuitively in developing an axiomatic basis for quantum mechanics. Statements about possibilities are also very significant for the interpretation of quantum mechanics.

## §1 Correspondence Rules

Let us regard a $\mathscr{P}\mathscr{T}$ as given in the following form, introduced at the beginning of II (and described in [3], [30]): $\mathscr{P}\mathscr{T}$ is composed of a mathematical theory $\mathscr{M}\mathscr{T}$, correspondence rules (—) and a reality domain $\mathscr{W}$.

The correspondence rules are prescriptions of how to translate into $\mathscr{M}\mathscr{T}$ such facts which can be detected in nature or on devices, or arise by technical procedures. Only facts in the fundamental domain $\mathscr{G}$ (a part of $\mathscr{W}$) shall be considered. It is important that the description of facts in $\mathscr{G}$ does not use the theory under consideration. This does not mean that we use no theory at all. But we may use only "pretheories" to describe the fundamental domain $\mathscr{G}$, that is theories already established before interpreting the theory $\mathscr{P}\mathscr{T}$ to be considered.

Clearly, the fundamental domain $\mathscr{G}$ must be restricted to those facts which may be translated into the language of $\mathscr{M}\mathscr{T}$. However, further restrictions of $\mathscr{G}$ are often necessary. Such restrictions can be formulated as normative axioms in $\mathscr{M}\mathscr{T}$. All those facts must be eliminated from the fundamental domain which translated by the correspondence rules contradict the normative axioms. Examples will be given below (§2.4 and §3.2).

The correspondence rules have the following form: Some facts in $\mathscr{G}$ are denoted by signs, say letters $a_1, a_2, \ldots$. In $\mathscr{MT}$ some sets are singled out as pictorial sets $E_1, E_2, \ldots, E_r$ and some relations $R_1, R_2, \ldots$ as pictorial relations. By virtue of the correspondence rules, the facts of $\mathscr{G}$ are translated into an additional text in $\mathscr{MT}$ (sometimes called "observational report"), which is of the form

$$(\text{—})_r(1): \quad a_1 \in E_{i_1}, a_2 \in E_{i_2}, \ldots; \tag{1.1}$$

$$(\text{—})_r(2): \quad R_{\mu_1}(a_{i_1}, a_{k_1}, \ldots, \alpha_{\mu_1}), R_{\mu_2}(a_{i_2}, a_{k_2}, \ldots, \alpha_{\mu_2}) \ldots, \qquad [\text{not } R_{\nu_1}(a_{u_1}, \ldots)], \ldots. \tag{1.2}$$

Here the $\alpha_\mu$ are real numbers which may be absent in some of the relations $R_\mu$. Let $(\text{—})_r$ designate the total text $(\text{—})_r(1)$ plus $(\text{—})_r(2)$. This $(\text{—})_r$ resembles a map of the facts into $\mathscr{MT}$; therefore in [2] the correspondence rules have been called "mapping principles".

The distinction between $(\text{—})_r(1)$ and $(\text{—})_r(2)$ has no fundamental significance. Instead of $(\text{—})_r(1)$ one may introduce relations $T_i(x)$ equivalent to $x \in E_i$. Then $(\text{—})_r(1)$ takes the same form

$$T_{i_1}(a_1), T_{i_2}(a_2), \ldots$$

as $(\text{—})_r(2)$. It is essential, however, that for every sign $a_1$ appearing in $(\text{—})_r(2)$ there is a formula $a_1 \in E_{i_1}$ in $(\text{—})_r(1)$.

The correspondence rules are just rules for the translation of propositions from common language or from the language of pretheories into the relations $(\text{—})_r$. How this translation from pretheories should be done is shown in §3.4.

The pictorial relations $R_\mu$ of $\mathscr{PT}$ can be represented by subsets

$$r_\mu \subset E_{i_1} \times E_{i_2} \times \ldots \times \mathbf{R} \tag{1.3}$$

– that is, by the set of all those $(y_1, y_2, \ldots, \alpha)$ for which $R_\mu(y_1, y_2, \ldots, \alpha)$ is valid. (In this and all analogous formulas the factor $\mathbf{R}$ may be absent.) For the set on the right side of (1.3) we shall write $S_\mu(E_1, E_2, \ldots, E_r, \mathbf{R})$. Then for $U = (r_1, r_2, \ldots)$ we have

$$U \in S(E_1, E_2, \ldots, E_r, \mathbf{R}) \tag{1.4}$$

with

$$S(E_1, E_2, \ldots, E_r, \mathbf{R}) = \mathscr{P}S_1(E_1, \ldots) \times \mathscr{P}S_2(E_1, \ldots) \times \ldots. \tag{1.5}$$

$\mathscr{MTA}$ will denote the mathematical theory $\mathscr{MT}$ completed by the observational report $(\text{—})_r$.

We say, that the theory $\mathscr{PT}$ does not contradict experience if no contradiction occurs in $\mathscr{MTA}$. We do not claim as Popper [62] that this "no contradiction with experience" is the one and only criterion to accept a $\mathscr{PT}$.

We know that in general the relations $(\text{—})_r$ are not suitable to describe the facts in $\mathscr{G}$ because of imprecisions. Therefore we must generalize $(\text{—})_r(2)$ by introducing imprecision sets (see also [40]). We presume uniform structures of physical imprecision for the pictorial sets $E_1, \ldots, E_r$. These can be extended canonically as uniform structures for the sets $S_\mu(E_1, \ldots)$. With an imprecision set $n_\mu \subset S_\mu(E_1, \ldots) \times S_\mu(E_1, \ldots)$ as an element of the uniform structure for $S_\mu(E_1, \ldots)$, we introduce the smeared relations

$$\tilde{r}_\mu = \{x \mid \text{there is a } y \in r_\mu \text{ with } (x, y) \in n_\mu\},$$
$$\tilde{r}'_\mu = \{x \mid \text{there is a } y \notin r_\mu \text{ with } (x, y) \in n_\mu\}. \tag{1.6}$$

Instead of $(—)_r(2)$ we then take

$$(—)_r(2): \quad (a_{i_1}, a_{k_1}, \ldots, \alpha_{\mu_1}) \in \tilde{r}_{\mu_1}, (a_{i_2}, a_{k_2}, \ldots, \alpha_{\mu_2}) \in \tilde{r}_{\mu_2}, \ldots, (a_{u_1}, \ldots) \in \tilde{r}'_{v_1}, \ldots . \tag{1.7}$$

There can be two sources for imprecision:

(1) Imprecisions in the pretheories, called imprecisions of measurements; (2) an imprecision of the theory $\mathscr{PT}$ itself, in the sense that in $\mathscr{MTA}$ we get contradictions if we take too small imprecision sets $n_\mu$; we say, that $\mathscr{PT}$ itself is an imprecise picture of reality (we know no physical theory which is a precise picture of reality).

When the imprecision of a measurement is much smaller than the imprecision of the theory itself, this is an encouragement to find a "better" theory (a more comprehensive theory in the sense of §3).

The pictorial sets and relations in the case of the "usual" quantum mechanics will not be defined here because we will discuss this problem in §3.3, denoting the usual quantum mechanics by $\mathscr{PT}_5$. We will discuss this problem in a special section, since one usually formulates the connection between experience and the mathematics of Hilbert space not precisely, so that errors are possible.

## §2 The Physical Contents of a Theory

Quantum mechanics in the usual form is a very instructive example for various and very different opinions about what is "physical" of the mathematical objects in Hilbert space. Many stories are told which give imaginations without real contents. Therefore we must look for a more rigorous method to elaborate the physical contents of $\mathscr{MT}$ in a $\mathscr{PT}$. The essential point for this method is to introduce an axiomatic basis. We have done this for quantum mechanics in the previous chapters and will employ this procedure to elaborate the physical content of quantum theory more precisely.

### §2.1 Species of Structure

Every mathematical theory $\mathscr{MT}$ as a part of a $\mathscr{PT}$ has the form called by Bourbaki [50] a "theory of the species of structure $\Sigma$". Let us denote it by $\mathscr{MT}_\Sigma$, a term already introduced in III §1 (see also [3] and [48]).

Now $\mathscr{MT}$ (without the label $\Sigma$) shall contain only set theory together with the theories of real and complex numbers, **R** and **C**. (Bourbaki considers as $\mathscr{MT}$ also theories stronger than the theory of sets.)

The additional ingredient in $\mathscr{MT}_\Sigma$ is the species of structure $\Sigma$ which is given by

(1) a collection of letters $y_1, \ldots, y_r$ called *principal base sets*,

(2) the terms **R** and (sometimes) **C** as auxiliary base sets,

(3) a typification $t \in S(y_1, \ldots, y_2, \mathbf{R}, \mathbf{C})$ where $S$ is an echelon construction scheme, and $t$ is called the *structural term*,

(4) a transportable relation $P(y_1, \ldots, y_r, t)$, called the *axiom* of the species of structure $\Sigma$.

Here an echelon construction scheme is a prescription of how to get the intended set by finitely many steps, where a step is the taking of a product set or a power set. A transportable relation is defined as follows:

If $f_i$ are mappings $y_i \to z_i$, then in a canonical way we define mappings

$$S(y_1, \ldots) \xrightarrow{\langle f_1, f_2, \ldots, f_r \rangle^S} S(z_1, \ldots),$$

where $S$ is an echelon construction scheme. If the $f_i$ are bijective, also $\langle f_1, \ldots, f_r \rangle^S$ is so.

A relation $P(y_1, \ldots, y_r, t)$ is called transportable if $f_i$ bijective implies $P(z_1, \ldots, z_r, u)$ with $u = \langle f_1, \ldots \rangle^S t$.

If another $\Sigma'$ is given by a text of the form (1) to (4) with $x_1, \ldots, x_n$ as base sets, $s \in S'(x_1, \ldots)$ as structural term and $P'(x_1, \ldots, x_n, s)$ as axiom, we call terms $E_1, \ldots, E_r, U$ in $\mathcal{M}\mathcal{T}_\Sigma$ a derivation of the species of structure $\Sigma'$ if:

(1) $U$ is a structure of species $\Sigma'$, that is, $U \in S'(E_1, \ldots, E_r, \mathbf{R}, \ldots)$ and $P'(E_1, \ldots, E_r, U)$ are theorems in $\mathcal{M}\mathcal{T}_\Sigma$,

(2) each of the terms $E_1, \ldots, E_r, U$ is an intrinsic term in $\mathcal{M}\mathcal{T}_\Sigma$.

A term $V(y_1, \ldots, y_r, t)$ is called intrinsic, if $V$ is an element of an echelon set on $y_1, \ldots, y_r, \mathbf{R}, \mathbf{C}$ and if $V(y_1, \ldots, y_r, t)$ is canonically mapped on $V(z_1, \ldots, z_r, u)$ with $u = \langle f_1, \ldots, f_r \rangle^S t$ for bijective mappings $y_i \xrightarrow{f_i} z_i$.

Two species of structure $\Sigma, \Sigma'$ on the *same* base sets $x_1, \ldots, x_n$ are called equivalent relative to the deduction procedures $U(x_1, \ldots, x_n, s)$ and $V(x_1, \ldots, x_n, t)$ if $U(\ldots)$ is a structure of species $\Sigma'$ and $V$ one of species $\Sigma$ and if

$$U(x_1, \ldots, x_n, V(x_1, \ldots, x_n, t)) = t,$$
$$V(x_1, \ldots, x_n, U(x_1, \ldots, x_n, s)) = s.$$

For $\mathcal{M}\mathcal{T}_\Sigma$ as a part of a physical theory, of still greater importance is a case more general than that of equivalent structures: Let $\Sigma, \Sigma'$ be defined as before. In addition let be given a deduction of $\Sigma$ in $\mathcal{M}\mathcal{T}_{\Sigma'}$ by intrinsic terms $E_1(x_1, \ldots, x_n, s)$, $E_2(x_1, \ldots, x_n, s), \ldots, E_r(x_1, \ldots, x_n, s)$ as base sets for a structure $U(x_1, \ldots, x_n, s)$ of species $\Sigma$ ($U(\ldots)$ also an intrinsic term).

It could be that there is a species of structure $\Sigma_1$ richer than $\Sigma$, such that $U(x_1, \ldots, x_n, s)$ is also a structure of species $\Sigma_1$. Here "$\Sigma_1$ richer than $\Sigma$" means that $\Sigma_1$ has the same base and the same typification but a stronger axiom. How can we exclude such richer $\Sigma_1$?

We introduce an additional condition. We call the structure $U(x_1, \ldots, x_n, s)$ of species $\Sigma$ deduced in $\mathcal{M}\mathcal{T}_{\Sigma'}$ a *representation* of $\Sigma$ *in* $\Sigma'$ if the relation

there are $x_1, \ldots, x_n$ and $s$ and bijective mappings
$y_i \xrightarrow{f_i} E_i(x_1, \ldots, x_n, s)$ such that $s \in S'(x_1, \ldots, x_n)$
and $P'(x_1, \ldots, x_n, s)$ and $\langle f_1, \ldots f_r \rangle^S t = U(x_1, \ldots, x_n, s)$ (2.1.1 a)

is a theorem in $\mathcal{M}\mathcal{T}_\Sigma$. This theorem says that every structure $t$ of species $\Sigma$ is isomorphic to $U(x_1, \ldots, x_n, s)$ for suitably chosen $x_1, \ldots, x_n$ and $s$. For brevity, we shall sometimes say that $U$ represents $\Sigma$ in $\Sigma'$.

In the following we will abbreviate: "there is $x$" by "$\exists x$" and "for every $x$" by $\forall x$. Then (2.1.1a) can also be written

$$\begin{aligned}&\exists x_1, \ldots, \exists x_n, \exists s, \exists f_1, \ldots, \exists f_r\\ &\{f_i \text{ are bijective mappings } y_i \xrightarrow{f_i} E_i(x_1, \ldots, x_n, s), s \in S'(x_1, \ldots, x_n),\\ &P'(x_1, \ldots, x_n, s) \text{ and } \langle f_1, \ldots, f_r \rangle^S t = U(x_1, \ldots x_n, s)\}.\end{aligned} \tag{2.1.1b}$$

If $U(x_1, \ldots)$ is a representation of $\Sigma$ in $\Sigma'$, then $U(x_1, \ldots)$ cannot be a structure of a species richer than $\Sigma$. Then if $R(y_1, \ldots, y_r, t)$ is a transportable relation and if $R(E_1, \ldots, E_r, U)$ is a theorem in $\mathcal{M}\mathcal{T}_{\Sigma'}$ this and (2.1.1) imply that $R(y_1, \ldots, y_r, t)$ is also a theorem in $\mathcal{M}\mathcal{T}_\Sigma$.

A familiar example for a representation of a species of structure $\Sigma$ is analytical geometry. As $\mathcal{M}\mathcal{T}_{\Sigma'}$ we choose $\mathcal{M}\mathcal{T}$ (i.e. we introduce no $\Sigma'$ at all). As a single $E_i$ we take $E = \mathbf{R}^3$ and as the structure $U \in \mathscr{P}(\mathbf{R}^3 \times \mathbf{R}^3 \times \mathbf{R})$ the set of all $(x_1, x_2, x_3; y_1, y_2, y_3, \alpha)$ with

$$\alpha = g(x_1\, x_2, x_3; y_1, y_2, y_3) = \sqrt{(x_1 - y_1)^2 + (x_2 - y_2)^2 + (x_3 - y_3)^2}.$$

We determine $\Sigma$ by a base term $y$, a structure term $t \in T(y, \mathbf{R})$ with $T(y, \mathbf{R}) = \mathscr{P}(y \times y \times \mathbf{R})$, and an axiom $P(y, t)$ not given explicitly. This $P(\ldots)$ is to say, that $t$ determines a function $y \times y \xrightarrow{d} \mathbf{R}_+$, and gives conditions for $d$ which make $y$ a three-dimensional Euclidean space.

Then $U$ is a structure of species $\Sigma$ on the base $E = \mathbf{R}^3$, but even a representation of $\Sigma$. The theorem (2.1.1) here becomes

$$\exists f \{f \text{ is a bijective mapping } y \xrightarrow{f} \mathbf{R}^3 \text{ with } d(z_1, z_2) = g(f(z_1), f(z_2))\}. \tag{2.1.2}$$

The proof of this theorem is nothing but the proof, that the Euclidean geometry $\Sigma$ can be represented by orthogonal coordinates (namely $y \xrightarrow{f} \mathbf{R}^3$).

Another, not so simple example is given in this book. We take $\Sigma$ as given by the base sets $y_1 = K$, $y_2 = L$ (with $K$, $L$ from IV) and by a structure term $t \in K \times L \times \mathbf{R}$. The axiom $P(y_1, y_2, t)$ determines $t$ as a function $K \times L \xrightarrow{\mu} \mathbf{R}$ (see also III T5.1.4) with

(i) $0 \leq \mu(w, g) \leq 1$.

(ii) $\mu(w_1, g) = \mu(w_2, g)$ for all $g \in L$ implies $w_1 = w_2$.

(iii) $\mu(w, g_1) = \mu(w, g_2)$ for all $w \in K$ implies $g_1 = g_2$.

(iv) There is a $g_0 \in L$ with $\mu(w, g_0) = 0$ for all $w \in K$.

(v) There is a $g_1 \in L$ with $\mu(w, g_1) = 1$ for all $w \in K$.

(vi) For each $g \in L$ there is a $g' \in L$ with $\mu(w, g) + \mu(w, g') = 1$ for all $w \in K$.

With $K$ and $L$ (instead of $\mathscr{K}$ and $\mathscr{L}$) we also construct the spaces $\mathscr{B}, \mathscr{B}'$ as in IV. Then $P(\ldots)$ claims that $K$ is convex, norm-complete and separable, while

$L$ is convex and $\sigma(\mathscr{B}', \mathscr{B})$-complete. $P(\ldots)$ further contains the axioms AV 1.1, A 1.2$s$, AV 2$f$, AV id, AV 3, AV 4$s$ (from VI), and AV 4$a$, AV 4$b$ (from VIII §4.1).

Let us specify $\Sigma'$ by a base set $x = \mathscr{H}$ and such structure terms and axioms that

$$\mathscr{H} = \bigcup_{\nu} \mathscr{H}_{\nu}$$

holds with (disjoint) Hilbert spaces $\mathscr{H}_{\nu}$ over the field $\mathbf{C}$ (see [2] A VI §15). Then we may define the intrinsic terms $E_1 = K(\mathscr{H}_1, \mathscr{H}_2, \ldots)$ and $E_2 = L(\mathscr{H}_1, \mathscr{H}_2, \ldots)$ where the elements $w \in K(\ldots)$ and $g \in L(\ldots)$ are given as sequences $w = (w_1, w_2, \ldots)$ resp. $g = (g_1, g_2, \ldots)$. Here the $w_{\nu}$ are selfadjoint operators in $\mathscr{H}_{\nu}$ with $w_{\nu} \geq 0$ and $\sum_{\nu} \mathrm{tr}(w_{\nu}) = 1$, while the $g_{\nu}$ are selfadjoint operators in $\mathscr{H}_{\nu}$ with $0 \leq g_{\nu} \leq 1$. As structure $U$ we introduce the real function $\mu(w, g) = \sum_{\nu} \mathrm{tr}(w_{\nu} g_{\nu})$.

$U$ is a structure of species $\Sigma$ over the sets $E_1, E_2$ (see XIII §4.2). That $U$ is a representation of $\Sigma$ (i.e. (2.1.1) a theorem in $\mathscr{M}\mathscr{T}_{\Sigma}$) is proved in VI through VIII.

## §2.2 Axiomatic Bases

In §1 we have seen, how the correspondence rules enable us to compare experiences with the mathematical theory $\mathscr{M}\mathscr{T}$, i.e. to establish $\mathscr{M}\mathscr{T}\mathscr{A}$. Now we presume that $\mathscr{M}\mathscr{T}$ is of the form $\mathscr{M}\mathscr{T}_{\Sigma'}$ (possibly there is no $\Sigma'$, i.e. $\mathscr{M}\mathscr{T}_{\Sigma'} = \mathscr{M}\mathscr{T}'$ is the theory of sets and real or complex numbers only). The pictorial sets $E_1, \ldots, E_r$ are supposed to be intrinsic terms in $\mathscr{M}\mathscr{T}_{\Sigma'}$. The observational report shall be denoted by $(\text{—})_r'$ and the physical theory by $\mathscr{P}\mathscr{T}'$.

There arise several questions:

1. How is the fundamental domain delimited? It could be that a contradiction in $\mathscr{M}\mathscr{T}_{\Sigma'}\mathscr{A}'$ is only caused by the circumstance that in $(\text{—})_r'$ we have noted facts not belonging to the fundamental domain $\mathscr{G}'$.

2. What of $\mathscr{M}\mathscr{T}_{\Sigma'}$ can in principle be rejected by experience? What of $\mathscr{M}\mathscr{T}_{\Sigma'}$ are purely mathematical ingredients without any significance to experiments? For instance, does the axiom $P'(\ldots)$ of $\Sigma'$ have any physical meaning?

The questions under 2. are burning if $\mathscr{M}\mathscr{T}'$ is only set theory (including the theory of real or complex numbers) and the pictorial sets $E_{\nu}$ as the relations $R_{\mu}$ are constructed by using only the set $\mathbf{R}$ of real numbers. In §2.1 we have encountered such an example, the analytical geometry. Even in applying a theory we mostly use a form where $\mathscr{M}\mathscr{T}'$ is only set theory; such a form allows to use the familar methods of dealing with real numbers. But if $\mathscr{M}\mathscr{T}'$ is only set theory, then all physical aspects of the theory must lie in the *definitions* of the pictorial sets $E_{\nu}$ and pictorial relations $R_{\mu}$. How can we exhibit the physical aspects of these definitions?

We try to answer all these questions by introducing an axiomatic basis for $\mathscr{P}\mathscr{T}'$.

The starting point for an axiomatic basis is the singling out of the pictorial sets and pictorial relations. Let $x_1, \ldots, x_n$ be the (principal) base sets and $s$ the structure term of $\Sigma'$. The $E_1, \ldots, E_{\nu}$ may be intrinsic terms $E_{\nu}(x_1, \ldots, x_n, s)$. We introduce the term $U(r_1, r_2, \ldots, \mathscr{N}_1, \ldots)$, where the $r_{\mu}$ from (1.3) represent the pictorial relations (see §1) and $\mathcal{N}_1, \ldots$ are uniform structures for the physical imprecision. Then we have

$$U \in S(E_1, \ldots, \mathbf{R}) \tag{2.2.1}$$

with

$$S(E_1, \ldots) = \mathcal{P}S_1(E_1, \ldots, \mathbf{R}) \times \mathcal{P}S_2(E_1, \ldots) \times \ldots \times \mathcal{P}(\mathcal{P}(X_1 \times X_1)) \times \ldots . \tag{2.2.2}$$

Here $\mathcal{P}(\mathcal{P}(X_1 \times X_1))$ gives the typification of the uniform structure $\mathcal{N}_1$, and $S_\mu$ is given by the right side of (1.3).

If $U$ is a structure of the species $\Sigma$ (with $y_1, \ldots, y_r$ the base terms and $t=(t_1, t_2, \ldots; \underline{\mathcal{N}}_1 \ldots)$ the structure term with $t_\mu(y_{i_1} \times y_{i_2} \times \ldots \times \mathbf{R}$ as in §2.1), we may construct a new $\mathcal{PT}$ from $\mathcal{PT}'$ as follows.

As fundamental domain $\mathcal{G}$ of $\mathcal{PT}$ we use the $\mathcal{G}'$ of $\mathcal{PT}'$. As mathematical theory of $\mathcal{PT}$ we choose $\mathcal{MT}_\Sigma$. The correspondence rules are transferred in obvious ways from $\mathcal{PT}'$:

1. Instead of $(-)'_r(1)$ given in (1.1) we write

$$(-)_r(1): \quad a_1 \in y_{i_1}, a_2 \in y_{i_2}, \ldots . \tag{2.2.3}$$

2. In $\mathcal{PT}'$ we had introduced the "smeared" relations $\tilde{r}_\mu, \tilde{r}'_\mu$ according to (1.6). Now we assume intrinsic terms $\underline{n}_\mu$ in $\mathcal{MT}_\Sigma$ which transported to $\mathcal{MT}_{\Sigma'}$ give the $n_\mu$. As smeared relations we define

$$\begin{aligned} \tilde{t}_\mu &= \{z \mid \text{there is a } z' \in t_\mu \text{ with } (z, z') \in \underline{n}_\mu\}, \\ \tilde{t}'_\mu &= \{z \mid \text{there is a } z' \notin t_\mu \text{ with } (z, z') \in \underline{n}_\mu\}. \end{aligned} \tag{2.2.4}$$

Then for $(-)'_r(2)$ in (1.7) we write

$$(-)_r(2): \quad (a_{i_1}, a_{k_2}, \ldots a_{\mu_1}) \in \tilde{t}_{\mu_1}, \ldots, (a_{u_1}, \ldots) \in \tilde{t}'_{v_1}, \ldots . \tag{2.2.5}$$

In this way, $\mathcal{PT}$ becomes a well defined physical theory. It can be proved that "$\mathcal{MT}_{\Sigma'}\mathcal{A}$ without contradiction" implies that also $\mathcal{MT}_\Sigma \mathcal{A}$ is without contradiction (see the end of this section and [3] §7.3).

A $\mathcal{PT}$ may be given (without recourse to $\mathcal{PT}'$!) in the form that the base terms $y_\nu$ of $\mathcal{MT}_\Sigma$ are the pictorial terms and the components $t_\mu$ of the structure term are the pictorial relations so that $(-)_r$ is given by (2.2.3) and (2.2.5). Then we say that $\mathcal{MT}_\Sigma$ is an axiomatic basis (of the first degree) of $\mathcal{PT}$. Sometimes we shortly call $\mathcal{PT}$ an axiomatic basis.

The theory $\mathcal{PT}$ constructed above is an axiomatic basis. Since we have only presumed that $U$ is a structure of species $\Sigma$, the theory $\mathcal{PT}'$ can be stronger than $\mathcal{PT}$. But if $U$ represents the species of structure $\Sigma$, the two theories $\mathcal{PT}'$ and $\mathcal{PT}$ are equivalent in the sense that $\mathcal{MT}_\Sigma \mathcal{A}$ is without contradiction if and only if $\mathcal{MT}_{\Sigma'}\mathcal{A}$ is so (proof at the end of this section and in [3] §7.3). Then we call $\mathcal{MT}_\Sigma$ and also $\mathcal{PT}$ axiomic bases (of the first degree) of $\mathcal{PT}'$.

The axiomatic basis for quantum mechanics (or more precisely: the axiomatic bases) developed in III through IX are not of the first degree and were not deduced from the "usual" quantum mechanics as $\mathcal{PT}'$. The reason was that the pretheories for the usual form of quantum mechanics are not well formulated. Only with very imprecise formulations in common language and much intuition the usual quantum mechanics can be applied. We will give a precise formulation of the usual quantum mechanics in §3.3.

First we want to extend the concept of an axiomatic basis beyond the first degree. Extensions can be made without any connection with the theory $\mathscr{PT}'$. In order not to do the same work twice, let us introduce the extensions in connection with $\mathscr{PT}'$ (whoever is not interested in such connections, may skip all those following developments which are related to $\mathscr{PT}'$).

Sometimes it is possible to eliminate some base terms $y_\nu$ by mappings. For some of the base terms in $\mathscr{MT}_\Sigma$ (which we denote by $y_{\nu_i}$), as intrinsic terms we assume injective mappings

$$y_{\nu_i} \xrightarrow{g_i} T_{\nu_i}(y_{\mu_1}, \ldots, \mathbf{R}). \tag{2.2.6}$$

Here the $T_{\nu_i}(y_{\mu_1}, \ldots, \mathbf{R})$ are echelon sets constructed only from those $y_{\mu_k}$ which do not belong to the family of the $y_{\nu_i}$. (In many cases there are mappings (2.2.6) where the $g_i$ are not injective. We will discuss these cases in §3.)

In the case of a connection between $\mathscr{PT}'$ and $\mathscr{PT}$ (discussed above where $U$ is a structure of species $\Sigma$ over the pictorial sets $E_\nu$), we assume that the mappings

$$E_{\nu_i} \xrightarrow{\hat{g}_i} T_{\nu_i}(E_{\mu_1}, \ldots, \mathbf{R}) \tag{2.2.7}$$

(the $g_i$ transported from $\mathscr{MT}_\Sigma$ to $\mathscr{MT}_{\Sigma'}$) are identical mappings of $E_{\nu_i}$. This implies

$$E_{\nu_i} \subset T_{\nu_i}(E_{\mu_1}, \ldots, \mathbf{R}). \tag{2.2.8}$$

Because of (2.2.6) we may eliminate the $y_{\nu_i}$ as base terms: Starting from $\Sigma$ we define the following species of structure $\Sigma^{(1)}$. As base terms of $\Sigma^{(1)}$ we take the $y_{\mu_k}$ only. Instead of the $y_{\nu_i}$ we introduce the sets $g(y_{\nu_i})$ as additional structure terms in $\Sigma^{(1)}$.

To make this rigorous we use new letters for the text $\Sigma^{(1)}$. The $y_{\mu_k}$ of $\Sigma$ are replaced by letters $z_k$, the $y_{\nu_i}$ by letters $t_i^{(1)}$. As typification of the $t_i^{(1)}$ we prescribe

$$t_i^{(1)} \subset T_{\nu_i}(z_1, \ldots, \mathbf{R}) \tag{2.2.9}$$

with the echelon construction schemes in (2.2.6). Besides the $t_i^{(1)}$ we retain the structure term $t$ of $\Sigma$ with the denotation $t_0^{(1)}$ and the same typification $S(y_1, \ldots)$ as in $\Sigma$. We only replace the $y_{\mu_k}$ by $z_k$ and the $y_{\nu_i}$ by the $T_{\nu_i}(\ldots)$ of (2.2.9):

$$t_0^{(1)} \in S(z_1, \ldots, T_{\nu_1}, \ldots). \tag{2.2.10}$$

The $t_i^{(1)}$ and $t_0^{(1)}$ may be comprised in one structure term $t^{(1)}$ of $\Sigma^{(1)}$.

As axiom $P^{(1)}(z_1, \ldots, t_1^{(1)}, \ldots, t_0^{(1)})$ of $\Sigma^{(1)}$ we introduce the following relations:

$$P_0^{(1)}(z_1, \ldots, t_1^{(1)}, \ldots, t_0^{(1)}) \quad \text{as the axiom} \quad P(y_1, \ldots, t) \quad \text{of } \Sigma$$

where we replace the $y_{\mu_k}$ by $z_k$, the $y_{\nu_i}$ by $t_i^{(1)}$ and $t$ by $t_0^{(1)}$. We add the relation $t_0^{(1)} \in S(z_1, \ldots, t_1^{(1)}, \ldots)$ which we get from the typification of $t$ in $\Sigma$ when $y_{\mu_k}$ is replaced by $z_k$ and $y_{\nu_i}$ by $t_i^{(1)}$. Then (2.2.10) as axiom for $\Sigma^{(1)}$ is not necessary. We get the last part of the axiom $P^{(1)}(\ldots)$ as follows:

Because of $P_0^{(1)}(\ldots)$, the mappings $g_i$ in (2.2.6) can be transported as injective mappings

$$t_i^{(1)} \xrightarrow{\hat{g}_i} T_{\nu_i}(z_1, \ldots) \tag{2.2.11}$$

(i.e. $\hat{g}_i$ intrinsic terms). As axiom we add that the $\hat{g}_i$ are identical maps of $t_i^{(1)}$ on itself (often this is already a theorem). Therefore $t_i^{(1)} \subset T_{v_i}(z_1, \ldots)$.

It is obvious that $t_0^{(1)}$ in $\mathcal{MT}_{\Sigma(1)}$ is a structure of species $\Sigma$ over the sets $z_1, \ldots, t_1^{(1)}, \ldots$. Moreover, (2.2.6) implies that $t_0^{(1)}$ represents $\Sigma$ in $\mathcal{MT}_{\Sigma(1)}$. To prove this we write the condition (2.2.1) for this case:

There are sets $z_1, \ldots, t_1^{(1)}, \ldots, t_0^{(1)}$ and bijective mappings $f_1, \ldots$ such that

$$t_i^{(1)} \in T_{v_i}(z_1, \ldots) \quad \text{and} \quad t_0^{(1)} \in S(z_1, \ldots, t_1^{(1)}, \ldots) \quad \text{and} \quad P_0^{(1)}(z_1, \ldots, t_1^{(1)} \ldots, t_0^{(1)}) \tag{2.2.12}$$

$$\text{and} \quad y_{v_i} \xrightarrow{f_{v_i}} t_i^{(1)}, \quad y_{\mu_k} \xrightarrow{f_{\mu_k}} z_k \quad \text{and} \quad \langle f_1, \ldots \rangle^S t = t_0^{(1)}.$$

To prove (2.2.12) in $\mathcal{MT}_\Sigma$, we only must take $z_k = y_{\mu_k}$, $f_{\mu_k}$ as the identical maps, $t_i^{(1)} = g_i(y_{v_i})$ and $f_{v_i} = g_i$: ($g_i$ in (2.2.6)) and $t_0^{(1)} = \langle f_1, \ldots \rangle^S t$.

There is something like a reciprocity: The terms $g_i(y_{v_i})$ and $\hat{t} = \langle f_1, \ldots \rangle^S t$ (where $f_{\mu_k}$ are identical maps and $f_{v_i} = g_i$) form a representation of $\Sigma^{(1)}$ in $\mathcal{MT}_\Sigma$. That these terms form a structure of species $\Sigma^{(1)}$ is equivalent to the following relation being a theorem in $\mathcal{MT}_\Sigma$:

$$g_i(y_{v_i}) \subset T_{v_i}(y_{\mu_1}, \ldots) \quad \text{and} \quad \hat{t} \in S(y_{\mu_1}, \ldots, g_{v_1}) \ldots) \tag{2.2.13}$$

and $P_0^{(1)}(y_{\mu_1}, \ldots, g_1(y_{v_1}), \ldots, \hat{t})$, and $\hat{g}_i$ identically maps $g_i(y_{v_i})$ on itself.

Here the $\hat{g}_i$ are defined as the mappings (2.2.6) transported by the mappings $f_{\mu_k}$ (as identical mappings) and $f_{v_i} = g_i$. Therefore $g(y_{v_i}) \xrightarrow{\hat{g}_i} g_i(y_{v_i})$ are identical mappings of $g_i(y_{v_i})$ in itself. The rest of (2.2.13) follows since the axiom $P(\ldots)$ of $\Sigma$ is transportable (§2.1).

To prove that the $g_i(y_{v_i}), \hat{t}$ form a representation of $\Sigma^{(1)}$, in $\mathcal{MT}_{\Sigma(1)}$ we must prove the theorem:

There are sets $y_1, \ldots, t$ and bijective mappings $f_1, \ldots$ such that $t \in S(y, \ldots)$ and $P(y_1, \ldots, t)$ and

$$z_k \xrightarrow{f_k} y_{\mu_k} \quad \text{and} \quad \langle f_1, \ldots \rangle \, T_{v_i} t_i^{(1)} = g_i(y_{v_i}) \quad \text{and} \quad \langle f_1, \ldots \rangle^S t_0^{(1)} = \hat{t}. \tag{2.2.14}$$

To prove this we take $y_{\mu_k} = z_k$, $y_{v_i} = t_i^{(1)}$, $t = t_0^{(1)}$ and $f_k$ as identical mappings. Then $g_i = \hat{g}_i$ ($\hat{g}_i$ in (2.2.11)) and $\hat{g}_i$ are identical mappings of $t_i^{(1)}$ into itself.

We easily define a $\mathcal{PT}^{(1)}$ with $\mathcal{MT}_{\Sigma(1)}$ as mathematical part. Instead of (2.2.3) we take

$$(\text{—})_r^{(1)}(1): \quad a_\alpha \in z_k, \quad a_\beta \in t_i^{(1)}, \ldots\ldots \tag{2.2.15}$$

where $z_k$ stands for $y_{\mu_k}$ and $t_i^{(1)}$ for $y_{v_i}$.

The uniform structures can be transported from $\mathcal{MT}_\Sigma$ to $\mathcal{MT}_{\Sigma(1)}$ and therefore a $\tilde{t}_\mu$ in (2.2.4) to a corresponding $\tilde{t}_{0\mu}^{(1)}$ as a smeared out component $t_{0\mu}^{(1)}$ of $t_0^{(1)}$. Then we replace (2.2.5) by

$$(\text{—})_r^{(1)}(2): \quad (a_{i_1}, a_{k_1}, \ldots, \alpha_{\mu_1}) \in \tilde{t}_{0\mu_1}^{(1)}, \ldots. \tag{2.2.16}$$

If we consider $\mathcal{PT}^{(1)}$ as a theory $\mathcal{PT}'$ (as needed above), the connection of $(\text{—})^{(1)}$ and $(\text{—})$ is such that $\mathcal{PT}$ is an axiomatic basis of the first degree of $\mathcal{PT}^{(1)}$.

In $\mathscr{P}\mathscr{T}^{(1)}$ all base terms $z_k$ of $\Sigma^{(1)}$ are pictorial sets. But there are other pictorial sets $t_i^{(1)}$ which according to (2.2.9) are subsets of echelon sets over the base sets $z_1, \dots, \mathbf{R}$. We say that $\mathscr{M}\mathscr{T}_{\Sigma^{(1)}}$ is of the $n$-th degree (relative to $\mathscr{P}\mathscr{T}$) if $n$ is the highest number of constituent power sets, i.e. the highest number of $\mathscr{P}$ in

$$\mathscr{P}(\dots \mathscr{P}(\dots \mathscr{P}(\dots \mathscr{P} \dots))).$$

We call also $\mathscr{M}\mathscr{T}_{\Sigma^{(1)}}$ an axiomatic basis in $\mathscr{P}\mathscr{T}^{(1)}$ and especially of the $n$-th degree. In this sense $\mathscr{P}\mathscr{T}^{(1)}$ is an axiomatic basis of the $n$-th degree of $\mathscr{P}\mathscr{T}$.

For some deductions it is practical to take the relations $a_\beta \in t_i^{(1)}$ from $(\text{—})_r^{(1)}(1)$ to $(\text{—})_r^{(1)}(2)$ and in (2.2.15) to write the relation $a_\beta \in T_{\nu_1}(z_1, \dots, \mathbf{R})$ instead of $a_\beta \in t_i^{(1)}$.

The axiomatic bases for quantum mechanics developed in III through IX are of the form $\mathscr{P}\mathscr{T}^{(1)}$. We will discuss this in § 3.3.

If there is a connection between $\mathscr{P}\mathscr{T}'$ and $\mathscr{P}\mathscr{T}$ as given in (2.2.7), (2.2.8), and $U$ is a structure of species $\Sigma$, we can collect the $E_{\nu_i}$ and $U$ to a structure term $V$. Then $V$ is a structure of species $\Sigma^{(1)}$ over the $E_{\mu_k}$. In $\mathscr{M}\mathscr{T}_{\Sigma'}$ we must prove:

$$\begin{aligned} &E_{\nu_i} \subset T_{\nu_i}(E_{\mu_1} \dots) \quad \text{and} \quad U \in S(E_1, \dots, E_r, \mathbf{R}) \\ &\text{and } P(E_1, \dots, E_r, U), \text{and } \tilde{g}_i \text{ are identical mappings of } E_{\nu_i}. \end{aligned} \tag{2.2.17}$$

This follows from (2.2.8), from (2.2.7) and the assumption formulated after (2.2.8) and from $U$ being a structure of species $\Sigma$ over $E_1, \dots, E_r$.

On the other hand, if $\{E_{\nu_i}, U\}$ is a structure of species $\Sigma^{(1)}$ over $E_{\mu_1} \dots$ then $U$ is a structure of species $\Sigma$ over $E_1, \dots, E_r$. In $\mathscr{M}\mathscr{T}_{\Sigma'}$ we must prove:

$$U \in S(E_1, \dots, E_r, \mathbf{R}) \quad \text{and} \quad P(E_1, \dots, E_r, U). \tag{2.2.18}$$

Since $\{E_\nu, U\}$ is a structure of species $\Sigma^{(1)}$, the theorem (2.2.17) holds in $\mathscr{M}\mathscr{T}_{\Sigma'}$ which implies (2.2.18).

If $U$ even represents $\Sigma$ in $\Sigma'$, then $V$ represents $\Sigma^{(1)}$ in $\Sigma'$. In $\mathscr{M}\mathscr{T}_{\Sigma^{(1)}}$ we must prove the following theorem:

There are sets $x_1, \dots, s$ and bijective mappings $h_1, \dots$ such that $s \in S'(x_1, \dots)$ and $P'(x_1, \dots, s)$ hold as well as

$$z_k \xrightarrow{h_k} E_{\mu_k}(x_1, \dots, s) \tag{2.2.19}$$

with

$$\langle h_1, \dots \rangle^{T_{\nu_i}} t_i^{(1)} = E_{\nu_i}(x_1, \dots) \quad \text{and} \quad \langle h_1, \dots \rangle^S t_0^{(1)} = U(x_1, \dots, s).$$

In $\mathscr{M}\mathscr{T}_\Sigma$ we find (2.1.1) as a theorem. Since $t_0^{(1)}$ is a structure of species $\Sigma$, in $\mathscr{M}\mathscr{T}_{\Sigma^{(1)}}$ we have the theorem:

There are sets $x_1, \dots, s$ and bijective mappings $f_1, \dots$ such that

$$s \in S'(x_1, \dots) \quad \text{and} \quad P'(x_1, \dots), \quad y_{\mu_k} \xrightarrow{f_{\mu_k}} E_{\mu_k}(x_1, \dots) \tag{2.2.20}$$

and

$$t_i^{(1)} \xrightarrow{f_{\nu_i}} E_{\nu_i}(x_1, \dots) \quad \text{and} \quad \langle f_1, \dots \rangle^S t_0^{(1)} = U(x_1, \dots).$$

If in (2.2.20) we replace the letters $f_{\mu_k}$ by $h_k$ and the $f_{\nu_i}$ by $k_i$, we see that (2.2.19) is a theorem if in $\mathscr{M}\mathscr{T}_{\Sigma^{(1)}}$ we prove that (2.2.20) implies

$$k_i t_i^{(1)} = \langle h_1, \dots \rangle^{T_{\nu_i}} t_i^{(1)}. \tag{2.2.21}$$

Since the $g_i$ in (2.2.6) are intrinsic terms, for bijective mappings $f_{\nu_i}$ with $\tilde{g}_i$ in (2.2.7) we have the diagram

$$\begin{array}{ccc} E_{\nu_i} & \xrightarrow{\tilde{g}_i} & E_\nu \\ \uparrow f_{\nu_i} & & \uparrow \langle f_{\mu_1}, \ldots \rangle^{T_{\nu_i}} \\ y_{\nu_i} & \xrightarrow{g_i} & T_{\nu_i}(y_{\mu_1}, \ldots) \end{array}$$

that is

$$\tilde{g}_i\, f_{\nu_2} = \langle f_{\mu_1}, \ldots \rangle^{T_{\nu_i}} g_i.$$

Since the $\tilde{g}_i$ are identical mappings (see after (2.2.7)), we conclude

$$f_{\nu_i} = \langle f_{\mu_1}, \ldots \rangle^{T_{\nu_i}} g_i,$$

to be transported to $\mathcal{M}\mathcal{T}_{\Sigma^{(1)}}$ as

$$k_i = \langle h_1, \ldots \rangle^{T_{\nu_i}} \hat{g}_i.$$

Since the $\hat{g}_i$ are the identities on $t_i^{(1)}$, we finally get (2.2.21).

If vice versa $V = \{E_\nu, U\}$ represents $\Sigma^{(1)}$ in $\Sigma'$, then $U$ represents $\Sigma$ in $\Sigma'$. We must prove that the theorem (2.2.19) in $\mathcal{M}\mathcal{T}_{\Sigma^{(1)}}$ implies (2.1.1) as theorem in $\mathcal{M}\mathcal{T}_\Sigma$. Since $\{g_i(y_{\nu_i}), \hat{t}\}$ is a structure of species $\Sigma^{(1)}$ in $\mathcal{M}\mathcal{T}_\Sigma$, the mappings (2.2.19) imply:

There are sets $x_1, \ldots, s$ and bijective mappings $h_1, \ldots$ such that $s \in S'(x_1, \ldots)$ and $P'(x_1, \ldots, s)$ hold as well as

$$y_{\mu_k} \xrightarrow{h_k} E_{\mu_k}(x_1, \ldots) \quad \text{with} \quad \langle h_1, \ldots \rangle^{T_{\nu_i}} g_i(y_{\nu_i}) = E_{\nu_i}(x_1, \ldots)$$

and

$$\langle h_1, \ldots \rangle^S \hat{t} = U(x_1, \ldots).$$

With $f_{\nu_i} = \langle h_1, \ldots \rangle^{T_{\nu_i}} g_i$ and with the definition of $\hat{t}$ follows (2.1.1).

One easily sees, how to translate directly the observational report (1.1), (1.7) into (2.2.15), (2.2.16). Therefore $\mathcal{P}\mathcal{T}^{(1)}$ can be derived from $\mathcal{P}\mathcal{T}'$ without use of $\mathcal{P}\mathcal{T}$. We call $\mathcal{P}\mathcal{T}^{(1)}$ also an axiomatic basic of $\mathcal{P}\mathcal{T}'$ if $V = \{E_{\nu_i}, U\}$ represents $\Sigma^{(1)}$ in $\Sigma'$.

Which of the three theories $\mathcal{P}\mathcal{T}'$, $\mathcal{P}\mathcal{T}$ or $\mathcal{P}\mathcal{T}^{(1)}$ is used, is a matter of taste. Better said, it depends on the problems which one wants to solve. Often *special* physical problems can be solved much more easily in $\mathcal{P}\mathcal{T}'$ since we generally are more familiar with the mathematics of $\mathcal{P}\mathcal{T}'$. Yet $\mathcal{P}\mathcal{T}$ is that form which shows most clearly the logics of the physical theory (see §2.3 and §4). *General* mathematical deductions often can be done best in $\mathcal{P}\mathcal{T}^{(1)}$. In §3.3 we will find additional reasons for preferring $\mathcal{P}\mathcal{T}'$ or $\mathcal{P}\mathcal{T}$ or $\mathcal{P}\mathcal{T}^{(1)}$ in quantum mechanics.

We have claimed that the three theories $\mathcal{M}\mathcal{T}_{\Sigma'}\mathcal{A}$, $\mathcal{M}\mathcal{T}_\Sigma\mathcal{A}$ and $\mathcal{M}\mathcal{T}_{\Sigma^{(1)}}\mathcal{A}$ with the "same" observational report $\mathcal{A}$ (i.e. (1.1), (1.7), resp. (2.2.3), (2.2.5), resp. (2.2.15), (2.2.16)) are equivalent in the following sense: A contradiction in one of these three theories implies a contradiction in the others. Let us prove this claim.

The observational report $(\text{—})_r(1)$ consists of relations $a_i \in y_\nu$, while $(\text{—})'_r(1)$ is formed by corresponding relations $a_i \in E_\nu$, the second part $(\text{—})_2(2)$ by the relations

(2.2.5) and $(—)'_r(2)$ by the corresponding relations (1.7). Collecting all $a_i$ into a term $A=(a_1, a_2, \ldots)$, we may write $(—)_r(1)$ as

$$A \in \tilde{T}(y_1, \ldots, y_r). \tag{2.2.22}$$

The corresponding $(—)'_r(1)$ gets the form

$$A \in \tilde{T}(E_1, \ldots, E_r), \tag{2.2.23}$$

where $\tilde{T}$ is an echelon construction scheme. The relations $(—)_r(2)$ may be comprised in a transportable relation

$$\tilde{P}(y_1, \ldots, y_r, t, A, \mathbf{R}) \tag{2.2.24}$$

and the corresponding $(—)'_r(2)$ in

$$\tilde{P}(E_1, \ldots, E_r, U, A, \mathbf{R}). \tag{2.2.25}$$

Similarly (2.2.15) and (2.2.16) take the forms

$$A \in \tilde{T}(\ldots z_k, \ldots, t_i^{(1)}, \ldots) \tag{2.2.26}$$

and

$$\tilde{P}(\ldots z_k, \ldots, t_i^{(1)}, \ldots, t_0^{(1)}, A, \mathbf{R}). \tag{2.2.27}$$

Here the $y_{v_i}$ resp. $y_{\mu_k}$ from (2.2.22) and (2.2.24) have been replaced by $t_i^{(1)}$ resp. $z_k$.

Also the connection between $\mathscr{PT}'$ and $\mathscr{PT}^{(1)}$ can be given such a form. To do this we first change $(—)_r(1)$ and $(—)'_r$ a little. Instead of a relation $a_\beta \in t_i^{(1)}$, in (2.2.15) we write $a_\beta \in T_{v_i}(\ldots)$ with $T_{v_i}(\ldots)$ in (2.2.11) and take $a_\beta \in t_i^{(1)}$ in $(—)_r^1(2)$. Because of $t_i^{(1)} \subset T_{v_i}(\ldots)$ this does not change any mathematical conclusion. We do the same with $(—)'_r$, replacing all $a_\beta \in E_{v_i}$ ($E_{v_i}$ the family defined by (2.2.7)) by $a_\beta \in T_{v_i}(E_{\mu_1} \ldots)$ and taking $a_\beta \in E_{v_2}$ in $(—)'_r(2)$.

Then $(—)_r^{(1)}(1)$ takes the form

$$A \in \tilde{\tilde{T}}(z_1, \ldots) \tag{2.2.28}$$

and the corresponding $(—)'_r(1)$ the form

$$A \in \tilde{\tilde{T}}(E_{\mu_1}, \ldots). \tag{2.2.29}$$

Likewise, $(—)_r^{(1)}(2)$ and $(—)_r(2)$ become

$$\tilde{\tilde{P}}(z_1, \ldots, t^{(1)}, A, \mathbf{R}) \quad \text{with} \quad t^{(1)} = \{t_i^{(1)}, t_0^{(1)}\}, \tag{2.2.30}$$

and

$$\tilde{\tilde{P}}(E_{\mu_1}, \ldots, V, A, \mathbf{R}) \quad \text{with} \quad V = \{E_{v_i}, U\}. \tag{2.2.31}$$

Therefore we need to prove only for one pair (e.g. $\mathscr{M}\mathscr{T}_\Sigma\mathscr{A}$, $\mathscr{M}\mathscr{T}_{\Sigma'}\mathscr{A}$) that a contradiction in one of them implies a contradiction in the other.

We have used the same letters $a_i$ for the observational report of $\mathscr{PT}$ and $\mathscr{PT}'$. This can raise mathematical confusion if we compare the two theories $\mathscr{M}\mathscr{T}_\Sigma\mathscr{A}$ and $\mathscr{M}\mathscr{T}_{\Sigma'}\mathscr{A}$. Therefore using new letters in $\mathscr{M}\mathscr{T}_\Sigma$ we introduce the text:

$$w \in \tilde{T}(y_1, \ldots, y_r, \mathbf{R})$$

and

$$\tilde{P}(y_1, \ldots, y_r, t, w, \mathbf{R}).$$

The typification

$$(t, w) \in S(y_1, \ldots) \times \tilde{T}(y_1, \ldots) \tag{2.2.32}$$

and the relation

$$P(y_1, \ldots, y_r, t) \quad \text{and} \quad \tilde{P}(y_1, \ldots, y_r, t, w, \mathbf{R}) \tag{2.2.33}$$

define a species of structure $\langle\Sigma\mathscr{A}\rangle$; we call it the test $\mathscr{A}$ of $\Sigma$. The term $(U, A)$ in $\mathscr{M}\mathscr{T}_{\Sigma'}\mathscr{A}$ is then a structure of species $\langle\Sigma\mathscr{A}\rangle$ over $E_1, \ldots, E_r$.

$\mathscr{M}\mathscr{T}_\Sigma\mathscr{A}$ and $\mathscr{M}\mathscr{T}_{\langle\Sigma\mathscr{A}\rangle}$ are the same theory since they are written only with different letters. We must compare the two theories $\mathscr{M}\mathscr{T}_{\langle\Sigma\mathscr{A}\rangle}$ and $\mathscr{M}\mathscr{T}_{\Sigma'}\mathscr{A}$.

Since $\{U, A\}$ is a structure of species $\langle\Sigma\,\mathscr{A}\rangle$ in $\mathscr{M}\mathscr{T}_{\Sigma'}\mathscr{A}$, every theorem in $\mathscr{M}\mathscr{T}_{\langle\Sigma\mathscr{A}\rangle}$ leads to a corresponding theorem in $\mathscr{M}\mathscr{T}_{\Sigma'}\mathscr{A}$. Therefore every contradiction in $\mathscr{M}\mathscr{T}_{\langle\Sigma\mathscr{A}\rangle}$ leads to a contradiction in $\mathscr{M}\mathscr{T}_{\Sigma'}\mathscr{A}$. In this sense, $\mathscr{P}\mathscr{T}'$ is a stronger theory than $\mathscr{P}\mathscr{T}$. If $U$ represent $\Sigma$ in $\mathscr{M}\mathscr{T}_{\Sigma'}$, then also the reverse holds.

In $\mathscr{M}\mathscr{T}_{\langle\Sigma\mathscr{A}\rangle}$ we first have the theorem:

$$\{s \in S'(x_1, \ldots) \text{ and } P'(x_1, \ldots), \text{and } f_i \text{ bijective mappings} \\ y_i \xrightarrow{f_i} E_i(x_1, \ldots, s) \text{ and } \langle f_1, \ldots\rangle^S t = U(x_1, \ldots)\} \tag{2.2.34}$$

implies

$$\{\hat{w} = \langle f_1, \ldots\rangle^{\tilde{T}} w \in \tilde{T}(E_1, \ldots) \text{ and } \tilde{P}(E_1, \ldots, E_r, U, \hat{w})\}.$$

If a contradiction in $\mathscr{M}\mathscr{T}_{\Sigma'}\mathscr{A}$ can be deduced, in $\mathscr{M}\mathscr{T}_{\Sigma'}$ we have the theorem:

$$\hat{w} \in \tilde{T}(E_1, \ldots) \Rightarrow \text{not } \tilde{P}(E_1, \ldots, E_r, U, \hat{w}).$$

Equivalent to this is the following theorem in $\mathscr{M}\mathscr{T}$ (and therefore also in $\mathscr{M}\mathscr{T}_{\langle\Sigma\mathscr{A}\rangle}$!):

$$\{s \in S'(x_1, \ldots) \text{ and } P'(x_1, \ldots)\} \Rightarrow \{\hat{w} \in \tilde{T}(\ldots) \Rightarrow \text{not } \tilde{P}(\ldots)\}.$$

Since $\{\hat{w} \in \tilde{T}(\ldots) \Rightarrow \text{not } \tilde{P}(\ldots)\}$ is the negation of the relation $\{\hat{w} \in \hat{T}(\ldots) \text{ and } \tilde{P}(\ldots)\}$, the left side of (2.2.34) implies $\{\hat{w} \in \hat{T}(\ldots) \text{ and } \tilde{P}(\ldots)\}$ and its negation. Therefore the negation of (2.1.1) is a theorem in $\mathscr{M}\mathscr{T}_{\langle\Sigma\mathscr{A}\rangle}$. Since (2.1.1) is a theorem in $\mathscr{M}\mathscr{T}_\Sigma$($U$ represents $\Sigma$) and therefore in $\mathscr{M}\mathscr{T}_{\langle\Sigma\mathscr{A}\rangle}$, we have a contradiction in $\mathscr{M}\mathscr{T}_{\langle\Sigma\mathscr{A}\rangle}$.

Thus we have proven that the theories $\mathscr{P}\mathscr{T}'$, $\mathscr{P}\mathscr{T}$ and $\mathscr{P}\mathscr{T}^{(1)}$ are equivalent.

We had assumed that the observational report $(—)_r$ describes real facts of the fundamental domain $\mathscr{G}$. But we never used this "reality" of $(—)_r$ in the above deductions. If $(—)_r$ were only an "imagined" observational report, all about $\mathscr{M}\mathscr{T}_\Sigma\mathscr{A}$, etc. would persist. The only difference would be that a contradiction in $\mathscr{M}\mathscr{T}_\Sigma\mathscr{A}$ with an "imagined" observational report $\mathscr{A}$ would never oblige us to reject the theory $\mathscr{P}\mathscr{T}$. On the contrary one would interpret such an effect as saying that the "imagined" observational report encompasses something impossible in nature. It has been proposed that contradictions in $\mathscr{M}\mathscr{T}_\Sigma\mathscr{A}$ for a "real" observational report form the only criterion to reject a theory (see Popper [62]). We will see in §2.4 that this is not the only reason to accept or reject a theory.

If we have an "imagined" observational report, instead of $\mathscr{A}$ let us write the letter $\mathscr{H}$ and call this report a "hypothetical report". We presume that $\mathscr{H}$ has no contradiction in itself, since $(—)_r$ for a real observational report cannot contradict itself unless we made a mistake in writing it down.

## §2.3 Laws of Nature and Theoretical Terms

In §2.2 we have explained what is an axiomatic basis $\mathscr{M}\mathscr{T}_\Sigma$ (resp. $\mathscr{M}\mathscr{T}_{\Sigma(1)}$) of a $\mathscr{P}\mathscr{T}'$ (with $\mathscr{M}\mathscr{T}_{\Sigma'}$). But we have not answered the question whether there is an axiomatic basis for every $\mathscr{P}\mathscr{T}$!

The problem to find an axiomatic basis for a given $\mathscr{P}\mathscr{T}'$ can be solved in any case trivially. We only need to choose the relation (2.2.1) as axiom $P(y_1, \ldots, y_r, t)$ of $\Sigma$. We feel that (2.1.1) as axiom is not yet in the form of a physical law that physicists have in mind. $\mathscr{M}\mathscr{T}_\Sigma$ with (2.1.1) as axiom gives no new physical insight campared with $\mathscr{M}\mathscr{T}_{\Sigma'}$. Only the logical structure of $\mathscr{P}\mathscr{T}$ with $\mathscr{M}\mathscr{T}_\Sigma$ is simpler than that of $\mathscr{P}\mathscr{T}'$ with $\mathscr{M}\mathscr{T}_{\Sigma'}$; this will be of interest in §4.

As an example let us as $\mathscr{M}\mathscr{T}'$ take analytic geometry (see page 157). An axiomatic basis with (2.1.1) as axiom would look as follows. With $y$ as base term and $d(z_1, z_2)=\alpha$ as distance relation we postulate the further axiom (2.1.2) (additional to the axiom that $d(z_1, z_2)$ is a real function). Obviously this axiomatic basis does not make the physical significance of Euclidean geometry more evident than $\mathscr{P}\mathscr{T}'$ does.

Nevertheless let us denote the axiomatic relation $P(\ldots)$ in an axiomatic basis $\mathscr{M}\mathscr{T}_\Sigma$ as "the physical laws". If $P(\ldots)$ is only of the form (2.1.1), we do not get an insight into the physical significance of these laws. Hence we desire another form of $P(\ldots)$ than (2.1.1). Therefore let us first classify various forms of physical concepts.

It is essential for physics to introduce new physical concepts. How do we introduce such new concepts?

For the introduction of new concepts, the form $\mathscr{P}\mathscr{T}'$ is not very suitable. In $\mathscr{M}\mathscr{T}_{\Sigma'}$ are defined the picture sets and picture relations. In $\mathscr{M}\mathscr{T}_{\Sigma'}$ it is not simple to say what we mean by a definition of new physical concepts. For a logical problem such as the definition of new concepts by old ones, an axiomatic basis is much better, also if the axiomatic relation has e.g. the form (2.1.1).

Such terms in $P(\ldots)$ which are connected with an "existential quantifier" $\exists$ (e.g. $\exists x(A(x))$, i.e. there is an $x$ with $A(x)$) are called theoretical auxiliary terms. Here it has been assumed that in $P(\ldots)$ quantifiers (not $\forall$) are replaced by $\exists$ and (not $\exists$) by $\forall$. If e.g. $P(\ldots)$ is given by (2.1.1), all terms $x_1, \ldots, x_n, s, f_1, \ldots, f_r$ are theoretical auxiliary terms. Sometimes $s$ is a collection of various terms $s_\mu$; then all $s_\mu$ are also called theoretical auxiliary terms. In this sense the "rectangular coordinates" introduced in (2.1.2) by the mapping $y \xrightarrow{f} \mathbf{R}^3$ are theoretical auxiliary terms.

In this sense, $\mathscr{M}\mathscr{T}_{\Sigma'}$ from $\mathscr{P}\mathscr{T}'$ is characterized by the theoretical auxiliary terms $x_1, \ldots, x_n, s, f_1, \ldots, f_r$ from (2.1.1a, b). Here it is essential that also the $f_i$ appear as theoretical auxiliary terms. Only in $\mathscr{M}\mathscr{T}_\Sigma$ with $P(\ldots)$ from (2.1.1a, b), and not already in $\mathscr{M}\mathscr{T}_{\Sigma'}$, can we see clearly what theoretical auxiliary terms are necessary for formulating $\mathscr{P}\mathscr{T}'$.

It is usual to extend the concept of theoretical auxiliary terms. To do this rigorously, we want to extend $\mathscr{M}\mathscr{T}_\Sigma$ (with $P(y_1, \ldots, y_r, t)$ as axiomatic relation) to a theory $\mathscr{M}\mathscr{T}_{\text{ext}}$ defined as follows.

We replace at least some of the terms bound by existential quantifiers in $P(\ldots)$ by new terms (constants) $u_1, u_2, \ldots$ in addition to the $y_1, \ldots, y_r, t$. As axioms for $\mathscr{M}\mathscr{T}_{\text{ext}}$ we use the typification $t \in S(y_1, \ldots, y_r, \mathbf{R})$ from $\mathscr{M}\mathscr{T}_\Sigma$ and

$P_{ext}(u_1, u_2, \ldots; y_1, \ldots, y_r, t)$ where we get $P_{ext}(\ldots)$ from $P(y_1, \ldots, y_r, t)$ by everywhere replacing $(\exists u)\, R(u)$ by $R(u)$.

For the example of the relation (2.1.1 b) as $P(\ldots)$, this transcription provides

$$P_{ext}(x_1, \ldots, x_n, s, f_1, \ldots, f_r; y_1, \ldots, y_r, t): s \in S'(x_1, \ldots, x_n) \text{ and } P'(x_1, \ldots, x_n, s)$$

and the $f_i$ are bijective mappings $y_i \to E_i(x_1, \ldots, x_n\, s)$ with

$$\langle f_1, \ldots, f_r \rangle^S\, t = U(x_1, \ldots, x_n, s). \tag{2.3.1}$$

Instead of $u_1, u_2, \ldots$ we have used the letters $x_1, \ldots, x_n, s, f_1, \ldots, f_r$. We define $\mathscr{P}\mathscr{T}_{ext}$ by using the same correspondence rules as for $\mathscr{P}\mathscr{T}$! Thus (2.3.1) is determined by the theory $\mathscr{P}\mathscr{T}'$; and if we "identify" the $y_i$ with the $E_i$ and $t$ with $U$ by the bijective mappings $f_i$, we regain $\mathscr{P}\mathscr{T}'$. But for logical purposes, $\mathscr{P}\mathscr{T}_{ext}$ is more suitable than $\mathscr{P}\mathscr{T}'$.

As in this example, we assume that $\mathscr{P}\mathscr{T}_{ext}$ has the form $\mathscr{M}\mathscr{T}_{\Sigma_{ext}}$ with a species of structure $\Sigma_{ext}$. The base terms of $\Sigma_{ext}$ are $y_1, \ldots, y_r, u_{\alpha_1}, \ldots$ and the structure terms are $t, u_{\beta_1}, \ldots$. The $y_1, \ldots, y_r$ are the pictorial terms, and $t$ is the collection of the pictorial relations of $\mathscr{P}\mathscr{T}_{ext}$. Then we can introduce an extended concept of theoretical auxiliary terms: We call every intrinsic term in $\mathscr{M}\mathscr{T}_{\Sigma_{ext}}$ a theoretical auxiliary term. In the usuable forms of physical theories one applies many theoretical auxiliary terms.

Often there are different forms of $\mathscr{P}\mathscr{T}'$ and correspondingly of $\mathscr{P}\mathscr{T}_{ext}$. Hence it is not surprising that we have no systematics of such theoretical auxiliary terms. Such terms have rather been adopted in the historical development of physics. Sometimes such terms have been dropped later; an example is the "ether". This status of the theoretical auxiliary terms may recommend to avoid them. Is this possible?

Before we try to answer this question, let us give some examples of theoretical auxiliary terms in the usual form of quantum mechanics.

Let us return to the representation of quantum mechanics in [2]. When we take $\mathscr{K}$ and $\mathscr{L}$ as pictorial terms and $\mathscr{K} \times \mathscr{L} \xrightarrow{\mu} [0, 1]$ as pictorial relation, then axiom $AQ$ in [2] III §5 has the form of $P(\ldots)$ in (2.1.1) if we take $\beta$, $\gamma$ of $AQ$ as the mappings $f_i$ of (2.1.1) and if we think of $\mathscr{K}$ and $\mathscr{L}$ as completed to $K$, $L$ (as done in [2] III §3 or here in VI D 3.5 and D 3.6). The corresponding structure in the form $\mathscr{P}\mathscr{T}_{ext}$ looks like this: $x_1 = x = \mathscr{H}$ with $s$ and $P'$ such that $\mathscr{H}$ is a "Hilbert space". $y_1 = K$, $y_2 = L$, $t \in K \times L \times \mathbf{R}$; $E(x, s) =$ set of all selfadjoint operators $W$ with $W \geq 0$ and $\operatorname{tr}(W) = 1$. $E_2(x, s) =$ set of all selfadjoint operators $F$ with $0 \leq F \leq 1$. $U(x, s) \in E_1 \times E_2 \times \mathbf{R}$ is determined by the relation $\operatorname{tr}(WF) = \alpha$.

$y_1 = K$ is the pictorial set for the "ensembles", $y_2 = L$ is the pictorial set for the "effects" and $\operatorname{tr}(WF) = \alpha$ is the "probability for the effect $F$ in the ensemble $W$". Thus $f_1$ and $f_2$ are two bijective mappings $y_1 = K \xrightarrow{f_1} E_1$ and $y_2 = L \xrightarrow{f_2} E_2$.

Examples for theoretical auxiliary terms are: The Hilbert space $\mathscr{H}$; the set of all closed subspaces of $\mathscr{H}$; the set of all elements $\varphi \in \mathscr{H}$ with $\|\varphi\| = 1$ (often called "states", not in accordance with our definition III D 5.2.2). For all these terms, the definition lets open whether they have a "physical" meaning and what this meaning could be, and whether there is something "in reality" that corresponds to these terms. There is e.g. another example of an auxiliary term which has no "physical" meaning (as we shall see later): This term is the so-called "phase". This phase is a relation between two vectors $\varphi_1, \varphi_2 \in \mathscr{H}$ of the same direction with $\|\varphi_1\| = \|\varphi_2\| = 1$. The phase $\alpha$ is defined by $\varphi_2 = e^{i\alpha}\varphi_1$. The widespread opinion that this phase "has" a physical meaning (i.e. that $\varphi_1$ and $\varphi_2$ are "physically" distinguished for $\alpha \neq 0$) caused many mistakes. In fact $\varphi_1$ and $\varphi_2$ have the same physical meaning and therefore $\alpha$ has no physical meaning. But what should we understand by what we called "physical" meaning? This can be clarified very well in the context of an axiomatic basis.

Therefore we regress to $\mathscr{PT}$ with $\mathscr{MT}_\Sigma$ as axiomatic basis (but without any presumption for the form of $P(\ldots)$).

As "theoretical (not auxiliary) terms" let us denote all intrinsic terms in $\mathscr{MT}_\Sigma$. The physical meaning of such terms is exactly defined by their deduction in $\mathscr{MT}_\Sigma$ and by the previously established physical meaning of the base terms $y_1, \ldots, y_r$ (that are the pictorial terms) and of the structural term (characterizing the pictorial relations).

During the development of an axiomatic basis for quantum mechanics in III through IX, we have introduced new concepts only by theoretical terms, i.e. by intrinsic terms. We will study some examples of such new concepts in §§3 and 4, to see the importance of new concepts in the context of other problems.

Obviously, we can transport all intrinsic terms of $\mathscr{MT}_\Sigma$ to terms in $\mathscr{MT}_{\Sigma_{\text{ext}}}$. In this sense every theoretical term from $\mathscr{MT}_\Sigma$ is also a theoretical auxiliary term from $\mathscr{MT}_{\Sigma_{\text{ext}}}$ but not vice versa.

Some of the theoretical auxiliary terms can also be "declared" to be theoretical terms by bijective mappings $f$ from one theoretical term (coming from $\mathscr{MT}_\Sigma$) onto another term in $\mathscr{MT}_{\Sigma_{\text{ext}}}$. For instance, the $E_i(x_1, \ldots, x_n, s)$ and $U(x_1, \ldots, x_n, s)$ can be "identified" by isomorphic mappings $y_i \xrightarrow{f_i} E_i$ with $t \to U$. Exactly this was the starting point of our analysis: The physical interpretation of the $E_i$ as pictorial terms and of $U$ as the composition of the pictorial relations. It is not difficult to identify by the $f_i$ also all theoretical terms (i.e. intrinsic terms in $\mathscr{MT}_\Sigma$) with corresponding terms in $\mathscr{MT}_{\Sigma_{\text{ext}}}$. But this in general does not give any physical meaning to the base sets $x_i$ or to such intrinsic terms of $\mathscr{MT}_{\Sigma_{\text{ext}}}$ which are not $f_i$-pictures of intrinsic terms of $\mathscr{MT}_\Sigma$.

In order to give the physical meaning of a theoretical term $X$ to another intrinsic term $V$ of $\mathscr{MT}_{\Sigma_{\text{ext}}}$, one may use a bijective mapping $g$ of $X$ on $V$. This $g$ should be defined as intrinsic term of $\mathscr{MT}_{\Sigma_{\text{ext}}}$, not necessarily as the canonical extension of the $f_i$. Then it is essential, however, not to forget the mapping $g$ and not to think that $V$ by itself has the physical meaning defined by $g$. For instance, possibly there is another bijective mapping $h$ of a theoretical term $Y$ onto $V$, so that $V$ based on $h$ has another physical interpretation than when based on $g$. Such a situation may cause great confusion if one forgets the mappings $g$ and $h$. Let us give an example from quantum mechanics.

As $\mathscr{MT}_{\Sigma_{\text{ext}}}$ we use the same theory as above, i.e. $y_1 = K$, $y_2 = L$, $E_1(x, s)$ the set of all selfadjoint operators $W$ with $W \geq 0$, $\text{tr}(W) = 1$, and $E_2(x, s)$ the set of all selfadjoint operators $F$ with $0 \leq F \leq 1$, with bijective mappings given by $y_1 = K \xrightarrow{f_1} E_1$, $y_2 = L \xrightarrow{f_2} E_2$. Then $X = \partial_e K$ is a theoretical term. The elements of $X$ shall be called extreme ensembles or extreme states. They are commonly called pure states or shortly states. We do not use these names to avoid misunderstandings.

$X$ is mapped by $f_1$ onto $\partial_e E_1(x,s)$, while $G=\partial_e L$ is a theoretical term, the set of decision effects.

Also $Y=\{$set of all atoms of the lattice $G\}$ is a theoretical term. The elements of $Y$ are often called the "finest" decision effects. $f_2$ maps $G$ on the set of the projection operators and especially $Y$ on the set of all $P_\varphi$ ($\|\varphi\|=1$) with $P_\varphi \psi = \varphi\langle\psi,\varphi\rangle$. The set of all $P_\varphi$ also equals $\partial_e E_1(x,s)$. Defining $V$ as this set of all $P_\varphi$ we have

$$X \xrightarrow{f_1} V \quad \text{and} \quad Y \xrightarrow{f_2} V.$$

Thus we get two different interpretations of $V$, one according to $f_1$ ($V$ as the set of extreme states), the other according to $f_2$ ($V$ as the set of finest decision effects). For instance, with $P_{\varphi_1}$ as extreme state and $P_{\varphi_2}$ as a finest decision effect we get $\mathrm{tr}(P_{\varphi_1} P_{\varphi_2}) = |\langle\varphi_1,\varphi_2\rangle|^2$ as the probability of the finest decision effect $P_{\varphi_2}$ in the extreme state $P_{\varphi_1}$.

Sometimes one replaces $V$ by the set of all $\varphi\in x=\mathscr{H}$ with $\|\varphi\|=1$. But there is no bijective mapping $P_\varphi\to\varphi$, since $\varphi$ and $\varphi'=e^{i\alpha}\varphi$ generate the same $P_\varphi=P_{\varphi'}$. If we define $\boldsymbol{\varphi}$ as the set of all $e^{i\alpha}\varphi$, then the bijective mapping $P_\varphi\to\boldsymbol{\varphi}$ allows to transport the physical interpretations of $V$ to te set of all $\boldsymbol{\varphi}$.

Since in $\mathscr{M}\mathscr{T}_{\Sigma_{\text{ext}}}$ there is no bijective mapping of an intrinsic term of $\mathscr{M}\mathscr{T}_\Sigma$ onto the set of all $\varphi\in\mathscr{H}$ with $\|\varphi\|=1$, the "phases" $\alpha$ cannot be interpreted.

As this example of the interpretation of $V$ by bijective mappings of theoretical terms $X$, $Y$ onto $V$, demonstrates, such transports of physical interpretations do not deepen the understanding of a theory. Why then are such transports used? The application of terms in $\mathscr{M}\mathscr{T}_{\Sigma_{\text{ext}}}$ can have advantages for practical calculations and for solving special problems. Then the language shortened by transported interpretations is very practical for the communication among experts (but can be dangerous for beginners or philosophers).

Since the aim of this book is not to find "short phrases", we will not pursue the introduction of theoretical auxiliary terms nor their possible interpretation by transports with the help of bijective mappings. On the contrary, we try to avoid any theoretical auxiliary term which is not a theoretical term.

The concept of theoretical terms appears to us still too broad since the set $\mathbf{R}$ of real numbers can enter the typification of an intrinsic term in $\mathscr{M}\mathscr{T}_\Sigma$ arbitrarily. For instance, in the above example of Euclidean geometry we denoted the base set of the points by $y$. Then the set of all bijective mappings $y\xrightarrow{f}\mathbf{R}^3$ with

$$d(z_1,z_2)=\sqrt{(f_1(z_1)-f_1(z_2))^2+(f_2(z_1)-f_2(z_2)^2+(f_3(z_1)-f_3(z_2))^2}$$

(i.e. the set of all rectangular coordinate systems) is an intrinsic term in $\mathscr{M}\mathscr{T}_\Sigma$, i.e. a theoretical term. The physical interpretation of this term is not immediately given by the pictorial term $y$ and the pictorial relation $d(z_1,z_2)=\alpha$. Surely also this theoretical term has a physical interpretation, which we will discuss in §4.

Let us now restrict the concept of theoretical terms, allowing the real numbers to enter the terms only in that manner in which their physical meaning is given by the pictorial relations. Therefore we define: A *physical term $B$* is an intrinsic term in $\mathscr{M}\mathscr{T}_\Sigma$ (i.e. a theoretical term) for which

$$B\in T(y_1,\ldots,y_r;t_1,t_2,\ldots) \tag{2.3.2}$$

is a theorem in $\mathcal{M}\mathcal{T}_\Sigma$. Here $T$ is an echelon construction scheme and the $t_\mu$ are the components of the structure term $t$, i.e. the pictorial relations. Thus **R** enters $B$ only through the $t_\mu$. Hence (2.3.2) can be fulfilled in a trivial form, i.e. so that (2.3.2) is contained in the definition of $B$.

The physical interpretation of such a term $B$ is given by the "logical" construction (2.3.2) from the pictorial sets and pictorial relations $t_\mu$, and by additional conditions which can be formulated intrinsically.

Let us denote an axiomatic relation $P(\ldots)$ in an axiomatic basis as *physically interpretable*, if in $P(\ldots)$ there are only such quantifiers $\exists z$ (and $\forall z$) for which a relation $z \in T(y_1, \ldots, y_r; t_1, t_2 \ldots)$ with $T$ as in (2.3.2) holds, e.g. in the form

$$\exists z[z \in T(y_r, \ldots, t_1, t_2 \ldots) \text{ and } \ldots]$$

or as a relation following from other relations in [...].

How such a physically interpretable relation $P(\ldots)$ can indeed be interpreted is said by the interpretation of the physical terms $z$ *and* by that of the quantifiers $\exists z$ (or $\forall z$). This last interpretation is *not* trivial and will be given in §2.4 and in a wider context in §4.

Therefore the relation (2.1.1 b) is not a physically interpretable axiom.

If we have an axiomatic basis $\mathcal{M}\mathcal{T}_\Sigma$ (as in §2.2), and if $P(\ldots)$ is physically interpretable we call $\mathcal{M}\mathcal{T}_\Sigma$ an axiomatic basis of the first degree with physically interpretable laws of nature. A $\mathcal{P}\mathcal{T}$ with such an axiomatic basis has exactly the desired form. Proceeding to $\mathcal{M}\mathcal{T}_{\Sigma(1)}$ as described in §2.2, we call $\mathcal{M}\mathcal{T}_{\Sigma(1)}$ an axiomatic basis of the $n$-th degree with physically interpretable laws of nature. Because of the fundamental significance of an axiomatic basis of the first degree with physically interpretable laws of nature let us briefly call such a basis a *simple* axiomatic basis.

There are philosophers of science who believe that there are physical theories (e.g. the quantum mechanics of atoms) for which there is no simple axiomatic basis, i.e. for which theoretical auxiliary terms are inevitable. Although we might demonstrate here that we have indeed given an axiomatic basis of higher than first *degree* with physically interpretable laws in III through IX, let us do so a little later because in §3 we shall develop a general scheme to compare various theories.

Having claimed a simple axiomatic basis as the "desired" form of a $\mathcal{P}\mathcal{T}$, we do not mean that other forms $\mathcal{M}\mathcal{T}_{\Sigma'}$ with theoretical auxiliary terms have no significance. In the contrary, a $\mathcal{M}\mathcal{T}_{\Sigma'}$ in which the species of structure $\Sigma$ of a simple axiomatic basis is represented (see §2.1), can be essential for the practical use of a $\mathcal{P}\mathcal{T}$. For instance, who would renounce analytic geometry and use only the Euclidean axioms?

## §2.4 Norms and Fundamental Domain

For the following discussions we assume $\mathcal{P}\mathcal{T}$ in the form of a *simple* axiomatic basis. The intent is to analyse the "physically interpretable" laws $P(\ldots)$ in such a form that one can distinguish between those parts of $P(\ldots)$ which are "pictures of real stuctures of the world" and those which are "descriptions of special concepts" or "prescriptions for actions". These last parts have to do with thinking and acting of human beings. Nevertheless we shall see that they reflect something of reality.

For instance, let us take a set $V$ with a relation $<$ and interpret $a<b$ as "$a$ is a part of $b$". Then axioms which make $V$ an ordered set are not coming from experience but from what we mean by "part of". An observational report cannot contradict the theory, unless we have made a "false application" of the concept of "to be a part of". We will see below another example from our axiomatic basis of quantum mechanics.

Such axioms which "define" (in the sense of $\mathcal{MT}$) a concept shall be called "conceptual norms". The "application" of such a concept must be such that the observational report does not contradict these axioms. In this sense the conceptual norms lay down the mode of application of the concept. It is obvious that such a concept cannot be applied everywhere. When it cannot be applied, we get no observational report corresponding to an application of the concept.

Often there are many other normative axioms in $\mathcal{MT}_\Sigma$, in contradictions to which we are "not interested". There may be facts which contradict these axioms. But we declare that such contradicting facts do not belong to the fundamental domain of $\mathcal{PT}$. Such norms in this sense confine the fundamental domain. Sometimes we say that these norms lay down how we should make the "right" experiments. Therefore let us call these axioms "action norms", not excluding the possibility that in nature we may find facts (i.e. no artifacts) which fulfill the action norms and therefore belong to the fundamental domain. Then we say that we may contemplate these natural facts "as if" they were produced by a "right" experiment.

Illustrative examples for conceptual norms and action norms are those axioms which in the study of probability are introduced (see AS 2.1 to AS 2.5) for the function

$$\mathcal{T}=\{(a,b)\,|\,a,b\in\mathcal{S}, a\supset b \text{ and } \alpha\neq\emptyset\}\xrightarrow{\lambda}[0,1].$$

These axioms concern a relation between $\mathcal{T}$ and $\mathbf{R}$. The real numbers are "pictures" of frequencies $N_+/N$. The axiom that $\lambda$ maps only into $[0, 1]$ is therefore a conceptual norm. The axiom that $\lambda$ is *mapping* $\mathcal{T}$ into $[0, 1]$, is an (idealized) action norm: This axiom says that we experiment in such a way that the frequences $N_+/N$ are "reproducible", i.e. that for the same pair $a$, $b$ we get (approximately) the same frequencies $N_+/N$. If not, then the experiment does not belong to the fundamental domain. Obviously, choosing $N$ large enough we find an approximate reproducibility for suitable $a$, $b$, as experimental physicists know.

It is easy to see ([2] II §3) that AS 2.1, AS 2.2, AS 2.3 are conceptual norms.

AS 2.4.1, AS 2.4.2 are pure idealizations. Below we will return to axioms which are pure idealizations.

AS 2.5 is an empirical law, as we will define such laws further below.

The normative axioms are connected with what protophysics calls norms (see [42]). The only difference seems to be that protophysics attaches great importance to instructions for actions which produce such facts that the corresponding observational report fulfills the normative axioms (obviously only if it is possible to follow the instructions). It would be a very interesting task, more extensively to compare the method of normative axiom and that of protophysics.

We say that the fundamental domain $\mathcal{G}$ of a $\mathcal{PT}$ is "intrinsically" defined if the norms alone confine $\mathcal{G}$. We will see in §3 that there also are external possibilities to confine $\mathcal{G}$. The physical content of the norms is precisely that "there is" a fundamental domain $\mathcal{G}$, i.e. that "it is possible to make" such experiments which do not contradict the norms. This "it is possible" will be discussed more rigorously in §4.

No $\mathcal{PT}$ is known which contains no norms. But in most of the $\mathcal{PT}$ the initially introduced normative axioms do not suffice to confine $\mathcal{G}$; i.e. $\mathcal{G}$ is not intrinsically defined. Not until after many applications and experiences with $\mathcal{PT}$ we learn to find additional restrictions for $\mathcal{G}$. Such additional restrictions are often formulated in common language; nevertheless we could formulate them as normative axioms. For Newton's mechanics we would add the "normative axiom" that all velocities should be smaller than $10^{-3}$ times the velocity of light. This example demonstrates that such additional normative axioms depend strongly on the imprecision sets used for the correspondence rules. Therefore we avoid "exact" axioms to confine $\mathcal{G}$. We want to permit different imprecision sets with different normative axioms to confine $\mathcal{G}$.

Obviously, experimental tests of a $\mathcal{PT}$ are only possible when we have defined $\mathcal{G}$. We do not know any $\mathcal{PT}$ which can be applied to "everything". It is an illusion that we can find a theory of the whole world. The reason is that the "physical method" cannot reach all realities, as we shall see later.

Before closing our general considerations concerning norms and going on to the special norms in quantum mechanics, we should remark about the "idealizations" we find in many axioms. Such an idealization (of the approximate reproducibility of frequencies) was the axiom that the relation between the elements of $\mathcal{T}$ and the real numbers is a *function* $\mathcal{T} \xrightarrow{\lambda} [0, 1]$. We have already seen in §1 that such idealizations are undone by using imprecision sets for the correspondence rules. Such a smearing by imprecision sets (as described in §1) sometimes causes great difficulties if the pictorial relations and the corresponding norms define a concept, i.e. if we have conceptual norms. One can get the impression that the concept is smeared. Not the concept is smeared but only the application of the concept to the facts (as illustrated in §1). Or to say it in another way: The concept rigorously defined in $\mathcal{MT}$ by conceptual norms is nowhere applicable *exactly*. Only if one applies the concept with a certain inaccuracy defined by an imprecision set, the concept is useful.

Similar is it with the action norms: The idealized action norms can only be fulfilled approximately; i.e. only by using imprecision sets one retains a fundamental domain on which the theory is applicable.

There is still another source of normative axioms. In $\mathcal{MT}_\Sigma$ some of the axioms can be imposed by pretheories, as we shall see in §3. In $\mathcal{PT}$ such axioms have the character of norms since an experimental test of these axioms is not intended; on the contrary it is a condition for the fundamental domain $\mathcal{G}$ of $\mathcal{PT}$ that the pretheories can be applied without contradictions.

Therefore for such axioms produced by pretheories we will be content with such forms as (2.1.1), since the physical significance of the axioms from pretheories is regarded as clarified in the pretheories.

To conclude let us try to get the axiom $P(\ldots)$ into such a form that we can separate the normative axioms (in most cases also including idealizations). The remaining axioms (as far as they are no pure idealizations such as AVid) are called "empirical" laws. These shall be investigated in the next section.

Let us list the norms introduced (in III through IX) to develop quantum mechanics.

AZ 1, AZ 2 are conceptual norms. AT 1 is produced by pretheories. APSZ 1 contains various axioms (see II §2.1, where AS 1.1, AS 1.2 are conceptual norms and AS 2.1 through AS 2.5 are already classified above). The same holds for APSZ 2 and APSZ 3, while APSZ 4.1, 2 are conceptual norms. APSZ 8.1, 2 are conceptual norms. APS 5.1 through APSZ 5.4 and APS 5.1.3, APS 5.1.4 are action norms which lay down how we should combine the systems for a "right" experiment. Also APS 6 (i.e. the reproducibility of the corresponding frequencies) is such a condition for the "right" combining. If an experiment contradicts APS 6 we do not say that APS 6 is defeated but that there are "disturbances" of the experiment. Only if we can remove the disturbances, we get an experiment of the fundamental domain. The same holds for APSZ 7.1.2.

APSZ 9 is a typical action norm, in this case also called selection norm. We should select for the fundamental domain only such experiments which do not contradict APSZ 9. Certainly we want to develop more comprehensive theories with greater fundamental domains, for instance in the way traced out in X and XI.

AP 1 and AR 1 are action norms. Before introducing them we have described possibilities to realize the demanded "direct mixtures".

AT 2 to AT 6 are also norms for "procedures" to measure trajectories. AT 7 is an empirical law (with idealizations; see the word "dense"), just so AT 8. Also AOb and APr are empirical laws.

All axioms AV ... in VI (except AVid, which is a pure idealization) are empirical laws. This claim seems to contradict the end of VI §7.1, where we declared the "microsystems" to form the fundamental domain of those physical systems for which the axioms AV 1.1, AV 1.2*s*, AVid, AV 3, AV 4 yield a usable theory. This sentence seems to claim that these axioms are norms for the fundamental domain; but it rather employs a language often used by physicists: The fundamental domain is that region of experience where the laws are valid. For others than "insiders" this sentence can indeed by misunderstood as if we make cyclic conclusions: By experiments of the fundamental domain we can test whether the empirical laws are good. If we have a contradiction, then we have facts that do not belong to the fundamental domain. Such a cyclic conclusion is not meant by the above sentence. This sentence shall only describe a *preliminary* situation in the development of a theory: We know that the stated empirical laws are not valid everywhere. But we do not yet know how to confine a suitable fundamental domain. We hope that this will be possible later. Our "present" ignorance shall be expressed by the above sentence: we hope that *it will be possible to define* a fundamental domain (the "microsystems") where the empirical laws are valid.

This has been done in IX, §2 and [2] VIII where we confined the fundamental domain to "atoms and molecules" and to processes of "not too high" energies (see above the analogous demand of not too high velocities in the case of Newton's mechanics).

The axioms AV 4a, AV 4b are empirical laws. Instead of characterizing the further axioms of IX resp. [2] now, we shall give remarks to this characterization in §3.

## §2.5 Empirical Laws and the Finiteness of Physics

Where are the empirical laws coming from? Is it possible to test empirical laws as we have described such tests in §2.2? Is it possible to deduce empirical laws from observational reports?

To discuss these questions we assume again $\mathscr{PT}$ in the form of a *simple* axiomatic basis and $P(\ldots)$ as split into the two parts $P_{\text{norm}}(\ldots)$ (the normative axioms) and $P_{\text{emp}}(\ldots)$ (the empirical axioms). Thus $P(\ldots)$ is equivalent to "$P_{\text{norm}}(\ldots)$ and $P_{\text{emp}}(\ldots)$". Let $\Sigma_{\text{norm}}$ be that species of structure where only $P_{\text{norm}}(\ldots)$ is required. This $\Sigma_{\text{norm}}$ is a "poorer" species of structure than $\Sigma$. The axioms of an observational report $(\text{—})_r(1)$ have according to (2.2.3) the form $a_i \in y_{v_i}$ and those of $(\text{—})_r(2)$ according to (2.2.5) the form

$$(a_{i_1}, a_{i_2}, \ldots, \alpha_\mu) \in \tilde{t}_\mu, \ldots, (a_{k_1}, a_{k_2}, \ldots, \alpha_\rho) \in \tilde{t}'_\rho .$$

Instead of a real observational report we can "invent" such a report, i.e. invent axioms of the same form. Let us write such invented axioms as $(\text{—})_h(1)$ and $(\text{—})_h(2)$, and call them "hypotheses" (of the first kind; in the sense of §4.1). For a hypothesis, in $(\text{—})_h(2)$ we also allow relations with $t_\mu$ instead of $\tilde{t}_\mu$. If such relations $(\text{—})_h(1)$ and $(\text{—})_h(2)$ are added to a $\mathscr{MT}$, we symbolize the hypothesis by $\mathscr{H}$ and denote the theory by $\mathscr{MTH}$.

If we form $\mathscr{MT}_\Sigma \mathscr{H}$ and $\mathscr{MT}_{\Sigma_{\text{norm}}} \mathscr{H}$ in this way, the letters $a_i$ in $(\text{—})_h$ are new constants of the theory.

Besides on $\Sigma_{\text{norm}}$ and $\Sigma$ we will also reflect on the following species of structure: Let $R_1$ and $R_2$ be two relations such that "$R_1$ and $R_2$" is a theorem in $\mathscr{MT}_\Sigma$. Then the relation "$P_{\text{norm}}$ and $R_1$ and $R_2$" is weaker than $P$. Let $\Sigma_{R_1}$ be the species of structure which differs from $\Sigma$ only in replacing $P$ by "$P_{\text{norm}}$ and $R_1$", and $\Sigma_{R_1 R_2}$ by "$P_{\text{norm}}$ and $R_1$ and $R_2$".

A relation $R_2$ is called "refutable" relative to $R_1$ if there is a hypothesis $\mathscr{H}$ such that $\mathscr{MT}_{\Sigma_{R_1}} \mathscr{H}$ is not contradictory while $\mathscr{MT}_{\Sigma_{R_1 R_2}} \mathscr{H}$ is contradictory. Equivalently, "not $R_2$" is a theorem in $\mathscr{MT}_{\Sigma_{R_1}} \mathscr{H}$.

An axiomatic basis is often developed by sharpening the axiomatic relation step by step, e.g. from $P_{\text{norm}}$ to "$P_{\text{norm}}$ and $R_1$" and from this to "$P_{\text{norm}}$ and $R_1$ and $R_2$", etc. Precisely this was the way we developed quantum mechanics (see e.g. VI). Therefore it is of high physical significance, whether a relation $R_2$ is refutable relative to $R_1$. If so, then $\mathscr{MT}_{\Sigma_{R_1 R_2}}$ genuinely confines the possible empirical facts relative to $\mathscr{MT}_{\Sigma_{R_1}}$, i.e. $R_2$ sharpens the descriptive power of the theory. In the preceding section we assumed that the action norms confine the fundamental domain. We can express this in the form that the action norms are refutable (if we take $R_1$ and $R_2$ as norms and especially $R_2$ as an action norm). Whereas for a good $\mathscr{PT}$ an empirical and refutable axiom $R_2$ must hold in the fundamental domain (i.e. for real observational reports), this need not be the case for an action norm; otherwise this norm would lose its meaning as "action" norm.

If there is a hypothesis such that $\mathscr{MT}_{\Sigma_{\text{norm}}} \mathscr{H}$ is not contradictory but $\mathscr{MT}_{\Sigma_{R_2}} \mathscr{H}$ is contradictory, we say that $R_2$ is refutable. For instance, we may ask whether $P_{\text{emp}}$ is refutable. The well known contention of Popper says that the physical content of a $\mathscr{PT}$ is that $P_{\text{emp}}$ can be refuted by empirical facts. But this is only one half the contents. There are many "empirical laws" in physical theories which are not refutable!

Before Popper there was another opinion: The physical laws can be deduced from experience. In our formal description: There is a hypothesis $\mathscr{H}$ such that $P_{\text{emp}}$ is a theorem in $\mathscr{M}\mathscr{T}_{\Sigma_{\text{norm}}}\mathscr{H}$, or more strongly: There is an observational report $\mathscr{A}$ for which $P_{\text{emp}}$ is a theorem in $\mathscr{M}\mathscr{T}_{\Sigma_{\text{norm}}}\mathscr{A}$.

That $R_2$ is a theorem in $\mathscr{M}\mathscr{T}_{\Sigma_{R_1}}\mathscr{H}$ is equivalent with: $\mathscr{M}\mathscr{T}_{\Sigma_{R_1(\text{not } R_2)}}\mathscr{H}$ is contradictory, i.e. (not $R_2$) is relative to $R_1$ refutable. Also if there is no observational report $\mathscr{A}$ which makes $R_2$ a theorem in $\mathscr{M}\mathscr{T}_{\Sigma_{R_1}}\mathscr{A}$, we say that $R_2$ is relative to $R_1$ empirically deducible if there is a hypothesis $\mathscr{H}$ which makes $\mathscr{M}\mathscr{T}_{\Sigma_{R_1(\text{not } R_2)}}\mathscr{H}$ contradictory.

Looking for empirically deducible laws in physical theories, one will find such laws only in very simple theories. Therefore the above contention of Popper was an important progress.

Not to make mistakes, we must emphasize that $R_2$ empirically deducible (relative to $R_1$) does not imply that $R_2$ is also refutable (relative to $R_1$). It may be that $R_2$ is empirically deducible (relative to $R_1$) and also refutable. But in physical theories we have many empirical laws $R_2$ which are neither refutable (relative to $R_1$) nor empirically deducible (relative to $R_1$). What is the physical content of such laws?

This can only be answered by returning to the problem of impresision and idealization.

There are only finitely many relations in $\mathscr{H}$ (resp. $\mathscr{A}$), and there are no quantifiers in $\mathscr{H}$. This we call the finiteness of physics. Why do we in spite of this finiteness use infinite sets in $\mathscr{M}\mathscr{T}_{\Sigma}$? All infinities in $\mathscr{M}\mathscr{T}_{\Sigma}$ are idealizations! Why do we use such idealizations? There are two purposes for idealizations: The mathematics of infinities is much simpler than a purely finite mathematics; and by extrapolations "to infinity" we may conceal our ignorance (e.g. how long or how precisely a special structure in nature can be revealed).

But if we have introduced infinities in $\mathscr{M}\mathscr{T}_{\Sigma}$, we must cancel this by admitting that for all practical purposes with imprecisions we can replace infinite sets by finite ones. How can we formulate this more rigorously? (See also [3] §9 and [40].)

For every infinite set $M$, we must introduce a uniform structure of physical imprecision $p$ so that $M_p$ ($M$ endowed with $p$) is a precompact and metrizable space. There are so many examples for such structures in the preceding chapters that we only want to add some remarks to one of them:

In III D5.1.1 and III D5.1.2 we had introduced the sets $\mathscr{K}$ and $\mathscr{L}$. These $\mathscr{K}$ and $\mathscr{L}$ are countable because the sets of preparation and registration procedures were postulated as countable. This "countable" is the "smallest" idealization of finite, if "one does not know" how many (but finitely many) ensembles and effects can be realized. Then in IV §2 we had discussed the uniform structures of physical imprecision and saw that the countablity of $\mathscr{K}$ and $\mathscr{L}$ makes these sets precompact and metrizable.

The uniform structures of physical imprecision on $\mathscr{K}$ (resp. $K$) and $\mathscr{L}$ (resp. $L$) can be described in the vector space description of IV §3 by the dual pair $\mathscr{B}$, $\mathscr{D}$ and the corresponding topologies $\sigma(\mathscr{B}, \mathscr{D})$ and $\sigma(\mathscr{D}, \mathscr{B})$. But we have not suceeded by physical arguments to display the subspace $\mathscr{D}$ of $\mathscr{B}'$ most generally; some arguments are given in [2] VII §8, [2] VIII §7 and [45].

It was even less possible to display a special metric. Only in X §3 we introduced a metric in the state spaces and correspondingly in the set of ensembles to describe finite imprecisions for the purpose of embedding. There we have seen that a metric $d$ introduced in $K$ is equivalent to a norm-precompact subset $\Lambda$ selected in $L \cap \mathscr{D}$ (resp. a metric in $L$ to a subset in $K$) such that

$$d(w_1, w_2) = \sup_{g \in \Lambda} |\mu(w_1 - w_2, g)|.$$

The selection of such a $\Lambda \subset L \cap \mathscr{D}$ (resp. of a subset of $K$) is physically connected with the question what effects (resp. ensembles) can "easily" be realized by devices, a problem discussed in XI §4 and XII §3. For physics, $\Lambda$ (resp. the corresponding subset of $K$) can be physically described as the set of all those effects (resp. ensembles) the realization of which is "not *too* difficult". The discussion in XI §4 and XII §3 showed that we are far from solving this problem. Therefore we have not used metrics in $K$ and $L$.

Let us return to the general case of a precompact $M_p$.

For every specific vicinity (imprecision set) $u$ there is a finite subset $\tilde{M}$ of $M$ such that for every $x \in M$ there is an $\tilde{x} \in \tilde{M}$ with $(x, \tilde{x}) \in u$. By means of $u$ we can smear the pictorial relations (or other deduced relations) as described in §1. In such a way, in $\mathscr{M}\mathscr{T}_{\Sigma_{R_1 R_2}}$ we can as a theorem deduce a relation $\tilde{R}_2$ which has the same form as $R_2$ with the only difference that every set $M$ is replaced by a finite set $\tilde{M}$ and every relation by a corresponding smeared relation. We say that $\tilde{R}_2$ is the finite kernel of $R_2$.

$\Sigma_{R_1 \tilde{R}_2}$ is therefore a weaker species of structure than $\Sigma_{R_1 R_2}$. The kernel $\tilde{R}_2$ of $R_2$ depends on the imprecision set $u \in p$. If $\forall u (u \in p$ and $\tilde{R}_2) \Leftrightarrow R_2$, then we say that $R_2$ is a normal idealization of its finite kernels. Otherwise one has to look for a *pure* idealization axiom $R_{2id}$ such that $\{\forall u (u \in p$ and $\tilde{R}_2)$ and $R_{2id}\} \Leftrightarrow R_2$.

It is very cumbersome to formulate $\tilde{R}_2$ (compared with $R_2$). Already in VI §1.1 we have remarked on the finite content of the axiom AV 1.1 as we introduced the relation $A_2$. Very shortly sketched, the finite kernel of AV 1.1 would look like this: First we must select a neighborhood $U$ of 0 for the $\sigma(\mathscr{B}', \mathscr{B})$ topology, then a finite subset $\tilde{L}$ of $L$ such that for every $g \in L$ there is a $g' \in \tilde{L}$ with $g - g' \in U$. With the imprecision belonging to $U$, we must smear the order relation $\leq$ into $\lesssim$ so that $g_1 \leq g_2$ implies $g'_1 \lesssim g'_2$ for $g_1 - g'_1 \in U$ and $g_2 - g'_2 \in U$. Similarly we have to smear the relation $\mu(w, g) = \alpha$ into $\mu(w, g) \sim \alpha$. The set $\tilde{L}$ specifies a vicinity of $\sigma(\mathscr{B}, \mathscr{D})$ corresponding to which we can introduce a finite subset $\tilde{K}$ of $K$ and in analogy to IV (5.1) define

$$\tilde{K}_0(g) = \{w \mid w \in \tilde{K} \text{ and } \mu(w, g) \sim 0\}.$$

From AV 1.1 we then can deduce the following finite kernel: For each pair $g_1, g_2 \in \tilde{L}$, there is a $g \in \tilde{L}$ with $g \gtrsim g_1, g_2$ and $\tilde{K}_0(g_1) \cap \tilde{K}_0(g_2) \sim \tilde{K}_0(g)$. Here the last sign $\sim$ means that the sets are equal within the imprecision in $\tilde{K}$. This example demonstrates why it is so enormously simpler to use infinite sets instead of only finite sets.

If we speak of $R_2$ as an empirical law, the word "empirical" indicates a property of the finite kernels $\tilde{R}_2$ (not of the additional idealizations contained in $R_2$).

What is the "physical content" of the finite kernels $\tilde{R}_2$ if $R_2$ is neither refutable nor empirically deducible? The content is that $\tilde{R}_2$ is empirically deducible (as we require for physically meaningful axioms!). (Since $\tilde{R}_2$ is weaker than $R_2$, $R_2$ not refutable also implies $\tilde{R}_2$ not refutable.)

It is easy to see that the finite kernel of AV 1.1 defined above is empirically deducible since one should only adopt such a hypothesis $\mathscr{H}$ that all effects $g \in \tilde{L}$ and all ensembles $w \in \tilde{K}$ are present in $\mathscr{H}$ and the frequencies (corresponding to $\mu(w, g)$) are such that the finite kernel of AV 1.1 can be "read" from $\mathscr{H}$ (the finite kernel is only a statement about finitely many elements!).

The known cases of empirical laws $R_2$ which are neither refutable nor empirically deducible are of the same kind as AV 1.1. Such an $R_2$ has the form

$$R_2\colon\quad \forall z\, \exists w [A(z, w)],$$

where $A$ is a relation. The negation of $R_2$ then is

$$\text{not } R_2\colon\quad \exists z\, \forall w [\text{not } A(z, w)].$$

If $z$ and $w$ are elements of infinite sets, neither $R_2$ nor [not $R_2$] can be refutable by finitely many relations in $\mathscr{H}$. But the negation of the finite kernel of $R_2$ may be refutable.

That $\tilde{R}_2$ is empirically deducible does not mean that we ever have a real observational report from which $\tilde{R}_2$ can be deduced. On the contrary, in almost all physical theories we have in $\tilde{R}_2$ finite sets of so many elements that there are not enough human beings and not enough time for experiments to get an observational report sufficient to deduce $\tilde{R}_2$. We can only say that we are "on the way" to an empirical deduction of $\tilde{R}_2$. As long as "on this way" we do not find an indication that we have made a $z$ for which there are fundamental difficulties in nature to make a $w$ with $A(z, w)$, we adopt $\tilde{R}_2$ (and its idealization $R_2$) as axiom. But we can never exclude that "on this way" we could detect a limitation of $\tilde{R}_2$ and thus a new physical law (see the general investigations in §4).

The best form of $P(\ldots)$ would allow us to separate norms, finite kernels and pure idealizations. We can now define more precisely what we mean by a pure idealization: $R_2$ is a pure idealization if $\tilde{R}_2$ is a theorem in $\mathscr{M}\mathscr{T}_\Sigma$ (for every imprecision set).

We have not reached this "best form" of $P(\ldots)$ for our axiomatic basis of quantum mechanics since we have not separated finite kernels from idealizations. We will try to give some remarks for such a separation below. Since it is "usual" in the axioms to mix physical contents with idealizations, one has often made mistakes by claiming something as a physical structure what indeed was only an idealization. For instance, it sometimes has been claimed that there are infinite structures in nature. This is amusing since one has introduced infinities in $\mathscr{M}\mathscr{T}_\Sigma$ only to conceal our ignorance. One has deceived oneself by transforming ignorance into knowledge. Honestly we must admit that either physics demonstrates the finiteness of a structure in nature, or physics cannot say how comprehensive this structure in nature really is.

Let us now try to characterize the empirical axioms in III to VIII. Not only AV 1.1 is a law with empirically deducible finite kernels. We have introduced many similar laws: AOb and APr, AV 1.2, AV 2 or AV 1.2*s* (APr on page 145 has a mistake; $\sigma(\mathscr{B}', \mathscr{B})$ must be replaced by $\sigma(\mathscr{B}, \mathscr{D})$).

AV 3 is refutable if it is independent of the foregoing axioms; this we have not proven.

Very interesting for the finiteness of physics is the quantization law AV 4 or AV 4*s*. This axiom is the only to contain the word "finite", namely "finite dimension".

We interpreted the introduction of infinities in the mathematical part $\mathscr{MT}$ of a $\mathscr{PT}$ as a clever method to hide unsolved physical problems. AV 4 is therefore nothing but the claim that we have solved, at least in one respect, a physical problem. We have not solved all problems hidden behind the infinity of the set $K$.

That we have indeed solved a physical problem can be seen best by comparing AV 4 with the second part of $AV_{kl}$ (VI §7.1). This says that each exposed face of $K$ is infinite dimensional. Thus we cannot find an end to demixing ensembles: Every ensemble can be demixed into two orthogonal (IV D5.1) ensembles with similar weights if we only make the precision high enough. AV 4 on the contrary says, that demixing ends in the following sense.

For every given imprecision, by experimental demixing we can find approximate faces of such finite dimensions that these dimensions are *not altered by increasing precision*, i.e. the demixing of ensembles of such faces ends, since for higher precision at most such mixing components can be found which can be neglected because of their "very small" weights.

In this sense, AV 4 is an empirically deducible axiom.

AV 4a and AV 4b are refutable axioms.

We have tried to give a short characterization of the axioms we used for the development of quantum mechanics. Obviously this characterization was not rigorous. There remains the task to elaborate rigorously all the indications we have given. Such an elaboration would be mainly interesting for the philosophy of science.

## §3 Intertheory Relations

Physics does not consist of *one* theory. On the contrary there are many theories, even though one finds the opinion that there should be one theory "behind" all these theories. But we do not in the least know this "one" theory. The belief in a single but not discoverable theory is fruitless. We rather should examine how the various theories are connected. In the philosophy of science this problem is well known as the problem of intertheory relations. (As addition to this section see [3] §8 and [48].)

A false view of intertheory relations has been the source of many false opinions concerning the "truth" of a physical theory. Thus the misconception has arisen that no theory is really "true" but that during the development of physics a later theory becoming valid (i.e. a paradigm) makes an older theory "untrue" (i.e. rejects it as paradigm by a revolutionary act).

Our aim in studying intertheory relations is to see better the relations among the various forms of quantum mechanics given in the previous chapters.

## §3.1 Restrictions

We start with a theory $\mathscr{PT}_1$ and wish to construct a theory $\mathscr{PT}$ which makes less restrictive assertions about possible observational reports. The theory $\mathscr{PT}_1$ is assumed given in the form of an axiomatic basis of the first degree (but not necessarily simple). It is easy to transport all considerations to an axiomatic basis of higher degree or to a general form of a theory, only these transported forms are not so lucid.

Let the species of structure $\Sigma_1$ of the mathematical part $\mathscr{MT}_{\Sigma_1}$ of $\mathscr{PT}_1$ be characterized by the base sets $x_1, \ldots, x_n$ (which are the pictorial sets) and the structural term $s$ (the components $s_\mu$ of $s$ being the pictorial relations). Further consider in $\mathscr{MT}_{\Sigma_1}$ intrinsic terms $E_1, \ldots, E_r$ of the following type:

(α) $E_\nu$ is a subset of a product set $x_{\nu_1} \times x_{\nu_2} \times \ldots \times x_{\nu_p}$, or

(β) $E_\nu$ is the range of a mapping $f$ from a set $F$ of type (α) into an echelon set over $x_1, \ldots, x_n$.

As the mathematical part of the new theory $\mathscr{PT}$ we simply take the same $\mathscr{MT}_{\Sigma_1}$ as for $\mathscr{PT}_1$. (Of course, this need not render $\mathscr{PT}$ an axiomatic basis.) As pictorial sets for $\mathscr{PT}$ we single out the sets $E_1, \ldots, E_r$. To obtain pictorial relations, we seek intrinsic terms $u_\mu(x_1, \ldots, x_n, s)$, $(\mu = 1, \ldots, m)$ which have the following typification over the pictorial sets:

$$u_\mu(x_1, \ldots, x_n, s) \subset E_{\mu_1} \times E_{\mu_2} \times \ldots \times \mathbf{R} \tag{3.1.1}$$

($\mathbf{R}$ may be absent in some of the $u_\mu$). These subsets $u_\mu$ represent the new pictorial relations.

We must add that uniform structures of physical imprecision should be transported canonically to the sets $E_\nu$, whereby the mapping $f$ in (β) must become uniformly continuous.

The last step in defining $\mathscr{PT}$ consists in transforming the correspondence rules of $\mathscr{PT}_1$ into the new ones of $\mathscr{PT}$, without changing (in a first step) the fundamental domain $\mathscr{G}_1$ of $\mathscr{PT}_1$. We start from a hypothesis $\mathscr{H}_1$ in $\mathscr{PT}_1$ as defined in §2.5. Thus $\mathscr{H}_1$ has the form

$$a_1 \in x_\nu, a_2 \in x_{\nu_2}, \ldots \tag{3.1.2}$$

and

$$(a_{i_1}, a_{i_2}, \ldots) \in s_{\mu_1}, \ldots. \tag{3.1.3}$$

We must define a Hypothesis $\mathscr{H}$ in $\mathscr{PT}$ corresponding to $\mathscr{H}_1$ in $\mathscr{PT}_1$, i.e. we must deduce a theory $\mathscr{MT}_{\Sigma_1}\mathscr{H}$ from $\mathscr{MT}_{\Sigma_1}\mathscr{H}_1$.

Let $E_\nu$ be a term of type (α). If

$$(a_{i_1}, \ldots, a_{i_p}) \in E_\nu \tag{3.1.4}$$

is a theorem in $\mathscr{MT}_{\Sigma_1}\mathscr{H}_1$ (i.e. if $\mathscr{H}_1 \Rightarrow$(3.1.4) is a theorem in $\mathscr{MT}_{\Sigma_1}$), we introduce a new sign $b_1$ as an abbreviation for the $p$-tuple $(a_{i_1}, \ldots, a_{i_p})$:

$$(a_{i_1}, \ldots, a_{i_p}) = b_1. \tag{3.1.5}$$

While (3.1.4) and (3.1.5) imply

$$b_1 \in E_\nu, \tag{3.1.6}$$

$b_1$ is called a new pictorial element of the new pictorial term $E_\nu$.

Now let $E_\nu$ be of type ($\beta$), say $E_\nu = f(F)$ and $F \subset x_{\nu_1} \times x_{\nu_2} \times \ldots \times x_{\nu_p}$. We are looking for sets of $p$-tuples

$$(a_{i_1}, \ldots, a_{i_p}) \in F \tag{3.1.7}$$

which are mapped under $f$ onto the same element $b_k$:

$$f(a_{i_1}, \ldots, a_{i_p}) = f(a_{j_1}, \ldots, a_{j_p}) = \ldots = b_k. \tag{3.1.8}$$

In this event we introduce the sign $b_k$ as a new pictorial element of the new pictorial term $E_\nu$:

$$b_k \in E_\nu. \tag{3.1.9}$$

The new hypothesis $\mathscr{H}$ will consist of statements of the form (3.1.6), (3.1.9) and of all relations

$$(b_{i_1}, b_{i_2}, \ldots) \in u_\mu, \ldots; (b_{k_1}, b_{k_2}, \ldots) \notin u_\lambda, \ldots \tag{3.1.10}$$

which are theorems in $\mathscr{M}\mathscr{T}_{\Sigma_1}\mathscr{H}_1$. This completes the definition of the new correspondence rules of $\mathscr{P}\mathscr{T}$.

In the case of imprecise correspondence rules where (3.1.3) takes the form

$$(a_{i_1}, a_{i_2}, \ldots) \in \tilde{s}_{\mu_1}; \ldots, \tag{3.1.11}$$

one weakens (3.1.6), (3.1.8), (3.1.9) and (3.1.10).

In the case ($\alpha$) we replace (3.1.3) by neighborhoods $\tilde{E}_\nu$ of $E_\nu$ in $x_{\nu_1} \times \ldots \times x_{\nu_p}$ (corresponding to the smearing of the $s_\mu$) and look for a theorem

$$b_\nu \in \tilde{E}_\nu. \tag{3.1.12}$$

In the case ($\beta$) we replace $E_\nu = f(F)$ by a neighborhood $\tilde{E}_\nu$ of $E_\nu$ in the corresponding echelon set.

Instead of (3.1.7) we only seek relations

$$(a_{i_1}, \ldots, a_{i_p}) \in \tilde{F} \tag{3.1.13}$$

with $\tilde{F}$ as a neighborhood of $F$. Then we take such a decomposition $\{\sigma_k\}$ of $\tilde{E}_\nu$ in disjoint sets that the elements of one $\sigma_k$ are vicinal in the sense of the used imprecision set. We take the same sign $b_k$ for those $p$-tuples which obey

$$f(a_{i_1}, \ldots, a_{i_p}) \in \sigma_k, f(a_{j_1}, \ldots, a_{j_p}) \in \sigma_k, \ldots \tag{3.1.14}$$

(instead of (3.1.8)) and get (3.1.12). With $\tilde{u}_\mu$ as neighborhoods of $u_\mu$ (i.e. as smeared relations) one looks for theorems

$$(b_{i_1}, b_{i_2}, \ldots) \in \tilde{u}_\mu, \ldots; (b_{k_1}, b_{k_2}, \ldots) \in \tilde{u}'_\lambda, \ldots. \tag{3.1.15}$$

The thus defined theory $\mathscr{P}\mathscr{T}$ will be called a *restriction* of $\mathscr{P}\mathscr{T}_1$, in symbols:

$$\mathscr{P}\mathscr{T}_1 \to \mathscr{P}\mathscr{T}. \tag{3.1.16}$$

If $\mathscr{M}\mathscr{T}_{\Sigma_1}\mathscr{H}_1$ is not contradictory, then also not the theory $\mathscr{M}\mathscr{T}_{\Sigma_1}\mathscr{H}_1\mathscr{H}$ which grows out of $\mathscr{M}\mathscr{T}_{\Sigma_1}\mathscr{H}_1$ by introducing the new constants $b_k$ via (3.1.5), (3.1.8) and by adding (3.1.6), (3.1.9), (3.1.10) resp. (3.1.12), (3.1.15). Since $\mathscr{M}\mathscr{T}_{\Sigma_1}\mathscr{H}$ is weaker than $\mathscr{M}\mathscr{T}_{\Sigma_1}\mathscr{H}_1\mathscr{H}$, a not contradictory $\mathscr{M}\mathscr{T}_{\Sigma_1}\mathscr{H}_1$ implies $\mathscr{M}\mathscr{T}_{\Sigma_1}\mathscr{H}$ not contradictory. Therefore we call $\mathscr{P}\mathscr{T}_1$ more comprehensive than $\mathscr{P}\mathscr{T}$.

As fundamental domain of $\mathscr{PT}$ we have initially used the $\mathscr{G}_1$ of $\mathscr{PT}_1$, just labeling the facts by other signs. The new labels collect some older designations (by (3.1.5) or (3.1.8) resp. (3.1.14)), and one then "forgets" these older designations. This forgetting can restrict also the fundamental domain. For instance all those facts must be eliminated from the fundamental domain the signs of which do not belong to the *subsets* $E_\nu$ (resp. $\tilde{E}_\nu$) according to ($\alpha$) or $F$ (resp. $\tilde{F}$) according to ($\beta$). In this sense the fundamental domain $\mathscr{G}$ of $\mathscr{PT}$ can be smaller than $\mathscr{G}_1$!

It should be emphasized that the definition of the new pictorial sets $E_\nu$ and the new pictorial relations $u_\mu$ is not "imprecise". Imprecise are only the observational reports.

Widespread is the following simple case of restrictions: The $E_\nu$ are some of the $x_\nu$ and the $u_\mu$ are some of the $s_\mu$. Then the observational report of $\mathscr{PT}$ is simply a part of the observational report of $\mathscr{PT}_1$. We say that we forget every $x_\nu$ which is no $E_\nu$ and every $s_\mu$ which is no $u_\mu$. Such a restriction shall be called a standard restriction, in symbols $\mathscr{PT}_1 \xrightarrow{s} \mathscr{PT}$.

Sometimes it will be necessary to generalize the concept of restriction in the following way. We consider a special hypothesis $\mathscr{H}_s$ in $\mathscr{MT}_{\Sigma_1}$. We now admit also new pictorial sets $E_i$ and pictorial relations $u_\mu$ which are only definable in the stronger theory $\mathscr{MT}_{\Sigma_1}\mathscr{H}_s$. If the restricted theory $\mathscr{PT}$ obtained in this way is applied, one always must assume a realization of $\mathscr{H}_s$. This realization presents such facts in the fundamental domain $\mathscr{G}_1$ of $\mathscr{PT}_1$ which furnish $\mathscr{MT}_{\Sigma_1}$ with an observational report $\mathscr{A}$ "equal" to $\mathscr{H}_s$ (when suitable signs in $\mathscr{G}_1$ are used; see §1). In the case thus described we call $\mathscr{PT}$ a restriction of $\mathscr{PT}_1$ relative to $\mathscr{H}_s$, in symbols: $\mathscr{PT}_1 \xrightarrow{\mathscr{H}_s} \mathscr{PT}$.

Restrictions relative to a hypothesis $\mathscr{H}_s$ are widespread in physics. Frequently $\mathscr{H}_s$ describes a situation in a laboratory relative to which the theory $\mathscr{PT}_1$ will be restricted. A fixed $\mathscr{H}_s$ can be realized by a variety of possible facts in $\mathscr{G}_1$, for example by many different situations "of the same kind" in different laboratories. As an example let us take for $\mathscr{PT}_1$ special relativity and for $\mathscr{H}_s$ the specification of an inertial frame. Then we could restrict the set of all inertial frames to the subset of those with a velocity less than $10^{-6}$ times the velocity of light relative to the frame $\mathscr{H}_s$ (the resulting restricted theory, however, is not yet Galilean relativity; see §3.2).

## §3.2 Embedding

Another important procedure for obtaining from $\mathscr{PT}$ a less comprehensive theory $\mathscr{PT}_2$ is an "embedding". The mathematical part $\mathscr{MT}_\Sigma$ of $\mathscr{PT}$ is not supposed to be an axiomatic basis, but rather to be given as $\mathscr{PT}$ in §3.1, i.e. by certain intrinsic terms $E_1, \ldots, E_r$ for pictorial sets and a term $U=(u_1, u_2 \ldots)$ representing the pictorial relations and thus satisfying

$$u_\mu \subset E_{i_1} \times E_{i_1} \times \ldots \times \mathbf{R} = T'_\mu(E_1, \ldots, E_r; \mathbf{R}). \tag{3.2.1}$$

Here $\mathbf{R}$ may be absent from some of the relations.

Now we consider a second species of structure $\Sigma_2$, with principal base sets $y_1, \ldots, y_p$ ($p \leq r$), typification $t \in T(y_1, \ldots, y_p; \mathbf{R})$ and axiom $P_2(y_1, \ldots, y_p; t)$. We suppose that $T$ has the precial form

$$T = \mathscr{P}T_1 \times \mathscr{P}T_2 \times \dots$$

with

$$T_\mu(y_1, \dots, y_p; \mathbf{R}) = y_{\mu_1} \times y_{\mu_2} \times \dots \times \mathbf{R}. \tag{3.2.2}$$

Hence

$$t = (t_1, t_2, \dots) \quad \text{and} \quad t_\mu \subset y_{\mu_1} \times y_{\mu_2} \times \dots \times \mathbf{R}. \tag{3.2.3}$$

Let $i \to v_i$ be a mapping of indices. Then we assume that the following relation is a theorem in $\mathscr{M}\mathscr{T}_\Sigma$:

$$\begin{aligned}&(\exists y_1)(\exists y_2)\dots(\exists t_1)(\exists t_2)\dots(\exists f_1)(\exists f_2)\dots[t_\mu \subset T_\mu(y_1, \dots, y_p; \mathbf{R}) \text{ for all } \mu \\ &\quad \text{and } P_2(y_1, \dots, y_p, t) \text{ and } E_i \xrightarrow{f_i} y_{v_i} \text{ for all } i \\ &\quad \text{and } \mu_\mu = [\langle f_1, \dots, f_r, 1\rangle^{T'_\mu}]^{-1} t_\mu].\end{aligned} \tag{3.2.4}$$

Here $[\langle f_1, \dots\rangle^{T'_\mu}]^{-1} t_\mu$ is the set of all $z$ with $\langle f_1, \dots\rangle^{T'_\mu} z \in t_\mu$. It may be that $\langle f_1, \dots\rangle^{T'_\mu} u_\mu \neq t_\mu$ since the range of $\langle f_1, \dots\rangle^{T'_\mu}$ is not necessarily the total set (3.2.2). $[\langle f_1, \dots\rangle^{T'_\mu}]^{-1} t_\mu = u_\mu$ implies $\langle f_1, \dots\rangle^{T'_\mu} u_\mu \subset t_\mu$ and $z \notin u_\mu \Rightarrow f(z) \notin t_\mu$.

In the following, (3.2.4) will be called the "embedding theorem". It is easy to transport this theorem to a representation of $\Sigma_2$ in another species of structure.

The construction of $\mathscr{P}\mathscr{T}_2$ is performed as follows: $\mathscr{M}\mathscr{T}_{\Sigma_2}$ is the mathematical part of $\mathscr{P}\mathscr{T}_2$. As fundamental domain $\mathscr{G}_2$ we take $\mathscr{G}$ endowed with the *same* signs as in $\mathscr{P}\mathscr{T}$. To obtain correspondence rules of $\mathscr{P}\mathscr{T}_2$ we replace every relation of the form $a_k \in E_i$ in the $(\text{—})_r(1)$ of $\mathscr{P}\mathscr{T}$ by

$$a_k \in y_{v_i} \tag{3.2.5}$$

and every relation of the form

$$\begin{aligned}&(a_{i_1}, a_{i_2}, \dots) \in u_\mu, \quad \text{resp. } \notin u_\mu \text{ in } (\text{—})_r(2) \\ \text{by} \quad &(a_{i_1}, a_{i_2}, \dots) \in t_\mu, \quad \text{resp. } \notin t_\mu.\end{aligned} \tag{3.2.6}$$

This completes the definition of $\mathscr{P}\mathscr{T}_2$. All this together is called an *embedding* of $\mathscr{P}\mathscr{T}$ into $\mathscr{P}\mathscr{T}_2$, in symbols $\mathscr{P}\mathscr{T} \rightsquigarrow \mathscr{P}\mathscr{T}_2$. Obviously $\mathscr{M}\mathscr{T}_{\Sigma_2}$ is an axiomatic basis of the first degree of $\mathscr{P}\mathscr{T}_2$. We speak of an embedding $\mathscr{P}\mathscr{T} \rightsquigarrow \mathscr{P}\mathscr{T}_2$ also then, when $\Sigma_2$ is represented in the mathematical part of $\mathscr{P}\mathscr{T}_2$ (see above). Then the $f_i$ map the $E_i$ into the pictorial terms and the $t_\mu$ into the pictorial relations of $\mathscr{P}\mathscr{T}_2$.

For $\mathscr{P}\mathscr{T}_2$ we have prescribed that $\mathscr{G}_2$ equals $\mathscr{G}$ with the same signs. This signifies that as hypothesis $\mathscr{H}_2$ in $\mathscr{P}\mathscr{T}_2$ we cannot take any relation of the form (3.2.5), (3.2.6). We must examine whether there is a corresponding hypothesis $\mathscr{H}$ in $\mathscr{P}\mathscr{T}$ which yields $\mathscr{H}_2$ in the designated way. This is only another formulation for a restriction of the fundamental domain of $\mathscr{P}\mathscr{T}_2$: We must not extend the fundamental domain $\mathscr{G}_2$ beyond $\mathscr{G}$. The limitation of this domain $\mathscr{G}_2$ is thus *not* completely defined *intrinsically*.

If we take care of this limitation of the fundamental domain, $\mathscr{P}\mathscr{T}$ is more comprehensive than $\mathscr{P}\mathscr{T}_2$ in the following sense. Let $\mathscr{H}_2$ be a hypothesis in $\mathscr{P}\mathscr{T}_2$ when $\mathscr{H}$ is a corresponding hypothesis in $\mathscr{P}\mathscr{T}$. If $\mathscr{M}\mathscr{T}_{\Sigma_2}\mathscr{H}_2$ is contradictory, then [not $\mathscr{H}_2$] is a theorem in $\mathscr{M}\mathscr{T}_{\Sigma_2}$ and therefore

$$\begin{aligned}&(\forall y_1)\dots(\forall t_1)\dots\{t_\mu \subset T_\mu(y_1, \dots) \text{ for all } \mu \\ &\quad \text{and } P_2(y_1, \dots) \Rightarrow (\forall a_1)(\forall a_2)\dots[\text{not } \mathscr{H}_2(y_1, \dots; a_1, a_2, \dots)]\}\end{aligned} \tag{3.2.7}$$

a theorem in $\mathcal{MT}_\Sigma$. On the other hand, in $\mathcal{MT}_\Sigma \mathcal{H}$ the embedding theorem implies the theorem:

$$(\exists y_1)\ldots(\exists t_1)\ldots(\exists f_1)\ldots[t_\mu \subset T_\mu(\ldots) \text{ for all } \mu \text{ and } P_2(\ldots)$$
$$\text{and } E_i \xrightarrow{f_i} y_{v_i} \text{ for all } i \text{ and } \mathcal{H}_2(y_1, \ldots; f(a_1), \ldots)]. \tag{3.2.8}$$

Since this contradicts (3.2.7), also $\mathcal{MT}_\Sigma \mathcal{H}$ is contradictory.

In physics one often encounters imprecise embeddings. We know that (3.2.4) can only be an idealization. If $\mathcal{PT}$ and $\mathcal{PT}_2$ do not harmonize relative to the idealizations, we expect that an embedding theorem of the form (3.2.4) is impossible but that an imprecise embedding theorem can be proven.

Starting from smeared $\tilde{t}_\mu, \tilde{t}'_\mu$ (see §1), let us try to prove the following weaker version of the embedding theorem:

$$(\exists y_1)\ldots(\exists t_1)\ldots(\exists f_1)\ldots[t_\mu \subset T_\mu(\ldots) \text{ for all } \mu$$
$$\text{and } P_2(\ldots) \text{ and } E_i \xrightarrow{f_i} y_{v_i} \text{ for all } i \tag{3.2.9}$$
$$\text{and } u_\mu \subset [\langle f_1, \ldots\rangle^{T'_\mu}]^{-1}\tilde{t}_\mu \text{ and } u'_\mu \subset [\langle f_1, \ldots\rangle^{T'_\mu}]^{-1}\tilde{t}'_\mu].$$

(Here $u'_\mu$ is the complement of $u_\mu$; for the definition of $[\langle f_1, \ldots\rangle^{T'_\mu}]^{-1}\tilde{t}_\mu$ see after (3.2.4).)

In (3.2.9) we must add that the $f_i$ must be uniformly continuous relative to the uniform structures of physical imprecisions. Then

$$\tilde{u}_\mu = [\langle f_1, \ldots\rangle^{T'_\mu}]^{-1}\tilde{t}_\mu, \tilde{u}'_\mu = [\langle f_1, \ldots\rangle^{T'_\mu}]^{-1}\tilde{t}'_\mu \tag{3.2.10}$$

is a possible smearing of $u_\mu$ in $\mathcal{PT}$.

The question concerning (3.2.9) is: If (3.2.4) (i.e. (3.2.9) for $\tilde{t}_\mu = t_\mu$) is not a theorem we have to ask for the smallest possible imprecisions for which (3.2.9) is a theorem. Such a problem has been discussed in X §§3.1 and 3.2.

Similarly as above we can prove that $\mathcal{PT}$ is more comprehensive than $\mathcal{PT}_2$. For this proof we must choose $\tilde{u}_\mu$ and $\tilde{u}'_\mu$ (defined in (3.2.10)) as the smeared pictorial relations of $\mathcal{PT}$. For such an imprecise embedding we use the same symbol $\mathcal{PT} \rightsquigarrow \mathcal{PT}_2$.

We shall discuss further only the case that (3.2.4) has $p=r$ and $i \to \mu_i$ is a bijection. Then we can choose such indices that $E_i \xrightarrow{f_i} y_i$.

If in addition the $f_i$ are bijections, then $U$ in $\mathcal{MT}_\Sigma$ is a structure of species $\Sigma_2$ over the pictorial sets $E_1, \ldots, E_r$, i.e. $P_2(E_1, \ldots; U)$ is a theorem in $\mathcal{MT}_\Sigma$. But also conversely: If $U$ is a structure of species $\Sigma_2$ in $\mathcal{MT}_\Sigma$ over $E_1, \ldots, E_r$, then the theorem (3.2.4) (with $p=r$ and $f_i$ bijections) follows. This last special case of embedding shall be called a standard embedding and be symbolized by $\overset{s}{\rightsquigarrow}$. We shall use this notation also for (3.2.9) with $p=r$ and bijections $f_i$. Then we say that $\Sigma_2$ is an approximation to $U$ (see also below).

But what if the $f_i$ are not bijective? If an $f_i$ is not injective, we can divide the set $E_i$ into classes of those elements which have the same image. Then a preceding restriction $\mathcal{PT}_1 \to \mathcal{PT}$ in (3.3.1) can be attained by choosing as pictorial sets in $\mathcal{MT}_\Sigma$ the set of classes in the $E_i$ (instead of these $E_i$ themselves). Therefore we shall only discuss the possibility that the $f_i$ are injective but not necessarily surjective.

Then $P_2(E_1, \ldots, U)$ need not to be a theorem in $\mathcal{MT}_\Sigma$! How can this happen although $\mathcal{PT}$ is a stronger theory than $\mathcal{PT}_2$? This is possible since only those pictures of facts are admitted which are not elements of $y_i \setminus f_i(E_i)$. This opens a possibility very decisive for the development of physics: $P_2(\ldots)$ only in the following sense contains less restricting axioms than $\mathcal{MT}_\Sigma$ does: The refutable axioms (in the sense of §2.5) contained in $P_2(\ldots)$ are also contained as theorems in $\mathcal{MT}_\Sigma$. For some of the empirically (in the sense of §2.5) deducible axioms $R(\ldots)$ contained in $P_2(\ldots)$, however, [not $R(E_1, \ldots)$] is a theorem in $\mathcal{MT}_\Sigma$ ([not $R(\ldots)$] is then refutable in $\mathcal{PT}$!). This means that the real possibilities in nature are restricted as compared with the possibilities postulated by some of the axioms in $P_2(\ldots)$. Precisely this is the meaning of the embedding introduced in X §2.2: The possibilities for preparing and registering postulated in $\mathcal{PT}_{q\,\exp}$ are unrealistic.

The restriction of the fundamental domain $\mathcal{G}_2$ (see also above) can be described by the subsets $f_i(E_i)$. These $f_i(E_i)$ are no intrinsic terms. Therefore, in order to use the less comprehensive theory $\mathcal{PT}_2$ one will look for additional normative axioms (to make $P_2(\ldots)$ stronger, equivalent to the postulate that the pictures are elements of $f_i(E_i)$).

As an example let us take the theory $\mathcal{PT}$ obtained at the end of §3.1 by restricting special relativity. This $\mathcal{PT}$ can be embedded imprecisely into the theory $\mathcal{PT}_2$ of Galilean relativity. Not all inertial frames will appear as embedding pictures, only those with velocities smaller than $10^{-6}$ times the velocity of light relative to a given inertial frame (symbolized by $\mathcal{H}_s$). This extra condition restricting the fundamental domain $\mathcal{G}_2$ remained unknown until the advent of special relativity. This is typical for the historical development of theories. One starts with a theory $\mathcal{PT}_2$ in which a too large fundamental domain is assumed. Then one learns by experience that only under certain restrictions of the fundamental domain the theory succeeds. Later on, $\mathcal{PT}_2$ is understood by means of a restriction $\mathcal{PT}_1 \rightarrow \mathcal{PT}$ followed by an embedding $\mathcal{PT} \rightsquigarrow \mathcal{PT}_2$. Then the embedding procedure yields a systematic derivation of the constraints formerly encountered only as empirical limitations (unless completely unknown).

We will return to the significance of such added normative contraints in §4.

Closing this section we must warn of a mistake. The embedding theorem alone does not prove that the theory $\mathcal{PT}$ is more comprehensive. In reality there can be the following situation between two theories $\mathcal{PT}$ and $\mathcal{PT}_3$:

$$\mathcal{PT} \rightsquigarrow \mathcal{PT}_2 \leftarrow \mathcal{PT}_3.$$

Then $\mathcal{PT}$ is not more comprehensive than $\mathcal{PT}_3$ because $\mathcal{PT}_2$ is only a restricted part of $\mathcal{PT}_3$! In our opinion, $\mathcal{PT}_{q\,\exp}$ contains unrealistic parts; but the embedding of $\mathcal{PT}_m$ can yield a theory $\mathcal{PT}'$ which contains less than necessary if the state space of $\mathcal{PT}_m$ is too "contracted" (X §2.6).

## §3.3 Networks of Theories and the Various Forms of Quantum Mechanics

The concept of theory-nets was introduced in [62], by emphasizing that many "physical theories" (in the intuitive sense) such as classical mechanics should be considered as a network of $\mathcal{PT}$'s. Let us show some aspects of this network for quantum mechanics (in the intuitive sense regarded as one theory).

Combining the two processes described in §3.1 and 3.2, we arrive at the following connection between three theories:

$$\mathscr{P}\mathscr{T}_1 \to \mathscr{P}\mathscr{T} \rightsquigarrow \mathscr{P}\mathscr{T}_2. \tag{3.3.1}$$

This represents a transition from the axiomatic basis $\mathscr{P}\mathscr{T}_1$ to the axiomatic basis $\mathscr{P}\mathscr{T}_2$, where $\mathscr{P}\mathscr{T}_2$ is less comprehensive than $\mathscr{P}\mathscr{T}_1$. In §3.1, 3.2 we have given an example where $\mathscr{P}\mathscr{T}_1$ as special relativity and $\mathscr{P}\mathscr{T}_2$ as Galilean relativity make

$$\mathscr{P}\mathscr{T}_1 \xrightarrow{\mathscr{H}_s} \mathscr{P}\mathscr{T} \rightsquigarrow \mathscr{P}\mathscr{T}_2.$$

Since in physics we often encounter the special connection (3.3.1) where the theory $\mathscr{P}\mathscr{T}$ only provides a rigorous explanation of the relation between $\mathscr{P}\mathscr{T}_1$ and $\mathscr{P}\mathscr{T}_2$, let us abbreviate (3.3.1) by

$$\mathscr{P}\mathscr{T}_1 \longrightarrow\!\!\!\rightsquigarrow \mathscr{P}\mathscr{T}_2. \tag{3.3.2}$$

Here the fundamental domain $\mathscr{G}_2$ of $\mathscr{P}\mathscr{T}_2$ is determined by a restriction from $\mathscr{G}_1$ to $\mathscr{G}$ (as described in §3.1) and a following embedding where $\mathscr{G}_2$ is the same domain as $\mathscr{G}$. In this way, $\mathscr{G}_2$ is defined by external conditions, i.e. by $\mathscr{G}$ (see §3.2). When we try to change $\mathscr{P}\mathscr{T}_2$ in such a way that $\mathscr{G}_2$ can be defined intrinsically by normative axioms, the embedding becomes a standard embedding:

$$\mathscr{P}\mathscr{T}_1 \overset{s}{\longrightarrow\!\!\!\rightsquigarrow} \mathscr{P}\mathscr{T}_2. \tag{3.3.3}$$

Here also the constraints for $\mathscr{G}_2$ are defined intrinsically. In (3.3.3) we have the most important relation between two theories.

We claim that all physical theories can be ordered in a network of relations of the form (3.3.2) (which should be changed into (3.3.3)). Thus by a network we mean a set of theories $\mathscr{P}\mathscr{T}_i$ and relations of the form (3.3.2) such that one can reach every theory $\mathscr{P}\mathscr{T}_k$ from any other $\mathscr{P}\mathscr{T}_i$ by going along arrows $\longrightarrow\!\!\!\rightsquigarrow$ (but possibly disregarding their directions). If one can reach $\mathscr{P}\mathscr{T}_3$ from $\mathscr{P}\mathscr{T}_i$ by going only in the direction of the arrows, we say that $\mathscr{P}\mathscr{T}_3$ is less comprehensive than $\mathscr{P}\mathscr{T}_i$.

The claimed existence of a network is the realistic counterpart of the utopian idea of a most comprehensive theory which comprises all others. During the development of physics, not always did all theories belong to the same network. There were disconnected parts of physics as e.g. mechanics, acoustics, optics (*and* chemistry as a very separated part, as it seemed at that time). Our claim is that today (and in the future) there is one and only one connected network of theories. It will be the task of this section to exhibit a subnetwork of theories which is called "quantum mechanics".

First there is a very simple link

$$\mathscr{P}\mathscr{T}_2 \overset{s\ \ s}{\longrightarrow\!\!\!\rightsquigarrow} \mathscr{P}\mathscr{T}_1, \tag{3.3.4}$$

where the restriction and the embedding are standard. The embedding shall be precise. The difference between $\mathscr{P}\mathscr{T}_2$ and $\mathscr{P}\mathscr{T}_1$ consists in "forgetting" some of the pictorial sets, pictorial relations and axioms. The reverse transition from $\mathscr{P}\mathscr{T}_1$ to $\mathscr{P}\mathscr{T}_2$ consists in adding to $\Sigma_1$ new base (pictorial) sets, new pictorial relations and new axioms. In this last direction proceeds the fundamental development of a theory. In this sense,

$$\mathscr{PT}_1 \overset{s\ \ s}{\leftsquigarrow\!\!-\!\!-} \mathscr{PT}_2 \overset{s\ \ s}{\leftsquigarrow\!\!-\!\!-} \mathscr{PT}_3 \ldots \overset{s\ \ s}{\leftsquigarrow\!\!-\!\!-} \mathscr{PT}_n \tag{3.3.5}$$

depicts a chain of theories developed from $\mathscr{PT}_1$ to $\mathscr{PT}_n$. One easily sees that we have given such "developments" in III through IX. Therefore we will not go here into details.

It must be emphasized that the way (3.3.5) from $\mathscr{PT}_1$ to $\mathscr{PT}_n$ of the development of an axiomatic basis is not a prescription for finding physical theories. It is allowed to seek theories in curious, even crazy ways: even contradictions are not forbidden. But if a theory has been established, then we should look for an axiomatic basis along (3.3.5).

We have already pointed out some links in the network of quantum mechanics in III §§4, 5.2 and 6.7. In the following we are not interested in a development chain like (3.3.5) but only in the "result" $\mathscr{PT}_n$ of such a development. Nevertheless there are various forms of quantum mechanics. The theory denoted in III §6.7 by $\mathscr{PT}_{1t}$ is that form of quantum mechanics which contains the most comprehensive description of the devices.

First we notice that $\mathscr{MT}_{\Sigma_{1t}}$ (the mathematical part of $\mathscr{PT}_{1t}$) is not of the first degree since there are pictorial sets (e.g. $\mathscr{Q}_1, \mathscr{Q}_2$) which are no base terms. We therefore go over to an axiomatic basis with a species of structure $\tilde{\Sigma}_{1t}$ of the first degree by a procedure presented in §2.2 (there we have described the transition from a structure $\Sigma$ of the first degree to a structure $\Sigma^{(1)}$ of higher degree).

As base terms of $\tilde{\Sigma}_{1t}$ we introduce the pictorial terms $M_1$, $M_2$, $\tilde{\mathscr{R}}_1$, $\tilde{\mathscr{R}}_2$, $Y_1$, $Y_2$, as this was done in III §1 and III §6.2. We add the following base terms (new as compared to $\Sigma_{1t}$): $\hat{\mathscr{Q}}_1, \hat{\mathscr{Q}}_2$, which are pictorial terms for the preparation procedures of the systems 1 resp. 2. Since $\hat{\mathscr{Q}}_1, \hat{\mathscr{Q}}_2$ are base terms, we must introduce a new relation $\beta_1(x_1, \tilde{a})$ as pictorial relation for "$x_1$ is prepared according to the procedure $\tilde{a}$". By

$$\tilde{a} \to a = \{x_1 \mid x_1 \in M_1 \text{ and } \beta_1(x_1, \tilde{a})\} \tag{3.3.6}$$

we define a mapping $\hat{\mathscr{Q}}_1 \to \mathscr{P}(M_1)$. The range of this mapping shall be denoted by $\mathscr{Q}_1$. We add the axiom (as conceptual norm) that the mapping (3.3.6) is injective. Thus $\hat{\mathscr{Q}}_1 \to \mathscr{Q}_1$ is bijective, while $\beta_1(x_1, \tilde{a})$ is equivalent to $x_1 \in \tilde{a}$. Then $\mathscr{Q}_1$ and the similarly defined $\mathscr{Q}_2$ are identified with the $\mathscr{Q}_1, \mathscr{Q}_2$ introduced in III §2.

The relations $\tilde{\mathscr{R}}_{10} \subset \tilde{\mathscr{R}}_1, \tilde{\mathscr{R}}_{20} \subset \tilde{\mathscr{R}}_2$ in AT 2 do not have the form desired for an axiomatic basis of the first degree. Therefore, instead of $\tilde{\mathscr{R}}_{i0}$ we introduce a pictorial relation $r_i(\tilde{b})$ with the interpretation: $\tilde{b}$ is a measuring method. Then we define $\tilde{\mathscr{R}}_{i0} = \{\tilde{b} \mid \tilde{b} \in \tilde{\mathscr{R}}_i \text{ and } r_i(\tilde{b})\}$ and have $\tilde{\mathscr{R}}_{i0} \subset \tilde{\mathscr{R}}_i$.

The main pictorial relations are the probability relations. $\lambda_{\text{Meas }1}, \lambda_{\text{Meas }2}$ contain only elements of base sets (see II, §3.1). This is not immediately the case for the other $\lambda$ introduced in III, §2. But we easily get the postulated form of pictorial relations by a procedure like the introduction of $\hat{\mathscr{Q}}_i$ above. For instance, instead of $\mathscr{S}_{12}$ from APSZ 6, in a first step we introduce a base term $\hat{\mathscr{S}}_{12}$ and a pictorial relation $\gamma(x, \tilde{c})$ for $x \in M$ and $\tilde{c} \in \hat{\mathscr{S}}_{12}$ with the interpretation: $x$ is selected according to the procedure $\tilde{c}$. By

$$\tilde{c} \to c = \{x \mid x \in M \text{ and } \gamma(x, \tilde{c})\}$$

we introduce a mapping $\hat{\mathscr{S}}_{12} \to \mathscr{P}(M)$, the range of which we denote by $\mathscr{S}_{12}$. We add the axiom that this mapping is injective and that $\mathscr{S}_{12}$ is generated by the $\theta_{12}$ defined in III (2.16).

In $\tilde{\Sigma}_{1t}$ we thus can introduce new base terms and define the probability relations as relations between elements of base terms (based on bijective mappings).

Obviously, this method of introducing additional base terms in $\tilde{\Sigma}_{1t}$ (as compared with $\Sigma_{1t}$) does not make the physical contents of $\mathscr{P}\mathscr{T}_{1t}$ clearer. Only if some logical problem seems not clear enough in $\Sigma_{1t}$, we should translate this problem to $\hat{\Sigma}_{1t}$.

If we presently skip the base terms and axioms of III, §7 and IX, we see that $\tilde{\Sigma}_{1t}$ obeys the postulates of a *simple* axiomatic basis. The axioms have been discussed already in §2.4. We now have the task to examine whether also the additional base sets and axioms of III §7 and IX fulfill the conditions for a simple axiomatic basis.

In III §7 we introduced the additional base term $\Delta$ and a pictorial relation. Now let us denote this relation by $\eta(a_2, \delta, a_2')$ with $a_2, a_2' \in \mathscr{Q}_2$, $\delta \in \Delta$ and interpret it as: $a_2'$ is the preparation procedure $a_2$ transported by $\delta$. By axioms we prescribe that $\eta$ defines a mapping $(\delta, a_2) \to a_2'$ which we denote by $a' = \delta a$, i.e. $\mathscr{Q}_2 \xrightarrow{\delta} \mathscr{Q}_2$. Additional relations and axioms lay down that $\Delta$ "is" the Galilei group. But $\Delta$ as Galilei group is a structure transported from pretheories. Therefore it is a familiar method (already discussed in §2.4), not to introduce all relations and axioms for this structure in the form of a simple axiomatic basis but to be content with the short formulation: $\Delta$ is the Galilei group. The relation $\eta(a_2, \delta, a_2')$ can also be viewed as $\eta(\tilde{a}_2, \delta, \tilde{a}_2')$ with $\tilde{a}_2, \tilde{a}_2' \in \hat{\mathscr{Q}}_2$. Thus the form of a simple axiomatic basis is attained.

In IX we have sketched additional structures (described in more detail in [2] through XVII). After discussing $\Delta$ and the consequences of the relation $\mathscr{Q}_2 \xrightarrow{\delta} \mathscr{Q}_2$ in [2] VII, in [2] VIII we introduced a new relation: "consisting of elementary systems". This is not a new pictorial relation, but a derived one (see [2] VIII §2 to 4) on the basis of an additional (empirical) axiom. In [2] VIII §2 this axiom is formulated as AZ with additional requirements imposed there.

This axiom is physically interpretable and therefore usable in $\tilde{\Sigma}_{1t}$. This definition of "composed of ..." should be complemented by a pictorial relation, in IX described by a mapping $\Pi \xrightarrow{\gamma} \mathscr{Q}'$ (also see [2] XVI §1). We easily see that this pictorial relation can be reduced to a relation between elements of base sets (by the procedure sketched above for the probability relations). Therefore also this relation is useable in $\tilde{\Sigma}_{1t}$.

In [2] VII §2 we have tried to give intuitive indications for this axiom *AZ*, but also expressed the desire to find a more transparent way for introducing the structure "composed of ...". The beginning of such a possibility is described in IX §2. There we first introduced the pictorial relation $\Pi \xrightarrow{\gamma} \mathscr{Q}'$ and additional axioms which permit us to get a bijective mapping (2.7). But this is not enough to describe the interaction of the parts in a composed system. We must add other axioms, e.g. *AZ*.

But all this discussion of the "best" way to introduce the structure "composed of ..." does not revoke the fact that $\tilde{\Sigma}_{1t}$ can be formulated as a simple axiomatic basis.

There is another feature of the presentation in [2] XI through XVI, which is all the more unsatisfactory: We introduced many additional "axioms" to describe "special" preparation procedures and measuring methods. These axioms are nothing but axiomatic relations for the mappings $\alpha$ and $\beta$ defined at the end of III §6.4: For special subsets of $\mathcal{Q}'_1 \times \psi_{1s}(\mathcal{F}_1)$ (resp. $\mathcal{Q}'_2 \times \psi_{2s}(\mathcal{F}_2)$) which can be described by the pretheories, the axioms give values $\alpha(a_1, k_1)$ (resp. $\beta(a_2, k_2)$). This method is clearly not elegant, but does not contradict the postulate of physically interpretable axioms.

In order to improve this method, one can take two ways. On both we begin with the same step already described in XI §7: We regard all these axiomatically introduced ensembles and observables as desired ones, i.e. we say nothing about the mappings $\alpha$ and $\beta$. The next step is to realize these desired ensembles and observables in the sense of V §8, resp. V §5. Then the first way goes over to the more comprehensive theory in XI, by which we can deduce the mappings $\alpha$, $\beta$ from the embedding in $\mathcal{PT}_{q\exp}$. The second way prefers to introduce very few observables by axioms concerning $\beta$ (e.g. for correlations between special $(a_2, k_2)$ and the impact observable in [2] XVI (6.1.13)). Then the theory from [2] XVII is used to find new observables and to construct ensembles by transpreparators. The actual way in experimental physics combines these two ways.

To formulate $\mathcal{MT}_{\tilde{\Sigma}_{1t}}$ we exclude the first way but leave open all possibilities of the second way. Thus we see how a theory $\mathcal{P}\tilde{\mathcal{T}}_{1t}$ with $\mathcal{MT}_{\tilde{\Sigma}_{1t}}$ as simple axiomatic basis can be formulated. $\mathcal{PT}_{1t}$ is equivalent to $\mathcal{P}\tilde{\mathcal{T}}_{1t}$ and has an axiomatic basis $\mathcal{MT}_{\Sigma_{1t}}$ of higher degree but also with physically interpretable axioms. The relation between $\mathcal{P}\tilde{\mathcal{T}}_{1t}$ and $\mathcal{PT}_{1t}$ is only a special case of that between $\mathcal{PT}$ and $\mathcal{PT}^{(1)}$ described in §2.2. With the concept of embedding we also can write this relation as

$$\mathcal{PT}_{1t} \rightleftarrows \mathcal{P}\tilde{\mathcal{T}}_{1t}. \tag{3.3.7}$$

The first embedding in (3.3.7) is nothing but the "identification" of the pictorial sets in $\mathcal{MT}_{\Sigma_{1t}}$ with pictorial sets in $\mathcal{MT}_{\tilde{\Sigma}_{1t}}$, as e.g. the range $\mathcal{Q}_1$ of $\hat{\mathcal{Q}}_1 \to \mathcal{P}(M_1)$ is identified with the pictorial set $\mathcal{Q}_1$ in $\mathcal{MT}_{\tilde{\Sigma}_{1t}}$. These identifications can also proceed in the opposite direction. Thus we get the second embedding in (3.3.7). It is obvious that $\mathcal{PT}_{1t}$ is easier to use than $\mathcal{P}\tilde{\mathcal{T}}_{1t}$.

Still simpler is $\mathcal{PT}_1$, as defined in III, §6.7. The relation of $\mathcal{PT}_{1t}$ and $\mathcal{PT}_1$ is fixed in III (6.7.1). We get $\mathcal{PT}_1'$ with the same mathematical part $\mathcal{MT}_{\Sigma_{1t}}$ of $\mathcal{PT}_{1t}$ if we omit $Y_1, Y_2, \tilde{\mathcal{R}}_1, \tilde{\mathcal{R}}_2$ as pictorial sets and use $\mathcal{R}_1, \mathcal{R}_2$ as new pictorial sets. The embedding in III (6.7.1) is a standard embedding by "identification" of the terms $\mathcal{R}_1, \mathcal{R}_2$ of $\mathcal{MT}_{\Sigma_{1t}}$ with the pictorial terms $\mathcal{R}_1, \mathcal{R}_2$ of $\mathcal{MT}_{\Sigma_1}$. By the transition from $\mathcal{PT}_{1t}$ to $\mathcal{PT}_1$ we have "forgotten" the description of the devices by trajectories.

$\mathcal{MT}_{\Sigma_1}$ is an axiomatic basis with physically interpretable axioms but not of the first degree. We easily get an axiomatic basis of the first degree by the procedure used above for the transition from $\Sigma_{1t}$ to $\tilde{\Sigma}_{1t}$. For instance, instead of $\mathcal{R}_1$, $\mathcal{R}_2$ we introduce base sets $\hat{\mathcal{R}}_1$, $\hat{\mathcal{R}}_2$ and relations $\alpha(x_i, \tilde{b}_i)$ with the interpretation: $x_i$ is registered by $\tilde{b}_i$. Then we introduce the mapping

$$\tilde{b}_i \to b_i = \{x_i \mid x_i \in M_i, \alpha(x_i, \tilde{b}_i)\}$$

and postulate that the mappings $\hat{\mathscr{R}}_i \to \mathscr{P}(M_i)$ are injective. $\mathscr{R}_i$ will then be defined by the range of this mapping. Are these base sets $\hat{\mathscr{R}}_i$ similar to the sets $\tilde{\mathscr{R}}_i$ of $\mathscr{M}\mathscr{T}_{\Sigma_{1t}}$? Indeed they are, except that an order structure of the sets $\hat{\mathscr{R}}_i$ does not appear (corresponding to the feature that $Y_1$, $Y_2$ do not appear in $\mathscr{M}\mathscr{T}_{\Sigma_1}$). Such an order structure of $\tilde{\mathscr{R}}_i$ was necessary to introduce the probabilities $\lambda_{\text{Meas}}$ which are unneeded for the formulation of $\mathscr{M}\mathscr{T}_{\Sigma_1}$.

A next step of "forgetting structures of the devices" is the transition from $\mathscr{P}\mathscr{T}_1$ to $\mathscr{P}\mathscr{T}_2$, described in III, §4 and symbolized by III (4.4). We get the restriction $\mathscr{P}\mathscr{T}_1'$ by using as pictorial sets only the terms $\mathscr{Q}$, $\mathscr{R}_0$, $\mathscr{R}$, $\mathscr{S}$. A standard embedding by identification of these terms with pictorial terms of $\Sigma_2$ yields $\mathscr{P}\mathscr{T}_2$. The axiomatic basis $\mathscr{M}\mathscr{T}_{\Sigma_2}$ is precisely that developed in [2], if we there use the axioms in [2] III §3 (not the axiom $AQ$ in [2] III §5!).

It is again easy to reformulate for $\mathscr{M}\mathscr{T}_{\Sigma_2}$ an axiomatic basis of the first degree. Therefore such a reformulation may be left to the interested reader.

We get a species of structure $\Sigma_{2Q}$ equivalent to $\Sigma_2$, if we replace the axioms in [2] III, §3 by the axiom $AQ$ in [2] III §5. Then $\mathscr{M}\mathscr{T}_{\Sigma_{2Q}}$ is again an axiomatic basis of $\mathscr{P}\mathscr{T}_2$, but not only with physically interpretable axioms since in $AQ$ we have introduced the theoretical auxiliary term of a sequence of Hilbert spaces.

In III §5.2, as a next step in the process of forgetting we considered the transition from $\mathscr{P}\mathscr{T}_2$ to a theory $\mathscr{P}\mathscr{T}_3$ symbolized by III (5.2.1). The restriction $\mathscr{P}\mathscr{T}_2'$ follows from $\mathscr{P}\mathscr{T}_2$ if as the only pictorial terms for $\mathscr{P}\mathscr{T}_2'$ we take the $\mathscr{K}$, $\mathscr{L}$ in $\mathscr{M}\mathscr{T}_{\Sigma_2}$ and the probability function $\mathscr{K} \times \mathscr{L} \xrightarrow{\mu} [0, 1]$ as single pictorial relation. $\Sigma_3$ is given by the base terms $\mathscr{K}$, $\mathscr{L}$ and the relation $\mu$ and additional axioms, which can be formulated by using $\mathscr{K}$, $\mathscr{L}$, $\mu$. The definition of the elements of $\Delta$ as transformations in $\mathscr{L}$ (resp. $\mathscr{K}$) can be transported from $\Sigma_2$ to $\Sigma_3$. The basis $\mathscr{M}\mathscr{T}_{\Sigma_3}$ is of the first degree.

The simple structure of $\mathscr{P}\mathscr{T}_3$ is attained at the expense of omitting from the fundamental domain some relations which are very important for the structure of the microsystems, namely the relations pictured by the concepts of observable and preparator. Obviously it is possible to add these concepts to $\mathscr{P}\mathscr{T}_3$ afterwards. But a foundation of these concepts similar to that given in V (or [2] IV) is not possible in $\mathscr{P}\mathscr{T}_3$ since we cannot establish the physical significance of the Boolean rings (with the help of which the concepts of observable and preparator are defined). To give these Boolean rings a physical interpretation it would be necessary to extend the species of structure $\Sigma_3$ by additional base terms $\mathscr{Q}$, $\mathscr{R}$ (endowed with structures similar to those of $\tilde{\mathscr{S}}$ in II §2.1) without an index set $M$. Such a theory $\mathscr{P}\mathscr{T}_{3\text{pr}}$ could serve as a step between $\mathscr{P}\mathscr{T}_2$ and $\mathscr{P}\mathscr{T}_3$ in the chain

$$\mathscr{P}\mathscr{T}_2 \to \mathscr{P}\mathscr{T}_2'' \xrightarrow{s} \mathscr{P}\mathscr{T}_{3\text{pr}} \to \mathscr{P}\mathscr{T}_{3\text{pr}}' \xrightarrow{s} \mathscr{P}\mathscr{T}_3.$$

Here $\mathscr{P}\mathscr{T}_2''$ would result from $\mathscr{P}\mathscr{T}_2$ by forgetting $M$ and the relations between $M$ and $\mathscr{Q}$, $\mathscr{R}$. Since $\Sigma_{3\text{pr}}$ contains $\mathscr{Q}$, $\mathscr{R}$ as base terms, $\mathscr{P}\mathscr{T}_{3\text{pr}}'$ results from $\mathscr{P}\mathscr{T}_{3\text{pr}}$ by forgetting besides $M$ also $\mathscr{Q}$, $\mathscr{R}$. But if we use $\mathscr{P}\mathscr{T}_{3\text{pr}}$ instead of $\mathscr{P}\mathscr{T}_3$ in order to define the important concepts of observable and preparator rigorously, then only a small step leads to the more transparent theory $\mathscr{P}\mathscr{T}_2$ developed in [2].

Neither $\mathscr{P}\mathscr{T}_2$ nor $\mathscr{P}\mathscr{T}_3$ coincides with the "usual" form of quantum mechanics taught in textbooks. Therefore let us also give the relation between $\mathscr{P}\mathscr{T}_2$ and this "usual" quantum mechanics to be called $\mathscr{P}\mathscr{T}_5$.

We introduce a species of structure $\Sigma_5$ by a Hilbert space $\mathscr{H}$ over the field **C** (of denumerable dimension). Let $\mathscr{L}_r(\mathscr{H})$ be the Banach space of all bounded self-adjoint operators (often enlarged to the set of all essentially selfadjoint operators; see [2] A IV § 10; who wants can in the following take $\mathscr{L}_r(\mathscr{H})$ as this set).

To describe the superselection rules we assume a decomposition $\mathscr{H} = \sum_\nu \oplus \mathscr{H}_\nu$. Let $\mathrm{Ob}(\mathscr{H})$ be the subset of all those operators in $\mathscr{L}_r(\mathscr{H})$ which map the subspaces $\mathscr{H}_\nu$ into themselves. This $\mathrm{Ob}(\mathscr{H})$ is singled out as pictorial set for what is commonly called an observable; but usually this concept is not completely clarified. To see this we need only read the deluge of words by which one tries to circumscribe an observable.

Let $K(\mathscr{H})$ be the subset of all those $W \in \mathscr{L}_r(\mathscr{H})$ which transform each $\mathscr{H}_\nu$ into itself and obey $W > 0$, $\mathrm{tr}(W) = 1$. This $K(\mathscr{H})$ is singled out as pictorial set for the ensembles (also called states). Again the meaning of ensemble (state) cannot be completely clarified.

As pictorial relation we use $\mathrm{tr}(WA) = \alpha$ where $W \in K(\mathscr{H})$ and $A \in \mathrm{Ob}(\mathscr{H})$. This $\mathrm{tr}(WA)$ is interpreted as the expectation value of $A$ in the ensemble $W$. The expectation value is the mathematical idealization of the mean value of the measured scale values for "many" measurements.

Special observables, e.g. position and momentum are introduced again without the possibility to clarify what these observables mean, e.g. what is the "position" observable. Similarly the physical significance of the time evolution in the Schrödinger- or Heisenberg picture ([2] X) is only interpreted by vague comparisons with classical mechanics, since one commonly does not use the correct interpretation of the Galileo transformations as transports of the measuring devices.

For all these unclear concepts, especially that of an observable, one had to pay dearly during the development of quantum mechanics; for instance all this unclearness caused an extensive and partly abstruse discussion of the "measuring process". All unclearness can be removed if we clarify the relation of $\mathscr{P}\mathscr{T}_5$ with $\mathscr{P}\mathscr{T}_2$. For $\mathscr{P}\mathscr{T}_2$ we will use the $\Sigma_{2Q}$ from above without changing the denotation of $\mathscr{P}\mathscr{T}_2$. We shall demonstrate the relation

$$\mathscr{P}\mathscr{T}_2 \to \mathscr{P}\mathscr{T}_4 \rightsquigarrow \mathscr{P}\mathscr{T}_5 \tag{3.3.8}$$

with a suitable restriction $\mathscr{P}\mathscr{T}_4$ constructed as follows.

As a first new pictorial set in $\mathscr{P}\mathscr{T}_4$ we use the set $K$ of ensembles. More difficult is it to define a new pictorial set formed by "scale observables". With this "interpretation" we do not immediately see an *intrinsic* term. A search for it reveals the more precise physical interpretation of "observable". A scale observable was defined as a decision observable with a scale. A scale for a Boolean ring is defined in V D 3.5.1. A scale observable is therefore a pair $\Sigma \xrightarrow{F} G$ and $y \in \mathscr{B}'(\Sigma)$ (see VII D 3.1). Since $\Sigma$ is not intrinsically defined we have to seek for an intrinsic analog of $\Sigma$. Such intrinsic Boolean rings are the sets $\mathscr{R}(b_0)$ with $b_0 \in \mathscr{R}_0$ (the observables were introduced as abstractions of these $\mathscr{R}(b_0) \xrightarrow{\psi_{b_0}} L$!).

We now can define a subset $\mathscr{R}_{0d}$ of $\mathscr{R}_0$ by those $b_0 \in \mathscr{R}_0$ for which $\mathscr{R}(b_0) \xrightarrow{\psi_{b_0}} L$

is nearly (in the sense of AOb in V §5) a decision observable. This "nearly" is typical for the physical applications of $\mathcal{PT}_5$, since in general there are no exact measuring methods for decision observables. Nevertheless we do not want to establish here a precise description of this "nearly" by imprecision sets. To simplify our considerations we will do as if the $b_0 \in \mathcal{R}_{0d}$ by $\mathcal{R}(b_0) \xrightarrow{\psi_{b_0}} G$ define decision observables.

$$b_{01}, b_{02} \in \mathcal{R}_{0d} \quad \text{and} \quad b_{01} \neq b_{02} \quad \text{imply} \quad \mathcal{R}(b_{01}) \cap \mathcal{R}(b_{02}) = \emptyset.$$

If $\mathcal{R}(b_{01}) \cap \mathcal{R}(b_{02}) \neq \emptyset$ then a $b \in \mathcal{R}(b_{01}) \cap \mathcal{R}(b_{02})$ would exist such that $b_0 = b_{01} \cap b_{02} \neq \emptyset$ with $b_0 \in \mathcal{R}_0$. Since $\psi(b_{01}, b_0) = \lambda 1$, $b_0 \neq \emptyset$ and $b_{01} \in \mathcal{R}_{0d}$, we get $\lambda = 1$, i.e. $b_0 = b_{01}$ and therefore $b_{01} = b_{02}$.

Thus the set $\mathcal{R}_d$ of all $b \in R$ with $b \subset b_0 \in \mathcal{R}_{0d}$ is partitioned into disjoint sets $\mathcal{R}(b_0)$ with $b_0 \in \mathcal{R}_{0d}$. It would be possible to define a restriction of $\mathcal{PT}_2$ by the new pictorial sets $K$, $\mathcal{R}_{0d}$ and $\mathcal{R}_d$ and the probability relation $\mu(w, \psi(b_0, b)) = \alpha$ with $w \in K$ and $b_0 \in \mathcal{R}_{0d}$, $b \in \mathcal{R}(b_0)$. To define $\mathcal{PT}_4$ we will use scales and the concept of expectation values (instead of $\mu(\ldots)$). The scales are only aids for representations of the Boolean rings $\mathcal{R}(b_0)$. This significance of scales was described in V §3.5 and [2] IV §2.5. Thus it is possible, for every $b_0$ to replace the Boolean ring $\mathcal{R}(b_0)$ by the set of scales $\mathcal{B}'(\mathcal{R}(b_0))$ (see V D3.5.1). $\mathcal{B}'(\mathcal{R}(b_0))$ is a theoretical, but not a physical term (in the sense of §2.3) since we have to use the real numbers (not only bound by the probability relation) for its definition. The scales play the same role for a Boolean ring as for instance coordinates for the Euclidean geometry.

We introduce the scales as aids for writing down the observational report in a form as clearly arranged as possible. How to do this?

As new pictorial sets use $\mathcal{R}_{0d}$ and the disjoint union $\bigcup_{b_0 \in \mathcal{R}_0} \mathcal{B}'(\mathcal{R}(b_0))$. The $b_0 \in \mathcal{R}_{0d}$ are given in the fundamental domain as registration devices. The scales $y \in \mathcal{B}'(\mathcal{R}(b_0))$ are arbitrarily "chosen" (for a given device $b_0$). They serve for a new description of the observational report. Instead of the probabilities for the various elements of $\mathcal{R}(b_0)$ we introduce the expectation values for the $y \in \mathcal{B}'(\mathcal{R}(b_0))$. This expectation value relation is given by

$$\mu(w, S' y) = \alpha. \tag{3.3.9}$$

Here $S'$ is the mapping $\mathcal{B}'(\mathcal{R}(b_0)) \xrightarrow{S'} \mathcal{B}'$ dual to the $\mathcal{B} \xrightarrow{S} \mathcal{B}(\mathcal{R}(b_0))$ defined in §3.2.

In the description of the observational report we must state the chosen scales $y$. From this observational report we can pass over to that of $\mathcal{PT}_4$ by forgetting $\mathcal{R}_{0d}$ and by using $K$ and $\mathrm{Ob}_4 = \bigcup_{b_0 \in \mathcal{R}_{0d}} S' \mathcal{B}'(\mathcal{R}(b_0))$ (the mapping $\bigcup_{b_0 \in \mathcal{R}_{0d}} \mathcal{B}'(\mathcal{R}(b_0)) \to \mathrm{Ob}_4$ by $S'$ is not necessarily injective!) as pictorial sets and (3.3.9) as pictorial relation. Then the axiom $AQ$ (see [2] III §5) of $\Sigma_{2Q}$ implies the embedding $\mathcal{PT}_4 \rightsquigarrow \mathcal{PT}_5$.

Also the fundamental domain of $\mathcal{PT}_4$ and thus of $\mathcal{PT}_5$ is very restricted compared with that of $\mathcal{PT}_2$. Only such devices ($b_0 \in \mathcal{R}_{0d}$) are admitted to the fundamental domain of $\mathcal{PT}_4$ which (nearly) "measure" decision observables. Since most measurements are not of this type, this restriction is at least very impractial. But very dubious is the imagination of some theorists that there is something real in nature that corresponds to the observables (in the usual sense of $\mathscr{P}\mathscr{T}_5$, i.e. to the elements of $\mathrm{Ob}(\mathscr{H})$) and that it is a question of experimental physics to seek for more or less precise measurements of these realities. These theorists feel pretty elevated above the work of the experimentalists. They think that theoretical physics has only to do with the clean reality, not dirtied by practical work, so clean that there is no such clean reality at all.

It cannot be the task of this book to discuss critically the deluge of literature concerning problems which solely arise from unclear and false ideas about the interpretation of the "usual" quantum mechanics $\mathscr{P}\mathscr{T}_5$. Therefore all those having problems with $\mathscr{P}\mathscr{T}_5$ should ask themselves critically whether or not these problems can be formulated within the framework of the theories $\mathscr{P}\mathscr{T}_{1t}, \mathscr{P}\mathscr{T}_1$ or $\mathscr{P}\mathscr{T}_2$. If not, then one must ask oneself on what these problems rest, apparently on some ideas additional to $\mathscr{P}\mathscr{T}_5$. But then, it is not more reasonable to abandon such additional ideas which are not compatible with quantum mechanics?

The reader may review the discussion in X §2.1 and 2.2 concerning the relations between various theories for macroscopic systems. The only difference is that there we assumed an embedding $\mathscr{P}\mathscr{T}_m \rightsquigarrow \mathscr{P}\mathscr{T}_{q\,\mathrm{exp}}$ in §2 to 3 to complete the theory $\mathscr{P}\mathscr{T}_m$ with respect to the dynamics. In this sense, the embedding postulate can be taken as an axiom of $\mathscr{P}\mathscr{T}_m$, an axiom of the form (2.1.1). Probably we have not yet found enough constraints for the embedding mappings. For this reason, until now we cannot say exactly *what* in the atomic structure of macrosystems is real and what not (see §4.8).

## §3.4 Pretheories

We have used the concept of pretheory very often but more intuitively than rigorously. We now are in the position to say rigorously what we understand by a pretheory. Already the discussions of restrictions and embedding revealed how the correspondence rules should be transferred. With $\mathscr{P}\mathscr{T}_1$ and $\mathscr{P}\mathscr{T}_2$ as the two forms of quantum mechanics one could have the impression that III (4.4) makes $\mathscr{P}\mathscr{T}_1$ a pretheory of $\mathscr{P}\mathscr{T}_2$ since that relation determines the correspondence rules of $\mathscr{P}\mathscr{T}_2$ by those of $\mathscr{P}\mathscr{T}_1$. There is something right in this impression.

We define: $\mathscr{P}\mathscr{T}_p$ is a pretheory of $\mathscr{P}\mathscr{T}$ if there is a standard restriction $\mathscr{P}\mathscr{T}'$ of $\mathscr{P}\mathscr{T}$ such that

$$\mathscr{P}\mathscr{T}_p \rightsquigarrow \mathscr{P}\mathscr{T}' \xleftarrow{\ s\ } \mathscr{P}\mathscr{T} \tag{3.4.1}$$

and the pictorial sets and relations of $\mathscr{P}\mathscr{T}$ which are "left" in $\mathscr{P}\mathscr{T}'$ are interpreted by $\mathscr{P}\mathscr{T}_p \rightsquigarrow \mathscr{P}\mathscr{T}'$.

As $\mathscr{P}\mathscr{T}_{1p}$ let us take the theory developed in III §1 to 3, as $\mathscr{P}\mathscr{T}_2'$ the standard restriction of the form $\mathscr{P}\mathscr{T}_2$ of quantum mechanics. Thus we forget all but $\mathscr{Q}$, $\mathscr{R}_0$, $\mathscr{R}$ and $\lambda_{\mathscr{Q}}$, $\lambda_{\mathscr{R}_0}$ and $\lambda_{\mathscr{S}}$ and the relations APS 1, APS 2, APS 3, APS 4, APS 5.1.1, APS 5.2, APS 6, APS 7 and APS 8 as axioms for $\Sigma_2$. Then we have

$$\mathscr{P}\mathscr{T}_{1p} \rightsquigarrow \mathscr{P}\mathscr{T}_2' \xleftarrow{\ s\ } \mathscr{P}\mathscr{T}_2 \tag{3.4.2}$$

where $\mathscr{Q}$, $\mathscr{R}_0$, $\mathscr{R}$ and $\lambda_{\mathscr{Q}}$, $\lambda_{\mathscr{R}_0}$, $\lambda_{\mathscr{S}}$ are interpreted by the "pretheory" $\mathscr{P}\mathscr{T}_{1p}$.

Obviously $\mathscr{PT}_{1p}$ is not the "total" theory $\mathscr{PT}_1$; but we have

$$\mathscr{PT}_1 \rightsquigarrow \mathscr{PT}_{1p}, \tag{3.4.3}$$

where we get $\mathscr{PT}_{1p}$ by forgetting all of $\mathscr{PT}_1$ except structures established in III §1 to 3.

(3.4.2), (3.4.3) and III (4.4) yield

$$\begin{array}{ccccc} & & \mathscr{PT}_1 & & \\ & \swarrow & & \searrow & \\ \mathscr{PT}_{1p} & \rightsquigarrow & \mathscr{PT}_2' & \xleftarrow{\ s\ } & \mathscr{PT}_2. \end{array} \tag{3.4.4}$$

This diagram pictures a desire emerging from a relation chain such as (3.4.1): to find a theory more comprehensive than $\mathscr{PT}_p$ *and* $\mathscr{PT}$. But this desire is not fulfilled in most of the situations where we use pretheories. For the form $\mathscr{PT}_{1t}$ of quantum mechanics, we have to use pretheories of the form described in II to interpret the base sets $Y_1$, $Y_2$ of $\Sigma_{1t}$. Our efforts in X and XI are nothing but steps in the direction to a more comprehensive theory than $\mathscr{PT}_{1t}$ *and* its pretheories.

In situations with more than one pretheory we have a diagram like

$$\begin{array}{ccccc} \mathscr{PT}_p' & \rightsquigarrow & \mathscr{PT}' & & \\ & & & \nwarrow^{s} & \\ \mathscr{PT}_p'' & \rightsquigarrow & \mathscr{PT}'' & \xleftarrow{\ s\ } & \mathscr{PT}. \\ \vdots & & \vdots & & \end{array} \tag{3.4.5}$$

There can be much more complicated situations in the network of physical theories than described here as examples. So far this network has been explicitly elaborated only in very few places. Nevertheless physicists find their way in this network like spiders in their webs.

## §4 Physically Possible, Physically Real Facts and Physically Open Questions

We have used words as "real" and "possible" not only in the preceding sections but also on many places in I to XII. But we have used these words only intuitively without defining rigorously what we mean. Obviously the interpretation of a $\mathscr{PT}$ is not exhausted by the correspondence rules introduced in §1. These rules are only the fundament on which we can build a more comprehensive interpretation language.

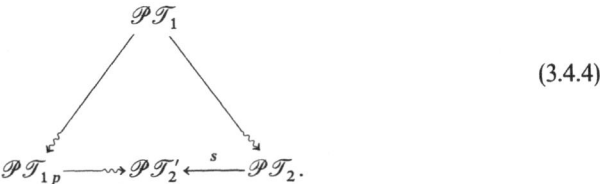

Certainly we adopt the observational report (1.1), (1.2) or (1.7) as statement of real facts formulated in the language of $\mathscr{PT}$. But we all know that we claim much more as real than is written down in the observational report. We use to infer other, not observed realities. For instance, we speak of real electrons although only interactions of macrosystems are observed. We speak of real atoms which compose macrosystems. We speak of real electrons in a semiconductor. Is all this correct? Or are electrons and atoms in macrosystems only imaginations suitable to explain some properties of the macrosystems? All this can only be clarified if we have a rigorous method to proceed from the observational report to other realities.

Much more mystical than the approach to realities seems to be that we speak of possible facts though there is no sign in $\mathscr{M}\mathscr{T}$ for such a logical category as "possible", i.e. there is no modal logic in $\mathscr{M}\mathscr{T}$. Obviously "possible" is a word of the interpreting language, and it is one of the most important words. In physics we mostly do not ask "what is?" but rather "what is possible?". Then the fundamental but not physical question arises "What of everything possible should we realize?"

The intuitive usage of the words "real", "there are", "this is possible", sometimes has led to errors. We need only mention questions such as: Has a single microsystem a "real state"? Have the microsystems "real properties"? Can hidden variables be real? Is there something like a "real propensity" for every possible process, if we describe processes by probabilities?

These and many other questions make it necessary to develop a rigorous method for introducing such words as "real" and "possible". The next sections are only preparations for such a development in §4.6 and §4.8.

## §4.1 Hypotheses in a $\mathscr{P}\mathscr{T}$

We have already introduced hypotheses in §2.5. We now want to extend this concept. For the following discussions it is suitable to take the mathematical part $\mathscr{M}\mathscr{T}_\Sigma$ as an axiomatic basis (not necessarily of the first degree and not necessarily simple). We will not change the theory $\mathscr{M}\mathscr{T}_\Sigma$ and therefore fix the base terms as $y_1, \ldots, y_r$, the structure terms by $t=(t_1, t_2, \ldots)$ and the axiomatic relation by $P(y_1, \ldots, t)$. We regard as given an observational report in the form (2.2.22) of (—)$_r$(1) and (2.2.24) of (—)$_r$(2). $\mathscr{M}\mathscr{T}_\Sigma\mathscr{A}$ is the theory with the additional constants $(a_1, a_2, \ldots)$ comprised in $A$ and the additional axioms (2.2.22), (2.2.24).

It is now possible, as in §2.5 to invent additional relations of the same form as the observational report. To distinguish the letters of the invented relations from those of the "real" observational report, let us denote them by $x_1, x_2, \ldots$ instead of $a_1, a_2, \ldots$. The additional invented report takes the form

$$X \in \tilde{T}_h(y_1, \ldots, y_r); \tag{4.1.1}$$

$$\tilde{P}_h(y_1, \ldots, y_r; t, A, X, \mathbf{R}). \tag{4.1.2}$$

Then $\mathscr{M}\mathscr{T}_\Sigma\mathscr{A}\mathscr{H}$ has the form $\mathscr{M}\mathscr{T}_\Sigma$ with the additional constants $a_1, a_2, \ldots;$ $x_1, x_2, \ldots$ and the additional axioms (2.2.22), (2.2.24), (4.1.1), (4.1.2). If there is no observational report, we again get $\mathscr{M}\mathscr{T}_\Sigma\mathscr{H}$ from §2.5, i.e. $\mathscr{M}\mathscr{T}_\Sigma$ with the only additional axioms (4.1.1) and (4.1.2). Without invented elements $x_1, x_2, \ldots,$ the theory $\mathscr{M}\mathscr{T}_\Sigma\mathscr{A}\mathscr{H}$ has the form $\mathscr{M}\mathscr{T}_\Sigma\mathscr{A}$ with the additional axiom

$$\tilde{P}_h(y_1, \ldots, y_r; t, A, \mathbf{R}) \tag{4.1.3}$$

which contains additional invented relations among the $a_i$. If there is no $A$ (i.e. no observational report), (4.1.3) has the form of an axiom additional to $\mathscr{M}\mathscr{T}_\Sigma$.

If $\tilde{T}_h$ and $\tilde{P}_h$ as above have the form of invented observational reports, we denote (4.1.1), (4.1.2) as a hypothesis of the *first kind*.

We will now extend this concept to more general hypotheses. Instead of $x_i$ which only are elements of the pictorial sets, we admit

$$x_1 \in T_{i_1}(y_1, \ldots, y_r, \mathbf{R}), \qquad x_2 \in T_{i_2}(y_1, \ldots, y_r, \mathbf{R}), \ldots$$

where the $T_k(\ldots)$ are *any* echelon construction schemes. Then $\tilde{T}_h(\ldots)$ in (4.1.1) has the more general form

$$\tilde{T}_h(y_1, \ldots, y_r, \mathbf{R}) = T_{i_1}(y_1, \ldots, y_r, \mathbf{R}) \times T_{i_2}(\ldots) \times \ldots.$$

We call the $T_k(\ldots)$ "extended pictorial sets". We also introduce "extended pictorial relations" by intrinsic terms $S_\eta(y_1, \ldots, y_r, \mathbf{R})$ of such a form that relations of the "extended" form

$$(x_\nu, \ldots, a_\mu, \ldots, \alpha) \in S_\eta(y_1, \ldots, y_r, \mathbf{R})$$

make sense. These extended relations can be collected into an axiom (4.1.2) with the only difference that $\tilde{P}_h(\ldots)$ can have a more generalized form.

A hypothesis of the described form which is not a hypothesis of the first kind, is called hypothesis of the *second kind*.

Whenever in the following we speak of hypotheses without addendum, all holds for hypotheses of the first and second kind.

For the following discussions it is advantageous to introduce the set

$$\tilde{E}_h(A) = \{X \mid X \in \tilde{T}_h \text{ and } \tilde{P}_h\}. \tag{4.1.4}$$

Then (4.1.1) and (4.1.2) can be comprised in

$$X \in \tilde{E}_h(A); \tag{4.1.5}$$

and $\mathcal{M}\mathcal{T}_\Sigma \mathcal{A}\mathcal{H}$ arises from $\mathcal{M}\mathcal{T}_\Sigma \mathcal{A}$ by adding (4.1.5) as axiom. If there are no invented elements $X$, we must replace (4.1.5) by (4.1.3).

Our definition of a hypothesis must be distinguished from a forecast. By a forecast we want to tell what will happen (or can happen) in the future. No physical theory contains a structure which tells us what is the moment "now". This should be so since physics is based on facts and on processes in devices; and there is no device which can tell us that the moment "now" is distinguished from moments in the past or future. Only we as human beings become aware in our consciousness of this "now". Only relative to this "now" we can say that something in our environs has happened or will happen. In the mathematical part $\mathcal{M}\mathcal{T}_\Sigma$ of a $\mathcal{P}\mathcal{T}$, there can be only a structure which characterizes the "direction" of time. But no structure in $\mathcal{M}\mathcal{T}_\Sigma$ tells us what has happened and what will happen, since there is no "now" in $\mathcal{M}\mathcal{T}_\Sigma$.

Nevertheless we know that the distinction of the future (what has not yet happened) from the past (what has happened and cannot be changed any more by our actions) is a decisive structure for our working. Where is that structure coming in?

It comes in by the observational report. Nothing in $\mathcal{P}\mathcal{T}$ tells us, *what* we can write down in $(\text{—})_r(1)$ and $(\text{—})_r(2)$. $\mathcal{P}\mathcal{T}$ tells us only *how* we must write down $(\text{—})_r(1)$ and $(\text{—})_r(2)$. The observational report can only be related to the past, not because

$\mathcal{PT}$ forbids us to write down something about the future, but because we as humans are unable to state facts of the future. Therefore the observational report contains only facts of the past; or more precisely: Past in physics is that about which we have an observational report.

A hypothesis of the first kind can be related to imagined facts in the past and in the future. A hypothesis is not restricted to the future. This will be important for the discussions in §4.6.

## §4.2 Classifications of Hypotheses

We presume that $\mathcal{M}\mathcal{T}_\Sigma\mathcal{A}$ is not contradictory; otherwise $\mathcal{PT}$ would be useless.

If $\mathcal{M}\mathcal{T}_\Sigma\mathcal{A}\mathcal{H}$ is contradictory we call $\mathcal{H}$ "false", otherwise "allowed".

$\mathcal{M}\mathcal{T}_\Sigma\mathcal{A}\mathcal{H}$ contradictory is equivalent to that [not $\mathcal{H}$], i.e. the negation of (4.1.5) resp. of (4.1.3) is a theorem in $\mathcal{M}\mathcal{T}_\Sigma\mathcal{A}$. The negation of (4.1.5) resp. of (4.1.3) has the form

$$X \notin \tilde{E}_h(A); \qquad \text{resp. [not } \tilde{P}_h(y_1, \ldots; t, A, \mathbf{R})]. \tag{4.2.1}$$

Equivalent to $X \notin \tilde{E}_h(A)$ is $\forall X(X \notin \tilde{E}_h(A))$, i.e.

$$\tilde{E}_h(A) = \emptyset. \tag{4.2.2}$$

$\mathcal{H}$ is false if and only if

$$\tilde{E}_h(A) \neq \emptyset \tag{4.2.3}$$

leads to a contradiction in $\mathcal{M}\mathcal{T}_\Sigma\mathcal{A}$, i.e. (4.2.2) is a theorem in $\mathcal{M}\mathcal{T}_\Sigma\mathcal{A}$. And vice versa: if (4.2.3) yields no contradiction in $\mathcal{M}\mathcal{T}_\Sigma\mathcal{A}$, then $\mathcal{H}$ is allowed.

Therefore "$\mathcal{H}$ allowed" is equivalent to: (4.2.3) can be added to $\mathcal{M}\mathcal{T}_\Sigma\mathcal{A}$ without contradiction.

It may be that (4.2.3) not only can be added to $\mathcal{M}\mathcal{T}_\Sigma\mathcal{A}$ without contradiction, but is a theorem in $\mathcal{M}\mathcal{T}_\Sigma\mathcal{A}$. Then we call $\mathcal{H}$ "theoretically existent".

If $\mathcal{H}$ is allowed and if

$$\tilde{E}_h(A) \neq \emptyset \Rightarrow \tilde{E}_h(A) \text{ has only one element} \tag{4.2.4}$$

is a theorem in $\mathcal{M}\mathcal{T}_\Sigma\mathcal{A}$, we call $\mathcal{H}$ allowed and determined.

If (4.2.3) and (4.2.4) are theorems in $\mathcal{M}\mathcal{T}_\Sigma\mathcal{A}$, i.e. if

$$\tilde{E}_h(A) \text{ has one and only one element}$$

is a theorem in $\mathcal{M}\mathcal{T}_\Sigma\mathcal{A}$, we call $\mathcal{H}$ theoretically existent and determined.

A hypothesis without $X$ may be called determined; i.e. allowed and determined if (4.1.3) can be added to $\mathcal{M}\mathcal{T}_\Sigma\mathcal{A}$ without contradiction, theoretically existent and determined if (4.1.3) is a theorem in $\mathcal{M}\mathcal{T}_\Sigma\mathcal{A}$.

Let us discuss the case that $\mathcal{H}$ is allowed but not theoretically existent. Then it must be possible, without contradiction to augment $\mathcal{M}\mathcal{T}_\Sigma\mathcal{A}$ by (4.2.3) or alternatively by the negation of (4.2.3), i.e. by (4.2.2). Therefore neither (4.2.3) nor (4.2.2) is a theorem in $\mathcal{M}\mathcal{T}_\Sigma\mathcal{A}$.

The relation (4.2.2) resp. [not $\tilde{P}_h(\ldots)$] may be called [neg $\mathcal{H}$]. As a hypothesis without invented elements $X$, this [neg $\mathcal{H}$] (i.e. $\tilde{E}_h = \emptyset$) must not be confused with [not $\mathcal{H}$] (i.e. $X \notin \tilde{E}_h$). Only if $\mathcal{H}$ itself has no invented elements $X$, then [neg $\mathcal{H}$] and [not $\mathcal{H}$] are equal, i.e. equal to [not $\tilde{P}_h$].

$\mathscr{H}$ is allowed but not theoretically existent if and only if [neg $\mathscr{H}$] is allowed but not theoretically existent. If [neg $\mathscr{H}$] is theoretically existent, then $\mathscr{H}$ is false.

In a next step let us try to express the classification of hypotheses by relations in $\mathscr{M}\mathscr{T}_\Sigma$. Obviously this step is only interesting if there is an observational report.

$\mathscr{H}$ is theoretically existent if (4.2.3) is a theorem in $\mathscr{M}\mathscr{T}_\Sigma\mathscr{A}$. This $\mathscr{M}\mathscr{T}_\Sigma\mathscr{A}$ arises from $\mathscr{M}\mathscr{T}_\Sigma$ by addition of (2.2.22), (2.2.24), i.e. of

$$A \in \tilde{T}(y_1, \ldots, y_r) \quad \text{and} \quad \tilde{P}(y_1, \ldots, y_r, t, A, \mathbf{R}) \tag{4.2.5}$$

as axiom. Therefore, (4.2.3) is a therorem in $\mathscr{M}\mathscr{T}_\Sigma\mathscr{A}$ if and only if

$$[A \in \tilde{T}(\ldots) \text{ and } \tilde{P}(\ldots)] \Rightarrow \tilde{E}_h(A) \neq \emptyset \tag{4.2.6}$$

is a theorem in $\mathscr{M}\mathscr{T}_\Sigma$. Equivalent to (4.2.6) is

$$\forall Z[(Z \in \tilde{T}(\ldots) \text{ and } \tilde{P}(\ldots)) \Rightarrow \tilde{E}_h(Z) \neq \emptyset]. \tag{4.2.7}$$

Therefore $\mathscr{H}$ is theoretically existent if and only if (4.2.7) is a theorem in $\mathscr{M}\mathscr{T}_\Sigma$.

Let us shorten (4.2.7) by introducing the set

$$\tilde{E}(y_1, \ldots, y_r, t) = \{Z \mid Z \in \tilde{T}(y_1, \ldots) \text{ and } \tilde{P}(y_1, \ldots)\}. \tag{4.2.8}$$

Then (4.2.7) takes the form

$$\forall Z[Z \in \tilde{E} \Rightarrow \tilde{E}_h(Z) \neq \emptyset]. \tag{4.2.9 a}$$

If $\mathscr{H}$ contains no invented elements $X$ we must replace (4.2.9 a) by

$$\forall Z[Z \in \tilde{E} \Rightarrow \tilde{P}_h(y_1, \ldots\ t, Z)]. \tag{4.2.9 b}$$

(4.2.9) can be read in the very clear form: *If* there is an observational report of the form $Z \in \tilde{E}$, it implies $\tilde{E}(Z) \neq \emptyset$ resp. $\tilde{P}_h(\ldots, Z)$.

In the same way we get the result: $\mathscr{H}$ is false if and only if

$$\forall Z[Z \in \tilde{E} \Rightarrow \tilde{E}_h(Z) = \emptyset] \tag{4.2.10 a}$$

resp.

$$\forall Z[Z \in \tilde{E} \Rightarrow \text{not } \tilde{P}_h(\ldots, Z)] \tag{4.2.10 b}$$

is a theorem in $\mathscr{M}\mathscr{T}_\Sigma$.

A hypothesis is allowed but not theoretically existent if (4.2.3) resp. (4.1.3) can be added as axiom to $\mathscr{M}\mathscr{T}_\Sigma\mathscr{A}$ without contradiction whereas (4.2.9) is not a theorem in $\mathscr{M}\mathscr{T}_\Sigma$. That (4.2.3) resp. (4.1.3) can be added as axiom to $\mathscr{M}\mathscr{T}_S\mathscr{A}$ is equivalent to:

$$\exists Z[Z \in \tilde{E} \text{ and } \tilde{E}_h(Z) \neq \emptyset] \tag{4.2.11 a}$$

resp.

$$\exists Z[Z \in \tilde{E} \text{ and } \tilde{P}_h(\ldots, Z)] \tag{4.2.11 b}$$

can be added as axiom to $\mathscr{M}\mathscr{T}_\Sigma$ without contradiction. That (4.2.9) is not a theorem in $\mathscr{M}\mathscr{T}_\Sigma$ is equivalent to that

$$\exists Z[Z \in \tilde{E} \text{ and } \tilde{E}_h(Z) = \emptyset] \tag{4.2.12 a}$$

resp.

$$\exists Z[Z \in \tilde{E} \text{ and not } \tilde{P}_h(\ldots, Z)] \tag{4.2.12 b}$$

can be added as axiom to $\mathscr{M}\mathscr{T}_\Sigma$ without contradiction; this is equivalent to [neg $\mathscr{H}$] being allowed.

Decisive for the classification of a hypothesis are the four relations (4.2.9), (4.2.10), (4.2.11) and (4.2.12). With the set

$$\tilde{E}_+ = \{Z \mid Z \in \tilde{E} \text{ and } \tilde{E}_h(Z) \neq \emptyset\} \tag{4.2.13a}$$

resp.

$$\tilde{E}_+ = \{Z \mid Z \in \tilde{E} \text{ and } \tilde{P}_h(\ldots, Z)\} \tag{4.2.13b}$$

the following equivalences hold:

$$\begin{aligned} &\tilde{E}_+ = \tilde{E} \Leftrightarrow (4.2.9),\\ &\tilde{E}_+ \neq \tilde{E} \Leftrightarrow (4.2.12),\\ &\tilde{E}_+ \neq \emptyset \Leftrightarrow (4.2.11),\\ &\tilde{E}_+ = \emptyset \Leftrightarrow (4.2.10). \end{aligned} \tag{4.2.14}$$

$\tilde{E} = \emptyset$ is no theorem in $\mathcal{M}\mathcal{T}_\Sigma$ since $\mathcal{M}\mathcal{T}_\Sigma \mathcal{A}$ is presumed to be not contradictory.

Thus we see that the following six cases are possible (we shorten: "it" for "is theorem in $\mathcal{M}\mathcal{T}_\Sigma$"; "ad" for "can be added as axiom to $\mathcal{M}\mathcal{T}_\Sigma$ without contradiction"):

$$\begin{aligned} &[+1] \quad \tilde{E}_+ = \tilde{E}\ \textit{it};\\ &[+] \quad \left\{\begin{matrix} \tilde{E} \neq \emptyset \Rightarrow \tilde{E}_+ \neq \emptyset\ \textit{it},\\ \tilde{E}_+ = \tilde{E}\ \textit{ad},\ \tilde{E} \neq \emptyset \Rightarrow \tilde{E}_+ \neq \tilde{E}\ \textit{ad}; \end{matrix}\right\}\\ &[0] \quad \left\{\begin{matrix} \tilde{E} \neq \emptyset \Rightarrow \tilde{E}_+ \neq \emptyset\ \textit{it},\\ \tilde{E} \neq 0 \Rightarrow \tilde{E}_+ \neq \tilde{E}\ \textit{it}; \end{matrix}\right\}\\ &[?] \quad \left\{\begin{matrix} \tilde{E}_+ = \tilde{E}\ \textit{ad},\ \tilde{E} \neq \emptyset \Rightarrow \tilde{E}_+ \neq \tilde{E}\ \textit{ad},\\ \tilde{E} \neq \emptyset \Rightarrow \tilde{E}_+ \neq 0\ \textit{ad}, \tilde{E}_+ = \emptyset\ \textit{ad}; \end{matrix}\right\}\\ &[-] \quad \left\{\begin{matrix} \tilde{E} \neq \emptyset \Rightarrow \tilde{E}_+ \neq \tilde{E}\ \textit{it},\\ \tilde{E} \neq \emptyset \Rightarrow \tilde{E}_+ \neq \emptyset\ \textit{ad}, \tilde{E}_+ = \emptyset\ \textit{ad}; \end{matrix}\right\}\\ &[-1] \quad \tilde{E}_+ = \emptyset\ \textit{it}. \end{aligned} \tag{4.2.15}$$

These six cases can be reduced if some special theorems hold, e.g. $E_+ \neq \emptyset \Rightarrow \tilde{E}_+ = \tilde{E}$. We will not discuss this here. For the case without observational report, $\mathcal{M}\mathcal{T}_\Sigma \mathcal{A}$ equals $\mathcal{M}\mathcal{T}_\Sigma$ and we have only the three cases

$$\begin{aligned} &[+1] \quad \tilde{E}_h \neq \emptyset\ \textit{it}; && \text{resp. } \tilde{P}_h\ \textit{it};\\ &[?] \quad \tilde{E}_h \neq \emptyset\ \textit{ad}, \tilde{E}_h = \emptyset\ \textit{ad}; && \text{resp. } \tilde{P}_h\ \textit{ad}, \text{not } \tilde{P}_h\ \textit{ad};\\ &[-1] \quad \tilde{E}_h = \emptyset\ \textit{it}; && \text{resp. not } \tilde{P}_h\ \textit{it}. \end{aligned} \tag{4.2.16}$$

Here $\tilde{E}_h = \{X \mid X \in \tilde{T}_h(y_1, \ldots, y_r, \mathbf{R})$ and $\tilde{P}_h(y_1, \ldots, y_r; t, \mathbf{R})\}$, i.e. $\tilde{E}_h$ and $\tilde{P}_h$ are identical with (4.1.4), resp. (4.1.3) but without $A$.

We see that $[+1]$ is equivalent to "$\mathcal{H}$ is theoretically existent", $[-1]$ to "$\mathcal{H}$ is false". In the other cases $[+]$, $[0]$, $[?]$, $[-]$, $\mathcal{H}$ is allowed. In the cases $[+]$ and $[0]$, $\tilde{E} \neq \emptyset \Rightarrow \tilde{E}_+ \neq \emptyset$ is a theorem in $\mathcal{M}\mathcal{T}_\Sigma$. In the cases $[?]$ and $[-]$, $\mathcal{M}\mathcal{T}_\Sigma$ does not decide between $\tilde{E}_+ = \emptyset$ and $\neq \emptyset$. Therefore, let us call $\mathcal{H}$ "strongly allowed" in the cases $[+]$, $[0]$, and "weakly allowed" in the other cases $[?]$, $[-]$. In the cases $[0]$ and $[-]$, $\tilde{E} \neq \emptyset \Rightarrow \tilde{E}_+ \neq \tilde{E}$ is a theorem in $\mathcal{M}\mathcal{T}_\Sigma$. This theorem says that there are elements $Z \in \tilde{E}$ for which $\tilde{E}_h(Z) = \emptyset$. Therefore it cannot be excluded that the observational report is such a $Z \in \tilde{E}$ that $\tilde{E}_h(Z) = \emptyset$. This may called "restrictively allowed". Thus the following classification of hypotheses $\mathscr{H}$ is possible:

Classification scheme

| | coarse | normal | fine |
|---|---|---|---|
| [+1] | allowed | strongly allowed | theoretically existent |
| [+] | | | unrestrictively strongly allowed |
| [0] | | | restrictively strongly allowed |
| [?] | | weakly allowed | unrestrictively weakly allowed |
| [−] | | | restrictively weakly allowed |
| [−1] | false | false | false |

$\mathscr{H}$ determined takes in $\mathscr{M}\mathscr{T}_\Sigma$ the form of the theorem:

$$\forall Z[Z \in \tilde{E}_+ \Rightarrow \tilde{E}_h(Z) \text{ has only one element}]. \tag{4.2.17}$$

The right side of (4.2.17) says that a mapping $\tilde{E}_+ \xrightarrow{f} \tilde{T}_h(\ldots)$ is defined by

$$Z \to X \in \tilde{E}_h(Z).$$

We have already encountered such mappings $f$ of a subset $\tilde{E}_+$ of $\tilde{E}$ into an echelon set $T_h(\ldots)$ in §3.1: $\tilde{E}_+$ is of type ($\alpha$) and therefore $f E_+ = \bigcup_{Z \in \tilde{E}_+} \tilde{E}_h(Z)$ of type ($\beta$). We shall see later that mappings $f$ of a subset of $\tilde{E}$ into echelon sets are decisive in physics.

## §4.3 Relations Between Various Hypotheses

In this section we presume for all hypotheses the same observational report, resp. no observational report.

We begin with the comparison of two hypotheses $\mathscr{H}_1$ and $\mathscr{H}_2$ which are made separately, i.e. we investigate the two mathematical theories $\mathscr{M}\mathscr{T}_\Sigma \mathscr{A} \mathscr{H}_1$ and $\mathscr{M}\mathscr{T}_\Sigma \mathscr{A} \mathscr{H}_2$.

We call $\mathscr{H}_1$ "sharper" than $\mathscr{H}_2$ if

$$\tilde{E}_h^{(1)}(Z) \subset \tilde{E}_h^{(2)}(Z) \quad \text{for all } Z \in \tilde{E} \tag{4.3.1 a}$$

is a theorem in $\mathscr{M}\mathscr{T}_\Sigma$. Then $\mathscr{H}_1$ and $\mathscr{H}_2$ have "the same" invented elements in the sense that (4.3.1 a) implies

$$T_h^{(1)}(\ldots) = T_h^{(2)}(\ldots). \tag{4.3.2}$$

Only the relation $X \in \tilde{E}_h^{(1)}(A)$ is sharper than the relation $X \in \tilde{E}_h^{(2)}(A)$. For hypotheses $\mathscr{H}_1$, $\mathscr{H}_2$ without invented elements, (4.3.1 a) must be replaced by

$$\tilde{P}_h^{(1)}(y_1, \ldots, y_r, t, Z) \Rightarrow \tilde{P}_h^{(2)}(y_1, \ldots, y_r, t, Z) \quad \text{for all } Z \in \tilde{E}. \tag{4.3.1 b}$$

If there is no observational report, (4.3.1 a) must be replaced by

$$\tilde{E}_h^{(1)} \subset \tilde{E}_h^{(2)} \tag{4.3.1 c}$$

resp.

$$\tilde{P}_h^{(1)}(y_1, \ldots, y_r, t) \Rightarrow \tilde{P}_h^{(2)}(y_1, \ldots, y_r, t). \tag{4.3.1 d}$$

We call $\mathscr{H}_1$ an extension of $\mathscr{H}_2$ for fixed observational report, if $\mathscr{H}_1$ has the form

$$X \in \tilde{E}_h^{(n)}(A) \quad \text{and} \quad X' \in \tilde{E}_h^{(e)}(A, X) \tag{4.3.3 a}$$

and if

$$\tilde{E}_h^{(n)} \subset \tilde{E}_h^{(2)}(Z) \quad \text{for all } Z \in \tilde{E} \tag{4.3.3 b}$$

is a theorem in $\mathscr{M}\mathscr{T}_\Sigma$. Then the hypothesis $X \in \tilde{E}_h^{(n)}(A)$ is sharper than $\mathscr{H}_2$. But $\mathscr{H}_1$ contains more invented elements than $\mathscr{H}_2$, namely the components of $X'$.

The set $E_h^{(1)}(Z)$ takes the form

$$\tilde{E}_h^{(1)}(Z) = \{X, X') \mid X \in \tilde{E}_h^{(n)}(Z) \text{ and } X' \in \tilde{E}_h^{(e)}(Z, X)\}.$$

With

$$\tilde{E}_h^{(n,e)}(Z) = \{X \mid X \in \tilde{E}_h^{(n)}(Z) \text{ and } \tilde{E}_h^{(e)}(Z, X) \neq \emptyset\} \subset \tilde{E}_h^{(n)}(Z) \subset \tilde{E}_h^{(2)}(Z)$$

we obtain

$$\begin{aligned} \tilde{E}_h^{(1)}(Z) &= \{Z \mid Z \in \tilde{E} \text{ and } \tilde{E}_h^{(1)}(Z) \neq \emptyset\} \\ &= \{Z \mid Z \in \tilde{E} \text{ and } \tilde{E}_h^{(n,e)}(Z) \neq \emptyset\} \\ &\subset \{Z \mid Z \in \tilde{E} \text{ and } \tilde{E}_h^{(2)}(Z) \neq \emptyset\} \end{aligned}$$

and thus

$$\tilde{E}_+^{(1)} \subset \tilde{E}_+^{(2)}. \tag{4.3.4 a}$$

If $\mathscr{H}_2$ has no invented elements, we must replace (4.3.3) by

$$\tilde{P}_h^{(n)}(y_1, \ldots, y_r, t, A) \quad \text{and} \quad X' \in \tilde{E}_h^{(e)}(A) \tag{4.3.5 a}$$

and

$$\tilde{P}_h^{(n)}(y_1, \ldots, y_r, t, Z) \Rightarrow \tilde{P}_h^{(2)}(y_1, \ldots, y_r, t, Z) \quad \text{for all } Z \in \tilde{E}. \tag{4.3.5 b}$$

For $\mathscr{H}_1$ we then have

$$\tilde{T}_h^{(1)}(\ldots) = \tilde{T}_h^{(e)}(\ldots)$$

and

$$\tilde{P}_h^{(1)}(\ldots): \tilde{P}_h^{(n)}(\ldots) \text{ and } \tilde{P}_h^{(e)}(\ldots).$$

This implies

$$\tilde{E}_h^{(1)}(Z) = \{X' \mid X' \in \tilde{E}^{(e)}(Z) \text{ and } \tilde{P}_h^{(n)}(y_1, \ldots, Z)\},$$

and therefore

$$\begin{aligned} \tilde{E}_+^{(1)} &= \{Z \mid Z \in \tilde{E} \text{ and } \tilde{E}_h^{(1)}(Z) \neq \emptyset\} \\ &\subset \{Z \mid Z \in \tilde{E} \text{ and } \tilde{P}_h^{(n)}(\ldots, Z)\} \\ &\subset \{Z \mid Z \in \tilde{E} \text{ and } \tilde{P}_h^{(2)}(\ldots, Z)\} = \tilde{E}_+^{(2)}, \end{aligned}$$

i.e. (4.3.4 a).

If there is no observational report, (4.3.4 a) must be replaced by

$$\tilde{E}_h^{(1)} \neq \emptyset \Rightarrow \tilde{E}_h^{(2)} \neq \emptyset, \tag{4.3.4 b}$$

and if moreover $\mathscr{H}_2$ has no invented $X$, (4.3.4b) must be replaced by

$$\tilde{E}_h^{(1)} \neq \emptyset \Rightarrow \tilde{P}_h^{(2)}. \tag{4.3.4c}$$

(4.3.4) does not imply (4.3.3) resp. (4.3.5). We call $\mathscr{H}_1$ "more restrictive" than $\mathscr{H}_2$ if (4.3.4) is a theorem. "More restrictive" is therefore less than "extension for fixed observational report".

The most significant procedure is the composition of hypotheses. We define the composition of $\mathscr{H}_1$, $\mathscr{H}_2$ as the hypothesis $\mathscr{H}$ given by

$$X_1 \in \tilde{E}_h^{(1)}(A) \quad \text{and} \quad X_2 \in \tilde{E}_h^{(2)}(A). \tag{4.3.6a}$$

For $\mathscr{H}$ we thus have:

$$\tilde{E}_h(A) = \tilde{E}_h^{(1)}(A) \times \tilde{E}_h^{(2)}(A) \tag{4.3.7a}$$

and $X = (X_1, X_2)$; hence (4.3.6) takes the form

$$X = (X_1, X_2) \in \tilde{E}_h^{(1)}(A) \times \tilde{E}_h^{(2)}(A). \tag{4.3.6b}$$

This implies

$$\begin{aligned} \tilde{E}_+ &= \{Z \mid Z \in \tilde{E} \text{ and } \tilde{E}_h(Z) \neq \emptyset\} \\ &= \{Z \mid Z \in \tilde{E} \text{ and } \tilde{E}_h^{(1)}(Z) \times \tilde{E}_h^{(2)}(Z) \neq \emptyset\} \\ &= \{Z \mid Z \in \tilde{E} \text{ and } \tilde{E}_h^{(1)}(Z) \neq \emptyset \text{ and } \tilde{E}_h^{(2)}(Z) \neq \emptyset\} \end{aligned}$$

and thus

$$\tilde{E}_+ = \tilde{E}_+^{(1)} \cap \tilde{E}_+^{(2)}. \tag{4.3.8a}$$

The same can simply be proved for hypotheses without invented $X_1$ or $X_2$.

If there is no observational report, we must omit the letter $A$ in (4.3.7a). If moreover there is no $X_1$, (4.3.7a) must be replaced by

$$\tilde{E}_h = \{X_2 \mid X_2 \in \tilde{E}_h^{(2)} \text{ and } \tilde{P}_h^{(1)}(y_1, \ldots, t)\}. \tag{4.3.7b}$$

If moreover $X_2$ is absent also, (4.3.7b) must be replaced by

$$\tilde{P}_h(y_1, \ldots, t) \colon \tilde{P}_h^{(1)}(y_1, \ldots, t) \text{ and } \tilde{P}_h^{(2)}(y_1, \ldots, t). \tag{4.3.7c}$$

We define $\mathscr{H}_1$, $\mathscr{H}_2$ as "compatible" if $\mathscr{H}_1$, $\mathscr{H}_2$ *and* the composite hypothesis $\mathscr{H}$ are at least allowed, i.e. if

$$\tilde{E}_+^{(1)} \cap \tilde{E}_+^{(2)} = \emptyset \tag{4.3.9a}$$

is no theorem in $\mathscr{M}\mathscr{T}_\Sigma$.

If there is no observational report, (4.3.9a) must be replaced by

$$\begin{aligned} &\tilde{E}_h^{(1)} = \emptyset \quad \text{or} \quad \tilde{E}_h^{(2)} = \emptyset; \\ \text{resp.:}\ &\text{not } \tilde{P}^{(1)} \quad \text{or} \quad E_h^{(2)} = \emptyset \\ \text{resp.:}\ &\text{not } P^{(1)} \quad \text{or} \quad \text{not } P^{(2)}. \end{aligned} \tag{4.3.9b}$$

If in the absence of an observational report $\mathscr{H}_1$ and $\mathscr{H}_2$ are of the first kind, they must contain invented $X_1$ and $X_2$. If $\mathscr{H}_1$ and $\mathscr{H}_2$ are allowed, neither $\tilde{E}_h^{(1)} = \emptyset$ nor $\tilde{E}_h^{(2)} = \emptyset$ are theorems in $\mathscr{M}\mathscr{T}_\Sigma$. Nevertheless "$\tilde{E}_h^{(1)} = \emptyset$ or $\tilde{E}_h^{(2)} = \emptyset$" can be a theorem in $\mathscr{M}\mathscr{T}_\Sigma$, i.e. $\mathscr{H}_1$ and $\mathscr{H}_2$ can be incompatible.

$\mathscr{M}\mathscr{T}_\Sigma$ is called "weakly complete" if every pair $\mathscr{H}_1$, $\mathscr{H}_2$ of the first kind *without* observational report is compatible. In §4.6 we will see why we desire at least a weakly complete theory. We emphasize that also for a weakly complete theory two $\mathscr{H}_1$, $\mathscr{H}_2$ *with* an observational report can be incompatible!

A hypothesis $\mathscr{H}$ is called "experimentally certain" if $\mathscr{H}$ is compatible with every allowed hypothesis of the first kind. For every allowed hypothesis $\mathscr{H}$ of the first kind, [neg $\mathscr{H}$] is not experimentally certain since $\mathscr{H}$ and [neg $\mathscr{H}$] are not compatible.

A hypothesis $\mathscr{H}$ is experimentally certain if and only if for every allowed hypothesis $\mathscr{H}_1$ of the first kind,

$$\tilde{E}_+ \cap \tilde{E}_+^{(1)} = \emptyset \tag{4.3.10a}$$

resp. in the case without observational report

$$\tilde{E}_h = \emptyset \quad \text{or} \quad \tilde{E}_h^{(1)} = \emptyset$$

and without $X$ (4.3.10b)

$$[\text{not } \tilde{P}_h] \quad \text{or} \quad \tilde{E}_h^{(1)} = \emptyset$$

is no theorem in $\mathscr{M}\mathscr{T}_\Sigma$.

$\mathscr{H}$ is experimentally uncertain if and only if there is an allowed hypothesis $\mathscr{H}_1$ of the first kind, for which (4.3.10) is a theorem. Because of $\tilde{E}_+^{(1)} \subset \tilde{E}$, the relation (4.3.10a) is equivalent to

$$\tilde{E}_+^{(1)} \subset \tilde{E} \setminus \tilde{E}_+, \tag{4.3.11a}$$

From the definition of [neg $\mathscr{H}$] follows that the set $\tilde{E}'_+$ (belonging to [neg $\mathscr{H}$]) equals $\tilde{E} \setminus \tilde{E}_+$.

Therefore (4.3.11a) is equivalent to $\mathscr{H}_1$ being more restrictive than [neg $\mathscr{H}$].

(4.3.10b) is equivalent to

$$\begin{aligned} &\tilde{E}_h^{(1)} \neq \emptyset \Rightarrow \tilde{E}_h = \emptyset, \\ \text{resp.} \quad &\tilde{E}_h^{(1)} \neq \emptyset \Rightarrow \text{not } \tilde{P}_h. \end{aligned} \tag{4.3.11b}$$

Thus we find again that $\mathscr{H}$ is experimentally uncertain if $\mathscr{H}_1$ is more restrictive than [neg $\mathscr{H}$].

[neg $\mathscr{H}$] is experimentally certain, if there is no allowed hypothesis $\mathscr{H}_1$ of the first kind for which

$$(\tilde{E} \setminus \tilde{E}_+) \cap \tilde{E}_+^{(1)} = \emptyset, \tag{4.3.12a}$$

resp. without observational report

$$\tilde{E}_h^{(1)} \neq \emptyset \Rightarrow \tilde{E}_h \neq \emptyset,$$

or without $X$ (4.3.12b)

$$\tilde{E}_h^{(1)} \neq \emptyset \Rightarrow \tilde{P}_h$$

are theorems in $\mathscr{M}\mathscr{T}_\Sigma$.

[neg $\mathscr{H}$] is experimentally uncertain, if there is an allowed hypothesis $\mathscr{H}_1$ of the first kind with (4.3.12), i.e. with

$$\tilde{E}_+^{(1)} \subset \tilde{E}_+ \tag{4.3.13a}$$

$$\begin{aligned} \text{resp.} \quad &\tilde{E}_h^{(1)} \neq \emptyset \Rightarrow \tilde{E}_h \neq \emptyset, \\ \text{resp.} \quad &\tilde{E}_h^{(1)} \neq \emptyset \Rightarrow \tilde{P}_h. \end{aligned} \tag{4.3.13b}$$

This is equivalent to $\mathscr{H}_1$ being more restrictive than $\mathscr{H}$.

Thus [neg $\mathscr{H}$] experimentally uncertain implies $\mathscr{H}$ allowed. We denote $\mathscr{H}$ for which [neg $\mathscr{H}$] is experimentally uncertain as ec-allowed. Every allowed $\mathscr{H}$ of the first kind is also ec-allowed, since such an $\mathscr{H}$ is more restrictive than itself. An allowed $\mathscr{H}$ is not ec-allowed if [neg $\mathscr{H}$] is experimentally certain.

If $\mathscr{H}$ is compatible with all experimentally certain hypotheses, then [neg $\mathscr{H}$] must be experimentally uncertain since $\mathscr{H}$ is not compatible with [neg $\mathscr{H}$]. Thus $\mathscr{H}$ must be ec-allowed.

If $\mathscr{H}$ is experimentally certain, then $\mathscr{H}$ is compatible with all ec-allowed hypotheses; this can be proved as follows: $\mathscr{H}_2$ ec-allowed implies that there is an $\mathscr{H}_1$ of the first kind and more restrictive than $\mathscr{H}_2$. If $\mathscr{H}_2$ were incompatible with $\mathscr{H}$, then $\mathscr{H}_2$ and therefore all the more $\mathscr{H}_1$ would be more restrictive than [neg $\mathscr{H}$] and thus $\mathscr{H}$ experimentally uncertain. Thus the ec-allowed hypotheses are precisely all the hypotheses compatible with all experimentally certain hypotheses. $\mathscr{H}$ compatible with all ec-allowed hypotheses implies that $\mathscr{H}$ is compatible with all allowed hypotheses of the first kind and therefore experimentally certain.

Thus we see a duality between experimentally certain and ec-allowed: $\mathscr{H}$ is ec-allowed if and only if $\mathscr{H}$ is compatible with all experimentally certain hypotheses; $\mathscr{H}$ is experimentally certain if and only if $\mathscr{H}$ is compatible with all ec-allowed hypotheses.

The concept of experimentally certain seems to us not restrictive enough since both $\mathscr{H}$ and [neg $\mathscr{H}$] can be experimentally certain! Then there are two experimentally certain hypotheses which are not compatible. We ask for all allowed $\mathscr{H}$ which are not compatible with all experimentally certain hypotheses. Such $\mathscr{H}$ cannot be ec-allowed, i.e. [neg $\mathscr{H}$] must be experimentally certain. To an experimentally certain $\mathscr{H}$ there is another experimentally certain $\mathscr{H}_1$ incompatible with $\mathscr{H}$ if and only if [neg $\mathscr{H}$] is experimentally certain.

We seek to sharpen the concept of certain in such a manner that all certain hypotheses are compatible. A first suggestive demand is:

$\mathscr{H}$ be called "almost certain" if $\mathscr{H}$ is experimentally certain *and* ec-allowed. $\mathscr{H}$ be denoted as "ac-allowed", if [neg $\mathscr{H}$] is not almost certain.

It follows that an ec-allowed hypothesis is also ac-allowed and that an allowed hypothesis of the first kind is also ac-allowed.

A hypothesis of the first kind which is experimentally certain is also almost certain, since every allowed hypothesis of the first kind is also ec-allowed (see above).

If $\mathscr{H}$ is compatible with all almost certain hypotheses, [neg $\mathscr{H}$] cannot be almost certain, since $\mathscr{H}$ is not compatible with [neg $\mathscr{H}$]. Therefore $\mathscr{H}$ is ac-allowed.

Let $\mathscr{H}$ vice versa be ac-allowed, i.e. [neg $\mathscr{H}$] not almost certain (i.e. [neg $\mathscr{H}$] not experimentally certain or not ec-allowed) then $\mathscr{H}$ is ec-allowed, or [neg [neg $\mathscr{H}$]] and thus $\mathscr{H}$ is experimentally certain. In both cases $\mathscr{H}$ is compatible with all almost certain hypotheses since these are experimentally certain *and* ec-allowed.

Let $\mathscr{H}$ be incompatible with an ac-allowed $\mathscr{H}_1$. Then $\mathscr{H}_1$ is more restrictive than [neg $\mathscr{H}$], and experimentally certain or ec-allowed. Therefore [neg $\mathscr{H}$] is experimentally certain or there is a hypothesis $\mathscr{H}_2$ of the first kind which is more restrictive than $\mathscr{H}_1$. Then $\mathscr{H}_2$ is also more restrictive than [neg $\mathscr{H}$], i.e. $\mathscr{H}$ is experimentally uncertain. Thus $\mathscr{H}$ is not almost certain.

There is again a duality: $\mathscr{H}$ is ac-allowed if and only if $\mathscr{H}$ is compatible with all almost certain hypotheses. $\mathscr{H}$ is almost certain if and only if $\mathscr{H}$ is incompatible with all ac-allowed hypotheses.

We have seen that $\mathscr{H}$ is almost certain if $\mathscr{H}$ is experimentally certain and if there is a $\mathscr{H}_1$ of the first kind which is more restrictive than $\mathscr{H}$. This suggests to sharpen the concept of almost certain as follows:

A hypothesis $\mathscr{H}$ be called "certain" if to every allowed $\mathscr{H}_2$ of the first kind there is an $\mathscr{H}_1$ of the first kind, compatible with $\mathscr{H}_2$ and more restrictive than $\mathscr{H}$.

Another hypothesis $\mathscr{H}$ be called "c-allowed", if [neg $\mathscr{H}$] is uncertain.

Let $\mathscr{H}$ be certain and $\mathscr{H}_2$ allowed and of the first kind. Then there is an $\mathscr{H}_1$ of the first kind, compatible with $\mathscr{H}_2$ and more restrictive than $\mathscr{H}$; thus $\mathscr{H}_2$ is all the more compatible with $\mathscr{H}$ and therefore $\mathscr{H}$ experimentally certain. Every experimentally certain $\mathscr{H}$ of the first kind is also certain since $\mathscr{H}_1$ in the definition of "certain" can be replaced by $\mathscr{H}$.

There follows at once that every ac-allowed $\mathscr{H}$, every ec-allowed $\mathscr{H}$ and every allowed $\mathscr{H}$ of the first kind are also c-allowed.

Let $\mathscr{H}$ be compatible with all certain hypotheses. Then [neg $\mathscr{H}$] must be uncertain and therefore $\mathscr{H}$ c-allowed. Let vice versa $\mathscr{H}_3$ be certain and not compatible with $\mathscr{H}$. Thus $\mathscr{H}_3$ is more restrictive than [neg $\mathscr{H}$]. Since $\mathscr{H}_3$ is certain, for every $\mathscr{H}_2$ of the first kind there is an $\mathscr{H}_1$ of the first kind, compatible with $\mathscr{H}_2$ and more restrictive than $\mathscr{H}_3$. Hence $\mathscr{H}_1$ is also more restrictive than [neg $\mathscr{H}$], i.e. [neg $\mathscr{H}$] is certain. Thus $\mathscr{H}$ compatible with all certain hypotheses is equivalent to [neg $\mathscr{H}$] uncertain.

Let $\mathscr{H}$ be compatible with all c-allowed hypotheses. Then [neg $\mathscr{H}$] is not c-allowed, i.e. [neg [neg $\mathscr{H}$]] must be certain and therefore also $\mathscr{H}$.

There is again a duality: $\mathscr{H}$ is c-allowed if and only if $\mathscr{H}$ is compatible with all certain hypotheses; $\mathscr{H}$ is certain if and only if $\mathscr{H}$ is compatible with all c-allowed hypotheses.

We can sharpen even more the concept of certain: $\mathscr{H}$ be called "perfectly certain" if there is an experimentally certain $\mathscr{H}_1$ of the first kind, which is more restrictive than $\mathscr{H}$.

Since one can choose $\mathscr{H}_1$ in the definition of "certain" equal to that in the definition of "perfectly certain", this last concept is sharper than "certain".

$\mathscr{H}$ be called "pc-allowed" if [neg $\mathscr{H}$] is not perfectly certain. We easily see that an experimentally certain $\mathscr{H}$ of the first kind is also perfectly certain and that an allowed $\mathscr{H}$ of the first kind is also pc-allowed.

We have again a duality: $\mathscr{H}$ is perfectly certain if and only if $\mathscr{H}$ is compatible with all pc-allowed hypotheses. $\mathscr{H}$ is pc-allowed if and only if $\mathscr{H}$ is compatible with all perfectly certain hypotheses.

If $\mathscr{H}$ is theoretically existent, it is compatible with all allowed hypotheses and therefore perfectly certain. Then $\mathscr{H}$ is all the more certain, almost certain and experimentally certain.

We repeat: For hypotheses of the first kind the concepts allowed, ac-allowed, c-allowed and pc-allowed are equivalent. The same is true for experimentally certain, almost certain, certain and perfectly certain.

The classification can be simplified for weakly closed theories $\mathscr{M}\mathscr{T}_\Sigma$, and for hypotheses which are not connected with the observational report. Here $\mathscr{H}$ not connected with the observational report means that $A$ is absent in (4.1.2) and thus also in (4.1.4).

Since $\mathcal{M}\mathcal{T}_\Sigma$ is weakly complete, every allowed hypothesis of the first kind which is not connected with the observational report, is experimentally certain. Therefore it is almost certain, certain and perfectly certain.

Let $\mathcal{H}$ be not connected with the observational report and almost certain. Then there is an $\mathcal{H}_1$ of the first kind which is more restrictive than $\mathcal{H}$. Such an $\mathcal{H}_1$ can be connected with the observational report. If we replace the letters $a_i$ of the observational report by invented elements $x_i'$, we get a hypothesis $\mathcal{H}_0'$ of the same form as the observational report. We define $\mathcal{H}_1'$ as the extension of $\mathcal{H}_0'$ which we get by the relations of $\mathcal{H}_1$ with the $a_i$ replaced by the $x_i$! Then $\mathcal{H}_1'$ is not connected with the observational report. Since the observational report together with $\mathcal{H}_1$ were more restrictive than $\mathcal{H}$, and $\mathcal{H}$ is not connected with the observational report, also $\mathcal{H}_1'$ must be more restrictive than $\mathcal{H}$. Then $\mathcal{H}_1'$ is experimentally certain since it is not connected with the observational report. Thus $\mathcal{H}$ is perfectly certain (and all the more certain).

If $\mathcal{M}\mathcal{T}_\Sigma$ is weakly complete, the concepts of almost certain, certain, and perfectly certain are equivalent for hypotheses not connected with the observational report. Therefore it is usual, for weakly complete $\mathcal{M}\mathcal{T}_\Sigma$ to "forget" the observational report for the discussion of hypotheses not connected with the observational report.

In the same way, for weakly complete $\mathcal{M}\mathcal{T}_\Sigma$ one deduces that we can forget all those parts of the observational report which by their letters are separated from that part of the observational report with which the interesting hypotheses are connected. A weakly complete theory is "so good" that we need not think of all experiments ever done if we "work with the theory".

Let $\mathcal{M}\mathcal{T}_\Sigma$ be weakly complete, and $\mathcal{H}$ a hypothesis without invented elements $X$ and not connected with the observational report, i.e. a relation $\tilde{P}_h(y_1, \ldots, y_r, t)$. If [not $\tilde{P}_h$] is experimentally uncertain, i.e. refutable in the sense of §2.5 (and therefore $\tilde{P}_h$ experimentally deducible), then $\tilde{P}_h$ is perfectly certain. To add $\tilde{P}_h$ as axiom to $\mathcal{M}\mathcal{T}_\Sigma$ does not change the physical contents of the theory. One only sharpens $\tilde{P}_h$ from perfectly certain to theoretically existent. We therefore in such cases add an axiom which makes $\tilde{P}_h$ theoretically existent if such an additional axiom makes the mathematical part of $\mathcal{P}\mathcal{T}$ more perspicuous.

Such an addition is a special case of a problem already treated in §2.5. Therefore let us here give some new aspects of this problem with the aid of the newly introduced concepts. For this purpose we do not presume $\mathcal{M}\mathcal{T}_\Sigma$ as weakly complete. If $\tilde{P}_h$ is experimentally uncertain, i.e. refutable but [not $\tilde{P}_h$] no theorem in $\mathcal{M}\mathcal{T}_\Sigma$, then the addition of $\tilde{P}_h$ as axiom would restrict the class of allowed hypotheses of the first kind and in this sense sharpen the physical theory. If $\tilde{P}_h$ is experimentally certain and [not $\tilde{P}_h$] uncertain, we can add $\tilde{P}_h$ as axiom without restricting the class of allowed hypotheses of the first kind.

If both $\tilde{P}_h$ and [not $\tilde{P}_h$] are experimentally certain, one could have the impression that we may add $\tilde{P}_h$ or [not $\tilde{P}_h$] as axiom. This judgement would be rash. If $\tilde{P}_h$ is of the form $\forall x(A(x))$ and thus [not $\tilde{P}_h$] of the form $\exists x(B(x))$ (with $B = \text{not } A$), it is for mathematical reasons important which of the two relations we add. That $\forall x(A(x))$ is experimentally certain says that there is no hypothesis of the first kind from which $\exists x(B(x))$ follows. An axiom $\forall x(A(x))$ is not only the better physical formulation of the experimental certainty of $\forall x(A(x))$, but also a mathematically more restrictive formulation.

If $\tilde{P}_h$ is experimentally certain and of the form $\forall x(A(x))$, and if we add $\tilde{P}_h$ as axiom, then $\tilde{P}_h$ becomes theoretically existent. Therefore we regard such an experimentally certain $\tilde{P}_h$ as almost equivalent to a theoretically existent one.

As an example we contemplate the relation (4.2.17). If this is a theorem, the hypothesis $\mathscr{H}$ is called determined. If (4.2.17) is experimentally certain (also if in addition its negation is experimentally certain), we call $\mathscr{H}$ "almost determined" and regard this "almost determined" as equally "good" as the sharper "determined".

In general the classes of almost certain, certain and perfectly certain hypotheses will not coincide. We prefer the concept certain. It appears to us as the least stringent concept which has the following additional property.

If $\mathscr{H}$ and $\mathscr{H}'$ are certain (perfectly certain) then the composite hypothesis $\mathscr{H}^*$ is also certain (perfectly certain).

Let us sketch the proof for certain hypotheses. Let $\tilde{E}_+$ and $\tilde{E}'_+$ be the sets belonging to $\mathscr{H}$ and $\mathscr{H}'$. Then $\tilde{E}_+ \cap \tilde{E}'_+$ belongs to $\mathscr{H}^*$.

Let an $\mathscr{H}_2$ of the first kind be given. Since $\mathscr{H}$ is certain, there is an $\mathscr{H}_1$ of the first kind, compatible with $\mathscr{H}_2$, such that $\tilde{E}_+^{(1)} \subset \tilde{E}_+$ is a theorem in $\mathscr{M}\mathscr{T}_\Sigma$. Since $\mathscr{H}_1$ and $\mathscr{H}_2$ are compatible, the hypothesis $\mathscr{H}_3$ composed of them is allowed and of the first kind with the set $\tilde{E}_+^{(3)} = \tilde{E}_+^{(1)} \cap \tilde{E}_+^{(2)}$. Since $\mathscr{H}'$ is certain there is an $\mathscr{H}_4$ of the first kind, compatible with $\mathscr{H}_3$, such that $\tilde{E}_+^{(4)} \subset \tilde{E}'_+$ is a theorem. Since $\mathscr{H}_3$ and $\mathscr{H}_4$ are compatible, $\tilde{E}_+^{(3)} \cap \tilde{E}_+^{(4)} = \tilde{E}_+^{(1)} = \tilde{E}_+^{(2)} \cap \tilde{E}_+^{(4)} = \emptyset$ is no theorem. Therefore the hypothesis $\mathscr{H}_5$ composed of $\mathscr{H}_1$ and $\mathscr{H}_4$ is compatible with $\mathscr{H}_2$ and

$$\tilde{E}_+^{(5)} = \tilde{E}_+^{(1)} \cap \tilde{E}_+^{(4)} \subset \tilde{E}_+ \cap \tilde{E}'_+,$$

i.e. $\mathscr{H}^*$ is certain.

The proof for hypotheses without observational report and the proof for perfectly certain hypotheses may be left to the reader.

If for $\mathscr{H}$ there is a theoretically existent $\mathscr{H}_1$ of the first kind which is more restrictive than $\mathscr{H}$, this $\mathscr{H}$ is also theoretically existent. The composition of two theoretically existent $\mathscr{H}$ and $\mathscr{H}'$ is also theoretically existent.

Let us apply these classifications and relations to an example important later. To a given $\mathscr{H}$ we define $\mathscr{H}_a$ by

$$A \in \tilde{E} \quad \text{and} \quad X \in \tilde{E}_+. \tag{4.3.14}$$

Here $\tilde{E}_+$ is given in (4.2.13). $\mathscr{H}_a$ is called "associated" to $\mathscr{H}$.

Because of $\tilde{E}_+ \subset \tilde{E}$, the invented $X$ in (4.3.14) are elements of the normal pictorial sets. Nevertheless the relation $X \in \tilde{E}_+$ need not to be of the first kind!

We first ask, which of the cases in (4.2.15) is possible for the associated hypothesis. For this hypothesis we have

$$\tilde{E}_+^{(a)} = \{Z \mid Z \in \tilde{E} \text{ and } \tilde{E}_+ \neq \emptyset\}, \qquad \tilde{E}^{(a)} = \tilde{E}. \tag{4.3.15}$$

Since $\tilde{E}_+$ does not depend on $Z$, we have the theorem

$$\tilde{E}_+^{(a)} = \tilde{E}^{(a)} \quad \text{or} \quad \tilde{E}_+^{(a)} = \emptyset.$$

Therefore we have only three possible cases:

$$\begin{array}{ll} [+1] & \tilde{E}^{(a)}=\tilde{E}_+^{(a)}\ it, \quad \text{i.e.} \quad \tilde{E}\neq\emptyset \Rightarrow \tilde{E}_+\neq\emptyset\ it. \\ [?] & \tilde{E}^{(a)}=\tilde{E}_+^{(a)}\ ad, \quad \tilde{E}_+^{(a)}=\emptyset\ ad; \quad \text{i.e.} \quad \tilde{E}\neq\emptyset \Rightarrow \tilde{E}_+\neq\emptyset\ ad; \\ & \tilde{E}\ \neq\emptyset \Rightarrow \tilde{E}_+=\emptyset\ ad. \\ [-1] & \tilde{E}_+^{(a)}=\emptyset\ it; \quad \text{i.e.} \quad \tilde{E}\neq\emptyset \Rightarrow \tilde{E}_+=\emptyset\ it. \end{array} \tag{4.3.16}$$

The case $[+1]$ for $\mathscr{H}_a$ (i.e. $\mathscr{H}_a$ theoretically existent) occurs if and only if one of the cases $[+1]$, $[+]$, $[0]$ occurs for $\mathscr{H}$. Thus $\mathscr{H}_a$ is false if and only if $\mathscr{H}$ is false.

Most interesting is the case that $\mathscr{H}_a$ is certain but not theoretically existent. We then have the case $[?]$ in (4.3.16). For $\mathscr{H}$ we have thus one of the cases $[?]$, $[-]$. $\mathscr{H}_a$ is certain if to every allowed $\mathscr{H}_2$ of the first kind there is an allowed $\mathscr{H}_1$ (of the first kind and compatible with $\mathscr{H}_2$) for which

$$\tilde{E}_+^{(1)} \subset \tilde{E}_+^{(a)}$$

is a theorem in $\mathscr{M}\mathscr{T}_\Sigma$. Thus $\mathscr{H}_a$ is certain if and only if to every allowed $\mathscr{H}_2$ of the first kind there is an allowed $\mathscr{H}_1$ of the first kind, compatible with $\mathscr{H}_2$, for which

$$\tilde{E}_+^{(1)} \neq \emptyset \Rightarrow \tilde{E}_+ \neq \emptyset \tag{4.3.17}$$

is a theorem in $\mathscr{M}\mathscr{T}_\Sigma$.

It follows that $\mathscr{H}$ certain implies $\mathscr{H}_a$ certain.

## §4.4 Behavior of Hypotheses under Extension of the Observational Report

Extending the observational report, we may get new letters $b_k$ added to the $a_i$ of the existing report. We comprise these into $(A, B)$ with $A=(a_1, a_2, \ldots)$ and $B=(b_1, b_2, \ldots)$. The extended observational report can be written

$$A \in \tilde{E}, \quad B \in \tilde{E}_h^{(r)}(A). \tag{4.4.1}$$

Here $\tilde{E}_h^{(r)}(A)$ is the set defined in (4.1.4), except that $\tilde{P}_h$ is not invented but "read" from real facts. The label $(r)$ shall point to this "reality" of $\tilde{P}_h$. As mathematical term, $\tilde{E}_h^{(r)}(Z)$ has the same structure as for a hypothesis of the first kind.

For the extended observational report, the set $\tilde{E}^{(\text{ex})}$ defined by (4.2.8) has by (4.4.1) the form

$$\tilde{E}^{(\text{ex})} = \{(Z, Z') \mid Z \in \tilde{E} \text{ and } Z' \in \tilde{E}_h^{(r)}(Z)\}. \tag{4.4.2}$$

With

$$\tilde{E}_+^{(r)} = \{Z \mid Z \in \tilde{E} \text{ and } \tilde{E}_h^{(r)}(Z) \neq \emptyset\} \tag{4.4.3}$$

we have

$$\tilde{E}^{(\text{ex})} = \{(Z, Z') \mid Z \in \tilde{E}_+^{(r)} \text{ and } Z' \in \tilde{E}_h^{(r)}(Z)\}. \tag{4.4.4}$$

The considered hypothesis $\mathscr{H}$ may be formulated already in $\mathscr{M}\mathscr{T}_\Sigma \mathscr{A}$, with $\tilde{E}_h(A)$ defined in (4.1.4). Therefore the additional elements $b_k$ do not enter $\mathscr{H}$. Thus the set $\tilde{E}_h^{(\text{ex})}(A, B)$, as defined in (4.1.4), for the *extended* observational report equals $\tilde{E}_h(A)$. Thus for $\mathscr{H}$ and the extended observational report we get:

$$\begin{aligned} \tilde{E}_+^{(\text{ex})} &= \{(Z, Z') \mid Z, Z') \in \tilde{E}^{(\text{ex})} \text{ and } \tilde{E}_h(Z) \neq \emptyset\} \\ &= \{(Z, Z') \mid Z \in \tilde{E} \text{ and } Z' \in \tilde{E}_h^{(r)}(Z) \text{ and } \tilde{E}_h(Z) \neq \emptyset\} \\ &= \{(Z, Z') \mid Z \in \tilde{E}_+ \text{ and } Z' \in \tilde{E}_h^{(r)}(Z)\}. \end{aligned} \tag{4.4.5a}$$

Here $\tilde{E}_+$ is defined in (4.2.13). Using (4.4.3) we find

$$\tilde{E}_+^{(ex)} = \{(Z, Z') \mid Z \in \tilde{E}_+ \cap \tilde{E}_+^{(r)} \text{ and } Z' \in \tilde{E}_h^{(r)}(Z)\}. \tag{4.4.5b}$$

By (4.4.4), (4.4.5b) we obtain the equivalences

$$\begin{aligned} &\tilde{E}_+^{(ex)} = \tilde{E}^{(ex)} \Leftrightarrow \tilde{E}_+^{(r)} = \tilde{E}_+ \cap \tilde{E}_+^{(r)} \Leftrightarrow \tilde{E}_+^{(r)} \subset \tilde{E}_+, \\ &\tilde{E}_+^{(ex)} = \emptyset \quad \Leftrightarrow \tilde{E}_+ \cap \tilde{E}_+^{(r)} = \emptyset. \end{aligned} \tag{4.4.6}$$

Using these relations we easily see, how the extension of the observational report can change the characterizations [+1] to [−1] of (4.2.15). Yet [+1] cannot change, i.e. a theoretically existent hypothesis remains theoretically existent. Similarly [−1] cannot change, i.e. a false hypothesis remains false. All other cases [+], [?], [0], [−] can change even into [−1], i.e. a not theoretically existent hypothesis can become false. This is not possible if the hypothesis is experimentally certain since then $\tilde{E}_+ \cap \tilde{E}_+^{(r)} = \emptyset$ cannot be a theorem in $\mathcal{M}\mathcal{T}_\Sigma$.

We even find that every experimentally certain $\mathcal{H}$ remains experimentally certain, every almost certain $\mathcal{H}$ remains almost certain, every certain $\mathcal{H}$ certain, and every perfectly certain $\mathcal{H}$ perfectly certain. We prove this only for "certain":

Let $\mathcal{H}_2$ be a hypothesis of the first kind for the extended observational report. $\mathcal{H}_2$ with this observational report can be given the form

$$A \in \tilde{E} \quad \text{and} \quad B \in \tilde{E}_h^{(r)}(A) \quad \text{and} \quad X' \in \tilde{E}_h^{(2)}(A, B). \tag{4.4.7a}$$

Formally this can be written as a hypothesis $\mathcal{H}_{20}$ for the unextended observational report:

$$A \in \tilde{E} \quad \text{and} \quad X \in \tilde{E}_h^{(r)}(A) \quad \text{and} \quad X' \in \tilde{E}_h^{(2)}(A, X). \tag{4.4.7b}$$

We must look for an $\mathcal{H}_1$ of the first kind:

$$A \in \tilde{E} \quad \text{and} \quad B \in \tilde{E}_h^{(r)}(A) \quad \text{and} \quad X' \in \tilde{E}_h^{(1)}(A, B),$$

which is compatible with $\mathcal{H}_2$ and more restrictive than $\mathcal{H}$; i.e. we have to prove that

$$\tilde{E}_+^{(1)} \subset \tilde{E}_+^{(ex)} \quad \text{and} \quad \tilde{E}_+^{(1)} \cap \tilde{E}_+^{(2)} \neq \emptyset \tag{4.4.8}$$

are theorems in $\mathcal{M}\mathcal{T}_\Sigma$.

Since $\mathcal{H}$ is certain relative to the unextended observational report, to $\mathcal{H}_{20}$ there must be a compatible $\mathcal{H}_{10}$ of the first kind which is more restrictive than $\mathcal{H}$. We write $\mathcal{H}_{10}$ in the form

$$A \in \tilde{E} \quad \text{and} \quad X'' \in \tilde{E}_h^{(10)}(A). \tag{4.4.9}$$

Since $\mathcal{H}_{10}$ is more restrictive than $\mathcal{H}$, we have

$$\tilde{E}_+^{(10)} \subset \tilde{E}_+ \tag{4.4.10}$$

as a theorem in $\mathcal{M}\mathcal{T}_\Sigma$. Since $\mathcal{H}_{10}$ is compatible with the $\mathcal{H}_{20}$ from (4.4.7b), it is all the more compatible with the hypothesis

$$A \in \tilde{E} \quad \text{and} \quad X \in \tilde{E}_h^{(r)}(A). \tag{4.4.11}$$

Thus the hypothesis composed of (4.4.9) and (4.4.11), i.e.

$$A \in \tilde{E} \quad \text{and} \quad X \in \tilde{E}_h^{(r)}(A) \quad \text{and} \quad X'' \in \tilde{E}_h^{(10)}(A), \tag{4.4.12}$$

is an allowed hypothesis of the first kind, from which we form an allowed $\mathscr{H}_1$ of the first kind by

$$A \in \tilde{E} \quad \text{and} \quad B \in \tilde{E}_h^{(r)}(A) \quad \text{and} \quad X'' \in \tilde{E}_h^{(10)}(A). \tag{4.4.13}$$

For this $\mathscr{H}_1$ let us prove (4.4.8).

From (4.4.13) follows

$$\begin{aligned} E_+^{(1)} &= \{(Z, Z') \mid (Z, Z') \in \tilde{E}^{(ex)} \text{ and } \tilde{E}_h^{(10)}(Z) \neq \emptyset\} \\ &= \{(Z, Z') \mid Z \in \tilde{E} \text{ and } Z' \in \tilde{E}_h^{(r)}(Z) \text{ and } \tilde{E}_h^{(10)}(Z) \neq \emptyset\} \\ &= \{(Z, Z') \mid Z \in \tilde{E}_+^{(10)} \text{ and } Z' \in \tilde{E}_h^{(r)}(Z)\}. \end{aligned}$$

This together with (4.4.10) and (4.4.5a) implies $\tilde{E}_+^{(1)} \subset \tilde{E}_+^{(ex)}$.

From (4.4.7a, b) follows

$$\begin{aligned} \tilde{E}_+^{(2)} &= \{(Z, Z') \mid (Z, Z') \in \tilde{E}^{(ex)} \text{ and } \tilde{E}_h^{(2)}(Z, Z') \neq \emptyset\} \\ &= \{(Z, Z') \mid Z \in \tilde{E} \text{ and } Z' \in \tilde{E}_h^{(r)}(Z) \text{ and } \tilde{E}_h^{(2)}(Z, Z') \neq \emptyset\}, \\ \tilde{E}_+^{(20)} &= \{Z \mid Z \in \tilde{E} \text{ and } \tilde{E}_h^{(20)}(Z) \neq \emptyset\}, \end{aligned}$$

where

$$\tilde{E}_h^{(20)}(Z) = \{(X, X') \mid X \in \tilde{E}_h^{(r)}(Z) \text{ and } X' \in \tilde{E}_h^{(2)}(Z, X)\}.$$

We define

$$F^{(2)}(Z) = \{X \mid X \in \tilde{E}_h^{(r)}(Z) \text{ and } \tilde{E}_h^{(2)}(Z, X) \neq \emptyset\}.$$

Then $\tilde{E}_h^{(20)}(Z) \neq \emptyset$ is equivalent to $F^{(2)}(Z) \neq \emptyset$. Therefore we have

$$\tilde{E}_+^{(20)} = \{Z \mid Z \in \tilde{E} \text{ and } F^{(2)}(Z) \neq \emptyset\}$$

and

$$\tilde{E}_+^{(10)} \cap \tilde{E}_+^{(20)} = \{Z \mid Z \in \tilde{E}_+^{(10)} \text{ and } F^{(2)}(Z) \neq \emptyset\}.$$

On the other hand we have

$$\tilde{E}_+^{(1)} \cap \tilde{E}_+^{(2)} = \{(Z, Z') \mid Z \in \tilde{E}_+^{(10)} \text{ and } Z' \in \tilde{E}_h^{(r)}(Z) \text{ and } \tilde{E}_h^{(2)}(Z, Z') \neq \emptyset\};$$

thus $\tilde{E}_+^{(1)} \cap \tilde{E}_+^{(2)} \neq \emptyset$ is equivalent to

$$\{Z \mid Z \in \tilde{E}_+^{(10)} \text{ and } F^{(2)}(Z) \neq \emptyset\} = \tilde{E}_+^{(10)} \cap \tilde{E}_+^{(20)} \neq \emptyset.$$

Therefore $\mathscr{H}_{10}$, $\mathscr{H}_{20}$ compatible, i.e. $\tilde{E}_+^{(10)} \cap \tilde{E}_+^{(20)} \neq \emptyset$ implies also $\mathscr{H}_1$, $\mathscr{H}_2$ compatible, i.e. $\tilde{E}_+^{(1)} \cap \tilde{E}_+^{(2)} \neq \emptyset$.

The hypothesis $\mathscr{H}_a$ associated with $\mathscr{H}$ changes its form under an extension of the observational report. While the unextended observational report has

$$\mathscr{H}_a\colon \quad A \in \tilde{E} \quad \text{and} \quad X \in \tilde{E}_+, \tag{4.4.14}$$

the extended observational report gives

$$\mathscr{H}_a^{(ex)}\colon \quad A \in \tilde{E} \quad \text{and} \quad B \in \tilde{E}_h^{(r)}(A) \quad \text{and} \quad (X, X') \in \tilde{E}_+^{(ex)}. \tag{4.4.15a}$$

From (4.4.5b) follows that this can be written

$$\mathscr{H}_a^{(ex)}\colon \quad A \in \tilde{E} \quad \text{and} \quad B \in \tilde{E}_h^{(r)}(A) \quad \text{and} \quad X \in \tilde{E}_+ \cap \tilde{E}_+^{(r)} \quad \text{and} \quad X' \in \tilde{E}_h^{(r)}(X). \tag{4.4.15b}$$

$\mathscr{H}_a^{(\text{ex})}$ differs from $\mathscr{H}_a$ for the extended observational report, which is

$$A \in \tilde{E} \quad \text{and} \quad B \in \tilde{E}_h^{(r)}(A) \quad \text{and} \quad X \in \tilde{E}_+. \tag{4.4.16}$$

Since a certain $\mathscr{H}$ remains certain (under an extension of the observational report), and since the hypothesis associated with a certain hypothesis is also certain, also $\mathscr{H}_a^{(\text{ex})}$ is certain. But if for instance $\mathscr{H}$ is not experimentally certain, $\tilde{E}_+ \neq \emptyset$ does not necessarily imply $\tilde{E}_+^{(\text{ex})} \neq \emptyset$ even if $\mathscr{H}_a$ is certain, i.e. $\mathscr{H}_a^{(\text{ex})}$ can become false although (4.4.16) is certain.

For later purposes we need the following definitions:

(1) $\mathscr{H}_1$ be called an "extension" of $\mathscr{H}_2$ if

$$\mathscr{H}_2: \quad A \in \tilde{E} \quad \text{and} \quad X \in \tilde{E}_h^{(2)}(A) \tag{4.4.17}$$

and

$$\mathscr{H}_1: \quad A \in \tilde{E} \quad \text{and} \quad B \in \tilde{E}_h^{(r)}(A) \quad \text{and} \quad X \in \tilde{E}_h^{(n)}(A) \quad \text{and} \quad X' \in \tilde{E}^{(e)}(A, B, X) \tag{4.4.18a}$$

with

$$E_h^{(n)}(Z) \subset \tilde{E}_h^{(2)}(Z) \qquad \text{for all } Z \in \tilde{E}. \tag{4.4.18b}$$

We see immediately that $\mathscr{H}_1$ in (4.3.3) indeed forms a special case of (4.4.18), namely if there is no $B$.

(2) $\mathscr{H}_1$ is called "more comprehensive" than $\mathscr{H}_2$ if (4.4.17), (4.4.18) hold except for the following changes: Some of the components $x_i$ of $(X, X')$ in (4.4.18a) can be replaced by some of the components $a_k$ or $b_k$ of $(A, B)$ if we have the theorem $x_i = a_k$ or $b_k$. Some components $x_l$, $x_k$ can be replaced by a single letter if $x_l = x_k$ holds.

We easily see that $\mathscr{H}_1$ more comprehensive than $\mathscr{H}_2$, and $\mathscr{H}_2$ more comprehensive than $\mathscr{H}_3$, imply $\mathscr{H}_1$ more comprehensive than $\mathscr{H}_3$. Thus "more comprehensive" defines an ordering in the field of hypotheses.

If an allowed hypothesis of the first kind is also called an "imaginable extension of the observational report", we can characterise:

A hypothesis $\mathscr{H}$ is experimentally certain if and only if there is no imaginable extension of the observational report for which $\mathscr{H}$ becomes false.

A hypothesis $\mathscr{H}$ is certain if and only if to every imaginable extension of the observational report there is another more comprehensive extension for which $\mathscr{H}$ is theoretically existent.

[neg $\mathscr{H}$] is certain if and only if $\mathscr{H}$ is experimentally uncertain for every imaginable extension of the observational report.

$\mathscr{H}$ is certain if and only if [neg $\mathscr{H}$] is experimentally uncertain for every imaginable extension of the observational report. [neg $\mathscr{H}$] is uncertain if and only if there is an imaginable extension of the observational report, for which [neg $\mathscr{H}$] is experimentally certain.

The associated hypothesis $\mathscr{H}_a$ is certain if and only if to every imaginable extension of the observational report there is another more comprehensive extension for which $\mathscr{H}$ belongs to one of the cases [+1], [+], [0], i.e. for which $\mathscr{H}$ is strongly allowed.

## § 4.5 The Mathematical Game

We have learned in the last sections 4.1 to 4.4 some procedures which are not customary in mathematics, not even in the theory of categories.

At some places we have used such concepts as categories without saying so. The "observables" for instance form such a categorial structure. But the theory of categories is not essential for physics, since an essential objective of this mathematical theory is to formulate higher infinities more rigorously than set theory can. In physics all is finite and infinities occur only as mathematical idealizations. For instance, for using a category of finitely many morphisms we do not need the great apparatus of category theory. This is no objection to using category theory for a better survey of the various mathematical theories used in physical theories, for instance for a general discussion of intertheory relations (see § 3).

The procedures described in § 4.1 to 4.4 are beyond the scope of usual mathematics since such concepts as the field of hypotheses, of certain, of c-allowed, etc. are no definitions in the scope of a mathematical theory. Also the "proofs" in this range are no mathematical proofs. For instance, in order to prove that a hypothesis $\mathscr{H}$ is certain we must give a method which to a $\mathscr{H}_2$ of the first kind can *construct* a compatible $\mathscr{H}_1$ of the first kind which is more restrictive than $\mathscr{H}$. Such construction proofs are necessary since the field of hypotheses is not a set in a mathematical theory. The hypotheses are not "existing", they are "made", made by us as humans applying the physical theory.

In this sense the mathematical framework of a physical theory is not a closed mathematical theory. Rather it is an open mathematical field within which we continually change the mathematical theories by observational reports and hypotheses. Only one part of all these theories is left unchanged (as long as we do not pass to a more or a less comprehensive theory): the axiomatic basis $\mathscr{M}\mathscr{T}_\Sigma$.

All this handling of mathematical theories within a $\mathscr{P}\mathscr{T}$ shall be called the "mathematical game" of $\mathscr{P}\mathscr{T}$. In §4.1 to 4.4 we have given the rules of this game. The introduced concepts are descriptions of situations of this game and not structures in $\mathscr{M}\mathscr{T}_\Sigma$.

A physical theory does not only describe (by the axioms of $\mathscr{M}\mathscr{T}_\Sigma$) "physical laws" which are "valid for ever" and in this sense an enduring structure in nature. It also contains a variable part. Some aspects of this variable part are given by the mathematical game. And the development of this game depends essentially on our actions in playing this game.

Although the axioms of $\mathscr{M}\mathscr{T}_\Sigma$ are not changed in playing this game, many of these axioms are already adjusted to this game. For instance, they determine whether a hypothesis is theoretically existent or not.

In practice the hypotheses are not formulated arbitrarily. One formulates part of the hypotheses as pure inventions and with the help of $\mathscr{M}\mathscr{T}_\Sigma$ tries to supply a hypothesis in such a way that it becomes theoretically existent or at least certain. For this purpose, today many computers are employed.

This mathematical game is not played for the sake of itself. This game is highly significant for physics, as we will discuss in the next section.

## 4.6 Possibility, Reality, Open Questions

At the beginning of §4 we have said that it may seem mystical how we can speak of "possible" facts though there are no corresponding logical signs in $\mathscr{MT}$. This is different in the mathematical game. An allowed hypothesis may also be called "possible", i.e. permitted to be added without contradiction. But we are not interested in "possible" moves of the purely mathematical game. We are interested in physics, i.e. in a physical interpretation of this game.

As well as a physical interpretation of a mathematical theory $\mathscr{MT}_\Sigma$ is not given by $\mathscr{MT}_\Sigma$ itself, also the physical interpretation of the mathematical game is not given by this game. The physical interpretation of $\mathscr{MT}_\Sigma$ was given by the correspondence rules, which permit us to write down the observational report in the language of $\mathscr{MT}_\Sigma$. We now wish to extend this interpretation to the mathematical game, using the classification of hypotheses from the preceding sections.

There first arises a difficulty: If we pass from one $\mathscr{PT}_1$ to a more comprehensive $\mathscr{PT}_2$, the classification of the hypotheses can change totally. This suggests not to rush to interpretations which could be refuted by a more comprehensive theory.

If the relation between $\mathscr{PT}_1$ and $\mathscr{PT}_2$ has the form

$$\mathscr{PT}_2 \overset{s\ \ s}{\leadsto} \mathscr{PT}_1 \tag{4.6.1}$$

the transition in (4.2.15) between the classifications $[+1]$, $[+]$, ..., $[-1]$ is not arbitrary. The following changes of the cases for a hypothesis by passing from $\mathscr{PT}_1$ to $\mathscr{PT}_2$ are possible (besides that every can be unchanged):

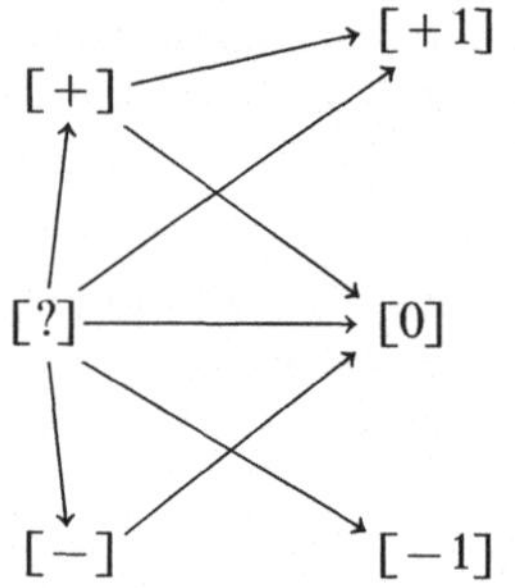


(4.6.2)

The diagram is based on the fact that theorems in $\mathscr{MT}_{\Sigma_1}$ go over into corresponding theorems in $\mathscr{MT}_{\Sigma_2}$ (see [3] § 10.3).

The step (4.6.1) occurs in the usual development of a theory as described by (3.3.5). The diagram (4.6.2) can be helpful to interpret the mathematical game at an early stage $\mathscr{PT}_i$ of the intended physical theory. We will not discuss such problems since we are only interested in the foundation of an interpretation going beyond that of the correspondence rules. A more technical elaboration for handling this interpretation must be left to future work.

Therefore we will only consider *one* theory $\mathscr{PT}$ and make such "presumptions" (concerning $\mathscr{PT}$) that the interpretation to be described makes sense. The presumptions will be better understood as we develop this interpretation.

Let us start by explaining what we mean by a "comparison of a hypothesis with experience".

Here we begin with the simplest but also fundamental case, that of an allowed hypothesis $\mathscr{H}$ of the *first* kind. We presume this for the next considerations until this presumption will be terminated explicitly.

The simplest case for a comparison is that where the extended observational report makes $\mathscr{H}$ false (it could not have been experimentally certain for the unextended observational report). We say that $\mathscr{H}$ is refuted by the experiment. This does not (!) imply that $\mathscr{H}$ will be refuted again if the experiment is repeated. What is meant by a repetition of an experiment in the context of the mathematical game will be defined rigorously below.

Contrary to a refutation of $\mathscr{H}$ is a "realization" of $\mathscr{H}$. What is meant by realization of $\mathscr{H}$? We consider the following change of $\mathscr{H}$ for the extended observational report: We try to replace the invented $x_i$ by letters $a_j$ (from the unextended observational report) and $b_k$ (from the extension) such that the new hypothesis (which has no invented elements!) is at least certain. Then we say to have "realized" $\mathscr{H}$. If $\mathscr{H}$ itself has no invented elements, we say that $\mathscr{H}$ is "realized" if $\mathscr{H}$ becomes at least certain for the extended observational report.

Let us now ask: under what circumstances is it "possible" to realize $\mathscr{H}$?

We begin with a simple but fundamental case: $\mathscr{H}$ has no observational report and may be allowed. Obviously the mathematical game itself cannot decide, whether $\mathscr{H}$ can be realized or not. If $\mathscr{P}\mathscr{T}$ is too weak, there can be many allowed hypotheses which cannot be realized. For instance, if we take thermodynamics without the second law, the mathematical game contains as allowed hypotheses so-called perpetuum-mobiles of the second kind. Thus we see again that the condition for a $\mathscr{P}\mathscr{T}$ to show no contradictions between $\mathscr{M}\mathscr{T}_\Sigma$ and the observational report, is too weak if we want to say whether a hypothesis can be realized. Without stronger requirements on $\mathscr{P}\mathscr{T}$ we cannot define what is meant by real and possible facts.

If $\mathscr{P}\mathscr{T}$ is not weakly closed, there are $\mathscr{H}_1$, $\mathscr{H}_2$ (of the first kind) without observational report, which are not compatible. If one of them were realized, it would be impossible to realize the other. $\mathscr{P}\mathscr{T}$ leaves unnecessarily open that $\mathscr{H}_2$ could perhaps be realizable although it is not since $\mathscr{H}_1$ was realized. Such a $\mathscr{P}\mathscr{T}$ is therefore too weak for our purposes. Therefore we assume that $\mathscr{P}\mathscr{T}$ is at least weakly closed. Then all allowed $\mathscr{H}$ without observational report are also certain.

Einstein's gravitation theory (general relativity) is not weakly closed. There are several cosmological models (i.e. hypotheses of the first kind without observational report) which are not compatible. But only one can be realized since there is only one cosmos from which we can get information, i.e. observational reports. In this case we try to decide between the models by observation, i.e. by asking: Which of the models is an allowed hypothesis *with* the observational report known until now?

A well known case of not weakly closed theories $\mathscr{P}\mathscr{T}$ is that where $\mathscr{P}\mathscr{T}$ contains adjustable parameters. These parameters are just determined by experiments, i.e. observational reports.

The presumption that $\mathscr{P}\mathscr{T}$ is weakly closed is not enough. It does not guarantee that all $\mathscr{H}$ without observational report are realizable. It may be, that we have forgotten a physical law what forbids some of these $\mathscr{H}$ as observational reports. Such an additional law would be a refutable law (§ 2.5).

We can pursue a similar consideration for a certain hypothesis *with* observational report:

$$\mathscr{H}:\quad A\in\tilde{E}\quad\text{and}\quad X'\in\tilde{E}_h(A).\tag{4.6.3}$$

We form the hypothesis $\mathscr{H}_0$ without observational reports:

$$X\in\tilde{E}\quad\text{and}\quad X'\in\tilde{E}_h(X).\tag{4.6.4}$$

Since $\mathscr{PT}$ was presumed weakly closed, $\mathscr{H}_0$ can be realized. Let

$$A'\in\tilde{E}\quad\text{and}\quad B\in\tilde{E}_h(A')\quad\text{and}\quad\ldots\tag{4.6.5}$$

be such a realizing observational report (here $B$ may contain some of the components of $A$). In general we do not know whether the observational report $A\in\tilde{E}$ can be extended to a form like (4.6.5). For an uncertain $\mathscr{H}$ we cannot exclude that the extension of $A\in\tilde{E}$ might produce an observational report contradicting $X'\in\tilde{E}(A)$, i.e. making $\mathscr{H}$ false. If $\mathscr{H}$ is certain, no such extension can occur; but this does not imply that there must be an extension realizing $\mathscr{H}$.

For a certain $\mathscr{H}$ our search for an extension of $A\in\tilde{E}$ which realizes $\mathscr{H}$ may appear hopeless. Then we should conjecture that we lack some law to tell us that some of the extensions (taken as hypotheses) are not compatible with $\mathscr{H}$, i.e. that $\mathscr{H}$ in a more comprehensive theory is not certain.

We denote $\mathscr{PT}$ as "g.$\mathscr{G}$-closed" if $\mathscr{PT}$ is weakly closed and all certain $\mathscr{H}$ of the first kind are realizable. (We use the same notation as in [3]; g stands for the German "gleich" what means "the same"; "g.$\mathscr{G}$" shall remind us that we have the *same* fundamental domain $\mathscr{G}$ with the *same* correspondence rules.) Is it a cyclic definition if we now claim that in a g.$\mathscr{G}$-closed $\mathscr{PT}$ a certain $\mathscr{H}$ can be realized, shortly, that $\mathscr{H}$ is "physically possible"? This is right. Such a cyclic situation can occur in the development of physical theories and in judging them. We can test whether there are contradictions between $\mathscr{MT}_\Sigma$ and observational reports, but we cannot experimentally deduce (see §2.5) all additional refutable laws, which are necessary to exclude not realizable $\mathscr{H}$. The work of physicists with a $\mathscr{PT}$ can suggest the conjecture that some new refutable law should be added. Thus a previous judgment that $\mathscr{PT}$ is g.$\mathscr{G}$-closed must perhaps be corrected. In this sense we are never absolutely sure that a $\mathscr{PT}$ is g.$\mathscr{G}$-closed.

If experience suggests the strong conjecture that a (not only certain but) theoretically existent $\mathscr{H}$ is not realizable, to $\mathscr{MT}_\Sigma$ we cannot add an axiom which makes $\mathscr{H}$ false resp. uncertain. Such an axiom must contradict $\mathscr{MT}_\Sigma$. In such a case we rather must replace some of the axioms of $\mathscr{MT}_\Sigma$ by others.

Just this is the case for $\mathscr{PT}_{q\exp}$. Instead of changing the axioms in the mathematical part $\mathscr{MT}_\Sigma$ of $\mathscr{PT}_{q\exp}$, we have used the procedure of embedding $\mathscr{PT}_m \rightsquigarrow \mathscr{PT}_{q\exp}$. This embedding procedure achieves two things: The axioms of $\mathscr{PT}_{q\exp}$ are partially replaced by those of $\mathscr{PT}_m$, and the relation between $\mathscr{PT}_m$ and the not g.$\mathscr{G}$-closed $\mathscr{PT}_{q\exp}$ is explicitly given.

We now presume $\mathscr{PT}$ to be g.$\mathscr{G}$-closed.

Let $\mathscr{H}$ be an allowed but not certain hypothesis of the form (4.6.3). Then a realization (4.6.5) of $\mathscr{H}_0$ should be possible. Now we cannot conclude that $A\in\tilde{E}$ must have an extension realizing $\mathscr{H}$. On the contrary there is an allowed

$$\mathscr{H}_1:\quad A\in\tilde{E}\quad\text{and}\quad X''\in\tilde{E}_h^{(1)}(A)\tag{4.6.6}$$

which is not compatible with $\mathscr{H}$ (for which $\tilde{E}_+ \cap \tilde{E}_+^{(1)} = \emptyset$ is a theorem in $\mathscr{M}\mathscr{T}_\Sigma$). If we had an observational report of the form

$$A \in \tilde{E} \quad \text{and} \quad B \in \tilde{E}_h^{(1)}(A) \quad \text{and} \quad \ldots,$$

$\mathscr{H}$ would become false and therefore could not be realized by any additional extension of that observational report. Nevertheless we say that $\mathscr{H}$ is physically possible "before" the observational report $A \in \tilde{E}$ is extended. What do we mean by such a proposition?

We just mean that $\mathscr{H}_0$ is realizable, i.e. that there may be experiments with an observational report of the form (4.6.5). If we have only the observational report $A \in \tilde{E}$ it is not yet decided whether an extension will or will not allow a realization of $\mathscr{H}$; but such a realization cannot be excluded.

To characterize this situation, we more precisely say: $\mathscr{H}$ is "conditionally physically possible".

For later we remark, that for $\mathscr{H}_0$ we have

$$\begin{aligned} E_h^{(0)} &= \{(X, X') \mid X \in \tilde{E} \quad \text{and} \quad X' \in \tilde{E}_h(X)\} \\ &= \{(X, X') \mid X \in \tilde{E}_+ \quad \text{and} \quad X' \in \tilde{E}_h(X)\} \end{aligned}$$

and for $\mathscr{H}_a$ associated to $\mathscr{H}$ obtain

$$\tilde{E}_h^{(a)} = \tilde{E}_+. \tag{4.6.8}$$

Obviously $\tilde{E}_h^{(0)} \neq \emptyset \Leftrightarrow \tilde{E}_h^{(a)} \neq \emptyset$, i.e. $\mathscr{H}_0$ certain (resp. theoretically existent) is equivalent to $\mathscr{H}_a$ certain (resp. theoretically existent). Also if $\mathscr{H}$ is only allowed but not certain, nevertheless $\mathscr{H}_a$ is certain (or just theoretically existent) since $\mathscr{H}_0$ is certain (resp. theoretically existent).

If we want to remark that the sharper condition "theoretically existent" is fulfilled (instead of only certain), we add the word "strongly".

Let $\mathscr{H}$ be not only certain but also determined. The observational report can be extended in such a form that $\mathscr{H}$ can be realized by replacing the $x_i$ by letters $a_i, b_k$ from the report. Since $\mathscr{H}$ is determined, there cannot be two different signs $a_i$ and $b_k$ by which one of the $x_i$ can be replaced, i.e. for every $x_i$ there can be only *one* fact in the fundamental domain corresponding to $x_i$. Since $\mathscr{H}$ is realizable, there must exist one fact, even if we have not "reported" it, i.e. if the observational report does not contain the signs corresponding to this fact. (We have not reported it because either it lies in the future or we have not noted it.) Therefore we say that $\mathscr{H}$ is "physically real".

We can weaken the condition "determined" to "almost determined". Then if in an extension of the observational report there would be *two* letters (e.g. $a_1, a_2$ with $a_1 \neq a_2$) by which *one* $x_k$ could be replaced, this would contradict the presumption that (4.2.17) is experimentally certain.

For the known physical theories (e.g. quantum mechanics) there are no determined (resp. almost determined) hypotheses *without* observational report. On the contrary, for these theories it is essential that a hypothesis $X \in \tilde{E}_h$ can be realized by "many" experiments $A_1 \in \tilde{E}_h$, $A_2 \in \tilde{E}_h$, .... Thus $\mathscr{M}\mathscr{T}_\Sigma$ does not say how the real world looks like. Only the mathematical game in connection with its physical interpretation gives us answers to the question what is real and what is possible.

It is an easily made mistake, to forget the normative axioms which define the fundamental domain. Thus one might judge a hypothesis as certain although it is false (i.e. not in the fundamental domain).

A not always remembered presumption for all questions about hypotheses is that also the observational report belongs to the fundamental domain. For instance, if by external influences (which may be excluded from the fundamental domain) the observational report leaves the fundamental domain, all conclusions about realizations of hypotheses may become senseless to the extent that the observational report has left the fundamental domain. Not only for the future but also for the past an observational report can be outside the fundamental domain (e.g. if the behaviour of a system before preparation does not belong to the fundamental domain); therefore conclusions about hypotheses in the past can also become senseless. Physicists are well aware of this presumption and say: Under the condition that ... (and here they formulate: that nothing "disturbs" the considered processes) this and that is possible or real.

For instance, we can say where the planets will be in the future under the condition that the planetary system will not be disturbed by other stars. We also can say where the planets were in the past, but not before they originated from clouds of matter.

If $\mathscr{H}$ is not certain we interpret $\mathscr{H}$ as "conditionally physically possible". Sometimes we can give an explicit condition for $\mathscr{H}$ to become possible. We may have to compatible hypotheses

$$\begin{aligned} \mathscr{H}_c\colon &\ A\in\tilde{E} \quad \text{and} \quad X'\in\tilde{E}_h^{(c)}(A), \\ \mathscr{H}\colon &\ A\in\tilde{E} \quad \text{and} \quad X\in\tilde{E}_h(A). \end{aligned}$$

For an extension

$$A\in\tilde{E} \quad \text{and} \quad B\in\tilde{E}_h^{(c)}(A)$$

of the observational report the hypothesis

$$A\in\tilde{E} \quad \text{and} \quad B\in\tilde{E}_h^{(c)}(A) \quad \text{and} \quad X\in\tilde{E}_h(A)$$

may be certain. Then we say that $\mathscr{H}$ is "physically possible under the condition $\mathscr{H}_c$", resp. (if $\mathscr{H}$ is determined) $\mathscr{H}$ is "physically real under the condition $\mathscr{H}_c$".

Thus we see how to reach more and more complicated structures of moves in the mathematical game and of physical interpretations of these moves.

To use only hypotheses of the first kind would make physics too clumsy. The fruitfulness of the physical language rests on the use of hypotheses of the first *and* second kind.

Let us consider a hypothesis $\mathscr{H}$ of the second kind:

$$A\in\tilde{E} \quad \text{and} \quad X\in\tilde{E}_h(A). \tag{4.6.9}$$

This $\mathscr{H}$ may be theoretically existent and determined. Then there is a mapping $\tilde{E}_+ \xrightarrow{f} \tilde{T}_h$ with $X = f(A)$, so that (4.6.9) implies

$$A \in \tilde{E} \Rightarrow f(A) \in \tilde{E}_h(A). \tag{4.6.10}$$

In this sense we say that the observational report $A \in \tilde{E}$ "realizes" the relation $X \in \tilde{E}_h(A)$ and that $\mathscr{H}$ is physically real.

We define a realization of $\mathscr{H}$ (not necessarily theoretically existent and determined) by an extension $\mathscr{H}_1$ of $\mathscr{H}$ for which $X \in \tilde{E}_h(A)$ is realized. Explicitly: There is given an extension

$$\mathscr{H}_1: \quad A \in \tilde{E} \quad \text{and} \quad B \in \tilde{E}_h^{(r)}(A) \quad \text{and} \quad X \in \tilde{E}_h^{(e)}(A, B) \tag{6.4.11}$$

where

$$[A \in \tilde{E} \text{ and } B \in \tilde{E}_h^{(r)}(A)] \Rightarrow \tilde{E}_h^{(e)}(A, B) \subset \tilde{E}_h(A). \tag{4.6.12}$$

To define the "realization" we first replace (as far as possible) some of the components $x_i$ of $X$ by letters from $A$ or $B$ so that the resulting hypothesis is theoretically existent and determined relative to the remaining $x_i$. (It is evident that only such $x_i$ can be replaced which are elements of the pictorial sets.)

Then (4.6.12) imples that already

$$A \in \tilde{E} \quad \text{and} \quad B \in \tilde{E}_h^{(r)}(A) \quad \text{and} \quad X \in \tilde{E}_h(A) \tag{4.6.13}$$

is theoretically existent. (4.6.12) implies that (4.6.11) is sharper than (4.6.13) in the sense of (4.3.1a). We have a realization if there is a "sharpening" of (4.6.13) and a suitable "replacing" so that a determined hypothesis results.

It would be suggestive to define $\mathscr{H}$ as physically possible if a realization of $\mathscr{H}$ is possible. But we want to weaken the definition of physically possible.

For this purpose we first consider $\mathscr{H}$ as theoretically existent, but not determined. We do not want to investigate whether there is a sharpening and replacing which makes the hypothesis determined. We will denote $\mathscr{H}$ as physically possible also without such an investigation, i.e. without knowing whether we can "select" one of the $X \in \tilde{E}_h(A)$. The word "possible" shall not express that we can give a method of realization but only that we are invited to look for such a method. Therefore a theoretically existent $\mathscr{H}$ shall be interpreted as physically possible.

If $\mathscr{H}$ is only certain, than to every extension of $A \in \tilde{E}$ there is a more comprehensive extension for which $\mathscr{H}$ is theoretically existent (see the end of §4.4). Hence in every case we can make such an extension that $\mathscr{H}$ becomes theoretically existent. Then we call $\mathscr{H}$ (weakly) possible, and (weakly) real if it is almost determined.

If $\mathscr{H}$ is not certain, the observational report can be extended in such a form that $\mathscr{H}$ becomes false. If we call $\mathscr{H}$ conditionally physically possible, we mean that there can be other experiments $A_1 \in \tilde{E}$, $A_2 \in \tilde{E}$, ... such that some of these provide an extension which makes $\mathscr{H}$ theoretically existent. This can be expressed by the associated hypothesis $\mathscr{H}_a$: An $\mathscr{H}_a$ being certain guarantees that there are experiments which make $\mathscr{H}_a$ and therefore also $\mathscr{H}$ theoretically existent.

Thus we reach the following scheme of interpretation, valid for hypotheses of the first and second kind ($\mathscr{P}\mathscr{T}$ presumed as g.$\mathscr{G}$-closed).

**Table 1.** Interpretation Scheme

| Classification of $\mathscr{H}$ | | Interpretation of $\mathscr{H}$ |
|---|---|---|
| theoretically existent and determined | | physically (strongly) real |
| theoretically existent | | physically (strongly) possible |
| certain and almost determined | | physically (weakly) real |
| certain | | physically (weakly) possible |
| c-allowed | associated hypothesis theoretically existent | conditionally physically (strongly) possible |
| c-allowed | associated hypothesis certain | conditionally physically (weakly) possible |
| [neg $\mathscr{H}$] certain | | physically (weakly) to exclude |
| false | | physically to exclude |

The mathematical game with this interpretation scheme becomes what we call the "physical game" of $\mathscr{PT}$. We see that the interpretation language of "real" and "possible" in this physical game depends decisively on a classification of hypotheses, which is not a purely mathematical question in the scope of $\mathscr{MT}_\Sigma$ (see §4.5).

Nevertheless there are many structures in $\mathscr{MT}_\Sigma$ which are adjusted to this game and therefore interpreted by corresponding moves in the game. For instance, (4.6.9) theoretically existent is equivalent to the theorem (or axiom) (4.2.9a) which is equivalent to

$$\forall Z[Z\in\tilde{E}\Rightarrow\exists X(X\in\tilde{E}_h(Z))]. \tag{4.6.14}$$

We therefore interpret (4.6.14) in the form: To every physically real $Z\in\tilde{E}$ it is physically possible to get an $X\in\tilde{E}_h(Z)$. An example for such an axiom is AV 1.1 in VI §1.1. We have already used (intuitively!) this interpreting language during the development of an axiomatic basis for quantum mechanics.

An example for a theoretically existent and determined hypothesis is given by the procedure of restriction $\mathscr{PT}_1\to\mathscr{PT}$ in §3.1. The "observational report" of $\mathscr{PT}$ is just a hypothesis which is theoretically existent and determined relative to the observational report of $\mathscr{PT}_1$ (see §4.8). Therefore the observational report of $\mathscr{PT}$ is physically real (on the basis of the observational report of $\mathscr{PT}_1$; see also §4.8).

By an embedding $\mathscr{PT}\rightsquigarrow\mathscr{PT}_2$ (see §3.2), the observational report is not changed. Thus the observational report of $\mathscr{PT}_2$ is physically as real as that of $\mathscr{PT}$.

Thus also such observational reports gained by pretheories (see §3.4) are physically real on the basis of the pretheories.

Obviously, the physical game of a $\mathscr{PT}$ is much simpler than it would be if we would include all the pretheories. Thus the physical game of quantum mechanics in the form denoted in §3.3 by $\mathscr{PT}_2$ (and presented in detail in [2]) is much simpler than that of $\mathscr{PT}_{1t}$ (notation as in §3.3). But in order to make experiments, we must retrace the chain of pretheories until the language and the work of craftsmen are reached. Only thus is it possible to get "realizations".

Two points must be emphasized:

There is no "pure" physics without technical applications. Such a pure physics would be a physics without physical game, i.e. only $\mathcal{M}\mathcal{T}_\Sigma$, i.e. only pure mathematics.

Most of the observational reports describe realities produced by human actions, i.e. artifacts. The only "naturally" given and "interesting" observational report seems to come from astronomy; and also this is not given without indirect measurements by highly technical devices. (For the concept of indirect measurement see § 4.8.)

The interpreting language of the physical game can be systematically developed from the simple propositions introduced above. One may introduce dialog games with the intention to formulate a logic for this interpreting language. Decisions in this dialog game are not only based on $\mathcal{M}\mathcal{T}_\Sigma$ but also on the observational reports. It is not our intent to develop here such a language and logic (see [2] IV § 8 and [64]). Only one fundamental decision about this language and logic is already made by our opinion what physics is, presented in this chapter XIII: Such a logic is a logic "a posteriori". It depends on the given structure of $\mathcal{M}\mathcal{T}_\Sigma$. The development of a $\mathcal{P}\mathcal{T}$ needs only a primitive logic ("and", "not") for the formulation of the observational report and the mathematical logic of $\mathcal{M}\mathcal{T}_\Sigma$. There is no "new" logic "a priori" which determines the structure of $\mathcal{M}\mathcal{T}_\Sigma$ and the formulation of the observational report. One often intends to develop a new logic "a priori" and to base quantum mechanics (i.e. some fundamental structures of $\mathcal{M}\mathcal{T}_\Sigma$, e.g. the Hilbert space structure) on this logic. From our point of view, this appears as if one would construct bridles and by this construction try to prove the existence of horses for which the bridles are suitable.

## § 4.7 Some Aspects of the Quantum Mechanical Game and the Role of Probability Theory in Physics

The role of probability theory in the physical game of a theory could be discussed in full generality (see [3]), since we have developed a general foundation of probability theory in the form of statistical selection procedures. But it is more elucidating to explain the role of probability for the example of quantum mechanics. Since we would lose control over the various hypotheses and their interpretations if we would consider different forms of quantum mechanics, let us only use the form denoted in § 3.3 by $\mathcal{P}\mathcal{T}_2$.

The pictorial sets are $M$, $\mathcal{Q}$, $\mathcal{R}$, $\mathcal{R}_0, \ldots$; one of the pictorial relations is the probability function $\lambda(a \cap b_0, a \cap b)$.

As a first example we adopt an observational report of the form

$$\begin{aligned} &a_i \in \mathcal{Q}' \quad (i=1, \ldots, 6); \quad b_{0k} \in \mathcal{R}_0, \quad b_k \in \mathcal{R}, \quad b_k \subset b_{0k} \quad (k=1, \ldots, 3); \\ &\lambda(a_i \cap b_{0k}, a_i \cap b_k) \approx 0 \qquad (i=1,2,3; k=1,2,3); \\ &\lambda(a_i \cap b_{0k}, a_i \cap b_k) \approx \alpha_{ik} \qquad (\alpha_{ik} \neq 0 \text{ for } i>3). \end{aligned} \tag{4.7.1}$$

We add the hypothetical part

$$\begin{aligned} &x_0 \in \mathcal{R}_0, \quad x \in \mathcal{R}, \quad x \subset x_0, \quad \lambda(a_i \cap x_0, a_i \cap x) \approx 0 \quad (i=1,2,3), \\ &\lambda(a_i \cap x_0, a_i \cap x) \gtrsim \alpha_{ik} \qquad (i>3, k=1,2,3). \end{aligned} \tag{4.7.2}$$

This hypothesis is theoretically existent by the axiom AV 1.1 in VI §1.1. Therefore (4.7.2) is physically possible. Since the hypothesis is not determined, (4.7.2) is not physically real until one has built a device $x_0$ with the demanded properties. The physical possibility of (4.7.2) does not tell us how to construct such a device. We will take up this question below.

Another example typical for quantum mechanics is given by an observational report of the form:

$$a\in\mathcal{Q}', \quad b_0\in\mathcal{R}_0, \quad b\in\mathcal{R}, \quad b\subset b_0,$$
$$\lambda(a\cap b_0, a\cap b)\approx\alpha, \quad m\in M, \quad m\in a, \quad m\in b_0. \tag{4.7.3}$$

As hypothetical part we take

$$m\in b. \tag{4.7.4}$$

According to (4.7.3), $m$ is a microsystem prepared by $a$ and measured by the device $b_0$. The hypothesis (4.7.4) concerns the question whether $m$ triggers the indication $b$.

We presume that $\alpha$ is physically distinguished from 0 and 1 (i.e. differs from 0 and 1 not only by its imprecision). Then the hypothesis (of the first kind without invented $x$) is not certain, but allowed. Therefore $m\in b$ is conditionally physically possible.

Here one often uses formulations which can easily be misunderstood.

One says "$m\in b$ is possible with the probability $\alpha$". If this is only a *short* formulation that the hypothesis (4.7.3), (4.7.4) is physically possible, nothing can be objected. But this formulation seems to be a relation between three things: $m$, $b$ and $\alpha$. Thus it can lead to mistakes or to imaginations which cannot be based on the theory.

One imagines e.g. that the system $m$ has the property to trigger $b$ with the probability $\alpha$, that there is something like a real propensity of $m$ to trigger $b$. The strength of this propensity is given by $\alpha$. This is an imagination added to the theory (compatible with the theory or not) since in the theory the probability is not a property of $m$ but of $(b_0, b)$ relative to $a$. Since (4.7.3) makes $m\in a\cap b_0$, a probability $\alpha\neq 0$ says that $m\in b$ is conditionally possible.

To assign a probability to $m$ relative to $b$ can cause unnecessary difficulties: If e.g. the observational report (4.7.3) is extended by the additional relations $\{a'\in\mathcal{Q}', \lambda(a'\cap b_0, a'\cap b)\approx\alpha', \lambda(a'\cap a\cap b_0, a'\cap a\cap b)\approx\beta$ and $m\in a'\}$ we get a new hypothesis, for which $m\in b$ is possible unless $\beta\approx 0$. But what probability should be assigned to $(m, b)$, $\alpha$ or $\alpha'$ or $\beta$? One could perhaps decide on $\beta$. But if one had forgotten to write down the additional relations, has one made a mistake when assigning on the basis of (4.7.3) the probability $\alpha$ to $(m, b)$? All these difficulties arise because of a logical mistake: One has done as if the relations $m\in a\cap b_0$ and $\lambda(a\cap b_0, a\cap b)\approx\alpha$ (between the five things $m$, $a$, $b_0$, $b$, $\alpha$) are only relations between $m$, $b$, $\alpha$.

But if one says that besides the probability introduced in the theory there is another "hidden" probability assigned to each pair $m\in M$ and $b\in\mathcal{R}$, one cannot forbid such "hidden variable theories" unless they are incompatible with the theory.

Sometimes another description of (4.7.3), (4.7.4) is added to our interpretation that $m\in b$ is conditionally physically possible.

This description calls (4.7.3) our "knowledge" and (4.7.4) a not yet "known" event (the triggering of $b$ by $m$). Then one calls $\alpha$ a measure of the "chance" that $m$ triggers $b$. (I use here the word chance instead of probability to distinguish between the probability function $\lambda$ and the intensity of subjective expectation.) In this sense of expectation, one restricts statements about the chance of events to events in the future (future indeed meant as the subjective future, defined relative to our consciousness).

Again such additional descriptions cannot be forbidden. They are indeed additional to what in §4.6 we called the physical game. We only object that such additional subjective descriptions are not necessary and not used in the every day work of physicists, which is described completely by the "physical game". In practice, only in one case one uses such additional "chance" descriptions, namely to persuade oneself or others to use a machine, argueing that the "chance" of an accident is very small.

There are some attempts to begin physics with the development of a language comprising concepts such as chance and of a corresponding logic. Such a language and logic shall establish fundamental structures of quantum mechnanics. This is not our understanding of physics. We object the same as at the end of §4.6.

We must emphasize that the hypothetical relation (4.7.4) is *not* necessarily related to the future! In connection with the EPR-Paradox we have already discussed in XII §2 the possibility that the indication $b$ can occur after the preparation is finished. What is then the meaning of $m \in b$ being conditionally physically possible? At the time when the observational report has been finished, $m \in b$ *has* already happened or not happened. Thus $m \in b$ is hypothetical not because it was not yet possible to detect $m \in b$, rather because we forgot to write down whether $m$ has triggered $b$, or because we gave someone the task to guess it. The possibility of $m \in b$ can therefore mean: It is possible that $m$ will trigger $b$; or it is possible that $m$ has triggered $b$. There are many examples in physics and technology where we are curious was has happened! In the physical game as described in §4.6, not only questions for the future but also for the past are possible.

We have discussed the hypothesis (4.7.3), (4.7.4) if $\alpha$ is physically distinguished from 0 and 1. For $\alpha=1$ the hypothesis is not only certain but even theoretically existent. Therefore $m \in b$ is physically real. It is physically real because $m$ has already triggered $b$ (which we perhaps forgot to report), or $m$ will trigger $b$. In the last case, $m \in b$ is real for the future. This case is similar to saying that the planets had a given position in the past or will have a given position in the future.

$\alpha=1$ is only a mathematical idealization. Physically, (i.e. in the observational report) we only have $\alpha \approx 1$ (not distinguishable from 1). Then we say that $m \in b$ is "practically certrain" and interpret such a hypothesis in the same way as a certain one.

Sometimes one reads that there is a difference between $\alpha=1$ and $\alpha \approx 1$ (e.g. $\alpha = 1\text{–}10^{-30}$) in this sense: for $\alpha=1$ it is perfectly certain that ..., and for $\alpha \approx 1$ it is only imperfectly certain that .... For $\alpha=1$ it is impossible, that not .... For $\alpha \approx 1$ it is possible, that not ..., although this can practically not be expected. But such a difference has nothing to do with physics since a probability $1\text{–}10^{-30}$ cannot be distinguished from 1 by any experiment. On the other hand, physics cannnot say that *anything* is perfectly certain, since physics gives no causes why anything should be perfectly certain. Only God can guarantee a perfect certainty and not we human beings in our human work named physical game.

For $\alpha=0$ the hypothesis (4.7.3), (4.7.4) is false and therefore $m \in b$ must physically be excluded. For $\alpha \approx 0$ we call the hypothesis practically false and $m \in b$ physically to be excluded as well.

As further examples let us take more complicated hypotheses, not completely writing down their observational reports. Such a report may contain the relations

$$\begin{aligned} &a \in \mathscr{Q}'; \quad b_{01}, b_{02} \in \mathscr{R}_0; \quad b_1, b_2 \in \mathscr{R}; \quad b_1 \subset b_{01}, \; b_2 \subset b_{02}; \\ &m \in M, \; m \in a \quad \text{and} \quad \lambda(a \cap b_{01}, a \cap b_1) \approx \alpha_1, \quad \lambda(a \cap b_{02}, a \cap b_2) \approx \alpha_2. \end{aligned} \tag{4.7.5}$$

But the observational report may contain results of many more experiments which imply $b_{01} \cap b_{02} = \emptyset$. (The reader may think of the case that $b_{01}$ is a device for measuring the momentum and $b_{02}$ one for the position at a time $t_0$; see also the discussion below.) Instead of writing down the observational report for all these experiments, to (4.7.5) we only add the relation $b_{01} \cap b_{02} = \emptyset$. The hypothesis may be defined by the additional hypothetic relation

$$m \in b_{01}. \tag{4.7.6}$$

This hypothesis of the first kind without invented elements is not certain, but allowed. Thus $m \in b_{01}$ is conditionally physically possible. In this case there is no probability which can be assigned to $m \in b_{01}$ since we can decide whether we take $b_{01}$ or $b_{02}$ as measuring devices.

In the same way we can discuss the hypothetical relation $m \in b_{02}$ instead of (4.7.6). If we take

$$m \in b_{01} \quad \text{and} \quad m \in b_{02}$$

instead of (4.7.6), we get a false hypothesis since $b_{01} \cap b_{02} = \emptyset$. A more interesting hypothesis arises if we replace (4.7.6) by

$$m \in b_1. \tag{4.7.7}$$

If $\alpha_1$ is distinguishable from 0 we have an allowed hypothesis (not certain, not even for $\alpha_1 = 1$!). Therefore (4.7.7) is conditionally physically possible. Here often the following *mistake* is made: One says that $m \in b_1$ is possible with the probabiity $\alpha_1$ (and impossible with the probability $(1-\alpha_1)$). This is a mistake since we can choose the device $b_{02}$, i.e. we can realize $m \in b_{02}$ which makes (4.7.7) false because of $b_{01} \cap b_{02} = \emptyset$. The choice of $b_{02}$ has made $m \in b_1$ to be physically excluded.

For instance for $m$ prepared by $a \in \mathscr{Q}'$ there is no probability for positions since one can measure the momentum and thus exclude any position possibility.

For the hypothesis (4.7.3), (4.7.4) we have a "constrained" possibility, constrained by probability. We cannot "make" $m \in b$, it "occurs". For (4.7.6) with (4.7.4) plus $b_{01} \cap b_{02} = \emptyset$ as observational report, we have a "free" possibility. We can dispose of (4.7.6). We therefore also call $m \in b_{01}$ "disposably possible". (4.7.7) is neither disposable nor constrained, it is partly disposably and partly constrained possible.

The disposable possibilities were always known to experimental physicists and engineers, but they played no role in theoretical discussions until the development of quantum mechanics. One was very astonished that such "practical" concepts should play a role in "pure physics" which should give us the structure of matter. Therefore one has made big efforts to eliminate the "actions of human beings" from the "pure physics", but without success. In this context it is very astonishing that in "pure physics" one introduced human consciousness instead of human actions. As we have described physics in §4.6 as a game, this cannot be separated from human actions. There does not exist something like a "pure physics" describing a "pure matter" which can be imagined as separated from all applications. Even to "observe" the matter in the universe we need modern devices (i.e. modern technology) to get effects from the universe in our laboratories.

We must make a short remark about our concept of "free" possibilities. If we apply Newton's mechanics to bullets, there are many free possibilities to prepare "initial values" of position and velocity. If we want to understand such preparations, we must leave the fundamental domain of Newton mechanics and pass to more comprehensive theories, e.g. theories of the chemical processes in a barrel. If we apply Newton's mechanics to big systems given in nature (no artifacts), e.g. the planetary systems, the "free" possibilities of the initial values are not free in the sense that we can make them. Yet they are "free" in the sense that they have been made by processes not belonging to the fundamental domain of Newton's mechanics, i.e. free "relative" to Newton's mechanics. The word "free" does not mean than we can make all this, but only that there are no constraints in the theory under consideration.

As a last example typical for quantum mechanics let us contemplate a desired observable (XI §7) as a hypothesis of the second kind. As desired observable we first take that defined by $\bar{F}(\mathscr{J},\mathscr{F})$ in [2] XVI (6.1.13), calling it the observable of the impact on a surface.

We define a hypothesis by an observational report $A \in \tilde{E}$ which expresses the construction of a laboratory reference frame. This frame makes it possible to define time intervals $\mathscr{J}$ and parts $\mathscr{F}$ of a plane surface.

It is typical for playing the physical game that one does not write down explicitly the whole observational report if it is gained by "well known" methods. We "know" how to get a laboratory frame. We do not know how to construct a device measuring the desired observable. Therefore we concentrate upon this question. To the observational report we add the hypothetical relations:

$$x_0 \in \mathscr{R}_0, \quad \mathscr{R}(x_0) \text{ is an approximation to } \Sigma \xrightarrow{\bar{F}} L \text{ with } \Sigma \text{ as the Boolean ring generated by the time intervals } \mathscr{J} \text{ and plane parts } \mathscr{F} \text{ mentioned above and by the mapping } \bar{F} \text{ from [2] XVI (6.1.13).} \tag{4.7.8}$$

The word "approximation" must be made concrete by specifying $\tilde{\Sigma}$ and $U$ as defined in AOb in V §5. $\tilde{\Sigma}$ can be defined with the help of the reference frame, i.e. by means of the observational report. To define $U$ we need finitely many $w \in K$. These must be given in explicit form, e.g. by finitely many "wave packets" $P_{\psi_i}$ with well defined $\psi_i$. Thus (4.7.8) has indeed the form $x_0 \in \tilde{E}_h(A)$.

If we have introduced AOb in V §5, the hypothesis (4.7.8) is theoretically existent. If we do not presume AOb, the hypothesis is certain. To prove this we must give a hypothesis of the first kind more restrictive than (4.7.8).

Since the $\psi_i$ are given, we can calculate the finitely many real numbers

$$\operatorname{tr}(P_{\psi_i}\,\bar{F}(\mathscr{I}_k,\sigma_k))=\langle\psi_i,\bar{F}(\mathscr{I}_k,\sigma_k)\psi_i\rangle=\alpha_{ik}$$

for the elements $(\mathscr{I}_k,\sigma_k)\in\tilde{\Sigma}$. We define a certain hypothesis of the first kind by

$$\begin{aligned}&x_0'\in\mathscr{R}_0,\quad x_k'\in\mathscr{R}(x_0)\text{ and the set }\{x_k'\}\text{ is a finite Boolean}\\&\text{ring isomorphic to }\tilde{\Sigma},\\&\qquad y_i\in\mathscr{Q}',\qquad \lambda(y_i\cap x_0'),\ y_i\cap x_k')=\alpha_{ik}\end{aligned}\tag{4.7.9}$$

with the same observational report $A\in\tilde{E}$. This hypothesis (4.7.9) is more restrictive than (4.7.8), as one easily sees.

Therefore, (4.7.8) is physically possible, (also without AOb) and in this case "freely possible". We can dispose of (4.7.8), i.e. we can construct a corresponding device.

The proof that (4.7.8) is certain can be repeated for any "given" observable. This is the meaning of the short remark after AOb in V §5.

The task to construct a device corresponding (4.7.8) cannot be solved when only the physical game of quantum mechanics in the form $\mathscr{P}\mathscr{T}_2$ (notation as in §3.3) is employed. In V §5 we have discussed this problem within $\mathscr{P}\mathscr{T}_{1t}$. But also this is not sufficient. It is necessary to pass to the theory presented in X and XI as an embedding of $\mathscr{P}\mathscr{T}_m$ in $\mathscr{P}\mathscr{T}_{q\exp}$. Only this theory can decide whether a constructed device fulfills the hypothesis (4.7.8).

But who tells us how to construct a device? No theory at all! Physical theories can only tell what is possible to do. They do not say which of many possible actions is the best to satisfy a given desire. Since no theory provides a systematic method for such a selection I must "invent" a construction of the desired device. This is a creative act and in this sense a great achievement of experimental physicists.

As another example of desired observables let us discuss the famous position – and momentum – observables (defined in [2] VII §4). Instead of (4.7.8) we take the hypothesis

$$\begin{aligned}&x_0\in\mathscr{R}_0,\quad \mathscr{R}(x_0)\text{ is an approximation to}\\&\text{the position- }\textit{and}\text{ momentum-observable.}\end{aligned}\tag{4.7.10}$$

Since the word "approximation" has been defined explicitly above, we need not explain it any more.

Whether (4.7.10) is a certain or a false hypothesis depends on the "approximation" required there. If the approximation is very close, the hypothesis becomes false, since the position- and momentum-observables are complementary ([2] IV D3.3). But there are finite approximations for which (4.7.10) becomes certain, resp. theoretically existent on the base of AOb, i.e. for which a corresponding device is possible.

To prove this we need only take an observable defined by the effects $\chi(\sigma)$ in X (3.4.6) with X (3.4.7) and $W$ in X (3.4.17). This observable $\Sigma\xrightarrow{\chi}L$ is an approximation to both, the position *and* momentum observable. Then the hypothesis (4.7.8) for $\Sigma\xrightarrow{\chi}L$ (instead of $\Sigma\xrightarrow{F}L$) is also one which fulfills (4.7.10).

Much more difficult is it to get a systematic survey of all "approximations" for which (4.7.10) is certain, i.e. realizable.

## §4.8 Real Facts and the Reality of Microsystems

In §4.6 we have interpreted a hypothesis

$$A \in \tilde{E} \quad \text{and} \quad X \in \tilde{E}_h(A) \tag{4.8.1}$$

as physically (strongly) possible if (4.8.1) is theoretically existent, i.e. if

$$\forall Z[Z \in \tilde{E} \Rightarrow \exists X(X \in \tilde{E}_h(Z))] \tag{4.8.2}$$

is a theorem. Therefore one interprets (4.8.2) in the form: The situation $X \in E(Z)$ is physically (strongly) possible under the condition $Z \in \tilde{E}$ (see the interpretation of (4.6.14)).

(4.8.2) has the form of well known mathematical "existence theorems". The proof of such theorems in $\mathscr{M}\mathscr{T}_\Sigma$ is therefore not only mathematically interesting, but in the context of $\mathscr{P}\mathscr{T}$ also of eminent physical importance. Many times physicists believe in such theorems even without proofs.

If besides (4.8.2) we have the theorem that $\tilde{E}_h(Z)$ has only one element, we interpret (4.8.2) in the form: The situation $X \in \tilde{E}_h(Z)$ is physically (strongly) real under the condition $Z \in \tilde{E}$. Theorems of the form "$\tilde{E}_h(Z)$ has only one element" are well known as uniqueness theorems. These theorems guarantee that the mathematical solution $X$ of a problem $X \in \tilde{E}_h(Z)$ describes physical reality under the condition that $Z \in \tilde{E}$ is real.

We may weaken the theoretical existence into certainty. A certain and determined hypothesis (4.8.1) is then also called physically (weakly) real. That (4.8.1) is determined implies: There is a mapping $\tilde{E} \xrightarrow{f} \tilde{T}_h(\ldots)$ (on physical grounds presumed uniformly continuous). When the range of $f$ is called $\tilde{E}_h^{(f)}$, we have

$$\tilde{E}_h^{(f)} = \bigcup_{Z \in \tilde{E}} \tilde{E}_h(Z). \tag{4.8.3}$$

There is a certain inversion. We start with the same definition as in ($\alpha$) and ($\beta$) of §3.1. The $E_\nu$ (as intrinsic terms) shall be subsets of products of pictorial sets. The $h_\nu$ (as intrinsic terms) shall be mappings $E \xrightarrow{h_\nu} T_\nu(\ldots)$ where the $T_\nu(\ldots)$ are echelon sets of the base sets of $\Sigma$. (For $h_\nu$ also the identical map $E_\nu \to E_\nu$ is allowed.) The range of $h_\nu$ shall be called $E_\nu^{(h)}$.

In addition there shall be a structure $U^{(h)}$ of the species $\Sigma^{(h)}$ over $E_1^{(h)}, E_2^{(h)}, \ldots$ as base sets (see §2.1).

Let $u_r^{(h)}$ be such subsets of product sets of the $T_\nu(\ldots)$ (and $\mathbf{R}$) that

$$(y_1, y_2, \ldots; \alpha) \in u_r^{(h)} \tag{4.8.4}$$

with $y_\lambda \in T_{\nu_\lambda}(\ldots)$ is a relation among $y_1, y_2, \ldots; \alpha$. Then $U^{(h)}$ shall have the form

$$U^{(h)} = (u_1^{(h)}, u_2^{(h)}, \ldots). \tag{4.8.5}$$

We will see that such a structure $U^{(h)}$ over $E_1^{(h)}, E_2^{(h)} \ldots$ can be interpreted as the picture of a real structure of real facts, labeled by the elements of $E_1^{(h)}, E_2^{(h)}, \ldots$.

Let us think of an observational report $A \in \tilde{E}$. It may be possible to find in $A$ finite sequences $(a_{i_1}, a_{i_2}, \ldots) \in E_\nu$. Then there are defined the elements $x_{i\nu} = h_\nu(a_{i_1}, \ldots) \in E_\nu^{(h)}$. With the collection $X = (\ldots x_{i\nu} \ldots)$, by $A \to X$ we can define a mapping $\tilde{E} \xrightarrow{f} \tilde{T}(\ldots)$ from $\tilde{E}$ into an echelon set.

Let it be possible to prove (in $\mathcal{M}\mathcal{T}_\Sigma \mathcal{A}$) some relations of the form (4.8.4) for the $x_{i\nu}$. These relations define a subset $\tilde{E}^{(u)} \subset \tilde{T}(\ldots)$ with

$$\tilde{E} \xrightarrow{f} \tilde{E}^{(u)}. \tag{4.8.6}$$

With the range $\tilde{E}^{(f)}$ of $\tilde{E}$, we therefore have

$$\tilde{E}^{(f)} \subset \tilde{E}^{(u)}. \tag{4.8.7}$$

With $\tilde{E}_h(A)$ as the set of the single element $f(A)$, the relation

$$A \subset \tilde{E} \quad \text{and} \quad X \in \tilde{E}_h(A) \tag{4.8.8}$$

is a theoretically existent and determined and therefore physically real hypothesis.

We have

$$\tilde{E}^f = \bigcup_{A \in \tilde{E}} \tilde{E}_h(A) \subset \tilde{E}^{(u)} \tag{4.8.9}$$

such that (4.8.8) implies

$$X \in \tilde{E}^{(u)}. \tag{4.8.10}$$

Thus (4.8.8) together with (4.8.10) is also a theoretically existent and determined hypothesis. Therefore it is physically real.

We summarize this to: By the measurement (observation) $A \in \tilde{E}$ we have also measured (observed) $X \in \tilde{E}^{(u)}$; or expressed more objectively: The real facts $A \in \tilde{E}$ imply the real facts $X \in \tilde{E}^{(u)}$.

This is the background for the following formulations: The $E_\nu^{(h)}$ are pictorial sets for real facts and the components $u_r^{(h)}$ of $U^{(h)}$ are pictorial relations for real relations, or shortly: The $E_\nu^{(h)}$ are sets of real facts with real relations $u_r^{(h)}$.

The facts represented by the observational report $A \in \tilde{E}$ guarantee the facts represented by $X \in \tilde{E}^{(u)}$. This last transition from $A \in \tilde{E}$ to $X \in \tilde{E}^{(u)}$ is exactly that from $\mathcal{P}\mathcal{T}_1$ to a restriction $\mathcal{P}\mathcal{T}$ (see §3.1).

We also say that by the facts (from the observational report) and by the structure laws (from the theory) we have "detected" the new realities (described by the structure $U^{(h)}$ of the species $\Sigma^{(h)}$ over the $E_\nu^{(h)}$). All physical detections are of this kind.

Another formulation is the following: We have "indirectly" measured $X \in \tilde{E}^{(u)}$ by the "direct" measurement $A \in \tilde{E}$ on the basis of the mapping $f$ in (4.8.6).

Sometimes we can measure indirectly what can also be measured directly. We can measure indirectly the position of the planets in two years by direct measurements made until now. In two years we can measure these positions also directly.

It may be confusing if we measure the same real situation indirectly with a higher precision than directly. This can happen if the mapping $f$ is magnifying, as e.g. the mapping of the object plane into the image plane of a microscope. In such cases one easily makes the mistake of defining a real situation by itself. Instead of this vicious circle, the theory provides a more precise extension of a not so precise measurement possibility (given by pretheories).

All observational reports finally should be reduced by pretheories to the language of craftsmen. Hence physics can detect only realities that can be defined by effects which can be stated by craftsmen and formulated in their language.

Therefore it is impossible, in physics to detect consciousness and all what I meet in my consciousness (red, green and yellow colors, beautiful sounds ect.). It is only possible physically to detect so-called "physical processes" in the brain. These are all what can be defined on the basis of theories by effects on devices.

Definitions for the contents of my consciousness are logically impossible since it is already known what I mean e.g. by colors. All concepts which are already known *before* physics (and which do not enter the language of craftsmen to describe the given facts on which physics is based) cannot be *defined* in physics. Hence such concepts all the more cannot be explained by physics.

In physics it is usual (similarly as in mathematics) *to give names* to sets $E_v^{(h)}$ of real facts endowed with a real structure $U^{(h)}$ of species $\Sigma^{(h)}$. We have done this often during the development of quantum mechanics in III through IX. Here let us only review the "detection of microsystems" in $\mathscr{P}\mathscr{T}_1$ (notation as in §3.3). With this detection we have connected the restriction from $\mathscr{P}\mathscr{T}_1$ to $\mathscr{P}\mathscr{T}_2$. The new pictorial sets are $M$, $\mathscr{Q}$, $\mathscr{R}_0$, $\mathscr{R}$ with the mappings $M \xrightarrow{h_1} M$ ($h_1$ = identity), $\mathscr{S}_1 \xrightarrow{h_2} \mathscr{Q}$, $\mathscr{S}_{20} \xrightarrow{h_3} \mathscr{R}$, $\mathscr{S}_2 \xrightarrow{h_4} \mathscr{R}$, where the last mappings are given by III (4.1) to (4.3). Structure terms are $\mathscr{S}$ and the probability function $\lambda_{\mathscr{S}}$ (see III D4.1 to APS 6). The characteristic relations for the new structure $\Sigma_2$ are given in III by APS 1 through APS 8. The definition III D4.2 is what we have called above "to give names". Thus the detection of real microsystems is justified.

Another not so difficult example of physical detections (in the realm of classical physics) is that of "electric charge", "electromagnetic fields", etc. In particle physics we see a growing family of newly detected microsystems, although the foundation of these detections is not yet as established as in such an "old" theory as quantum mechanics.

Much more questionable is the reality of the microsystems (atoms) as parts of a macrosystem. This problem is not yet completely solvable, since we lack a surely g.$\mathscr{G}$-closed theory for macrosystems. If $\mathscr{P}\mathscr{T}_{q\,\mathrm{exp}}$ were g.$\mathscr{G}$-closed, the question could be answered in the same way as for microsystems. But there are severe arguments that $\mathscr{P}\mathscr{T}_{q\,\mathrm{exp}}$ is *not* g.$\mathscr{G}$-closed. Hence we must not conclude from $\mathscr{P}\mathscr{T}_{q\,\mathrm{exp}}$ the reality of the microsystems as parts of the macrosystems.

The fundamental domain of most known theories $\mathscr{P}\mathscr{T}_m$ is too small to answer the question about the reality of those microscopic parts. Therefore we can only remark, how the solution of this problem could look like.

At the end of XI §6 we have considered microsystems which (as parts of a macrosystem) can be distinguished from the rest of the macrosystem. Such systems are "real" in the same sense as prepared microsystems (the reality of which we have explicitly deduced above). Nevertheless it can be that measurements of such "emitted" or "distinguished" microsystems give no new information about the macrosystem, if the state space is comprehensive enough: To an effect $g$ of the microsystems we can relate a trajectory effect $p(g)$ in XI (6.31).

But what about the reality of the many microsystems "inside" a macrosystem?

At the end of III §6.6 we have shortly described measurement possibilities for macrosystems by the scattering of microsystems. Such a scattering is extensively described in XI §1 and its embedding in $\mathscr{P}\mathscr{T}_{q\,\mathrm{exp}}$ in XI §6. By such scatterings we get information about the macrosystem. Formula XI (1.19) describes this additional information, going beyond $\langle \varphi_1^{(0)}(a_1), k_1 \rangle_1$ as the probability for the trajectory effect $k_1$ without scattering of microsystems ($\varphi_1^{(0)}(a_1)$ in XI (1.22)). Examples for such scatterings are well known; most famous is the scattering of $X$-rays to analyse the atomic structure of crystals.

XI (1.19) only then contains additional information about the macrosystem when $\Gamma(w^{(i)}, a_1)$ does not only depend on $\varphi_1^{(0)}(a_1)$, i.e. when $\tilde{\Gamma}$ cannot be introduced according to XI (1.24). We conjecture that for macrosystems there is a most comprehensive state space $Z$ such that XI (1.24) holds, i.e. that the preparation procedures for macrosystems cannot be better distinguished by scattering than by their trajectories in $Z$. This means: there are real microscopic structures of macrosystems; but these can be described in an objectivating manner.

If XI (1.24) is valid, i.e. $Z$ comprehensive enough, then (similarly as at the end of XI §6) the probability to get the trajectory effect $k_1$ and the effect $g$ of the scattered microsystems becomes

$$\langle \Gamma(w^{(i)}, a_1), k_1 g \rangle = \langle \tilde{\Gamma}(w^{(i)}, u_1), k_1 g \rangle = \langle u_1, k_1 p(g; w^{(i)}) \rangle. \tag{4.8.11}$$

Here we have $u_1 = \varphi_1^{(0)}(a_1)$, while $w^{(i)}$ is the ensemble of the impinging microsystems and $p(g; w^{(i)})$ describes a trajectory effect (depending on $g$ and $w^{(i)}$). Then (4.8.11) is the mathematical form for not getting information beyond that from the objectivating description in $Z$.

We know many microscopic but nevertheless objectivating descriptions of macrosystems, e.g. of semiconductors. But since these theories are not simple we will only use the theory of rarefied gases as described in X §3.5.

If we take as state space only that of aerodynamics, i.e. $(\boldsymbol{u}(\boldsymbol{r}), \rho(\boldsymbol{r}), T(\boldsymbol{r}))$ ($\boldsymbol{u}$ velocity field, $\rho$ density field, $T$ temperature field), this space is not sufficient to describe anything of the microstructure of the gas. This well known objection of the positivists against the reality of atoms is not valid, since this state space is too small to describe for instance scattering of microsystems on the gas. In our opinion, the state space of the Boltzmann distribution functions is comprehensive enough as far as the atoms can be approximated by systems without inner structures. The description by the Boltzmann distribution function gives us an imprecise (but objectivating!) description of the positions and momenta of the atoms. Thus the atoms are real, although not with precise positions and momenta.

The fundamental domain of this Boltzmann theory must be expanded if one wants to account for the inner structure of the atoms. Experiences show that an "accupation number" of the discrete energy levels probably is the right objectivating description.

Obviously, a general and systematic theory of the right state spaces describing also the microscopic structure of macrosystems is not yet developed. Such a theory would give precise answers to the question: What is for macrosystems physically possible or physically real?

At the end of this section we must warn of a mistake. This mistake rests on the false notion that "mathematical objects" must be pictures of physical objects. The denotation "mathematical objects" is only an intuitive one. In mathematics it is better to say "term" instead of object. The language of mathematics is characterized by "substantific" and "relational" signs [50] I §1.3. These signs are used to formulate "facts" in the mathematical language. The substantific signs are used to identify a fact by a sign, and the relational signs are used to make statements about factual relations among the facts designated by substantific signs (see §1).

This does not mean that these facts are physical objects. Also a trajectory (i.e. a dynamical process) can be designated by a substantific sign. We had physical objects defined in V §10. A one to one correspondence between substantific signs and physical objects is therefore a mistake.

As we have already discussed in §2.5, the mathematical set theory is only used for idealizations from finitely many signs to infinitely many. In any case, only finitely many signs are used for the observational reports.

Thus also the use of "elements of sets" as signs for physical facts does not mean that these facts are something rigid. As already emphasized, these facts can be processes or procedures as e.g. the preparation- and registration-procedures.

Also the new mathematical concept of categories does not change anything since we have again signs representing morphisms, the composition of morphisms as relations, etc. That all this does not bring a revolution follows from the fact that for *finitely* many signs the new mathematics is equivalent to the older one. Nevertheless, $\mathscr{M}\mathscr{T}_\Sigma$ as part of a physical theory and the observational reports have indeed something static. Neither $\mathscr{M}\mathscr{T}_\Sigma$ nor the once recorded observational reports can be changed (except one has made a mistake). But this is exactly the method of physics: Stated facts remain stated facts. Nevertheless there is a very dynamic process in physics. This process is not described by a mathematical theory but by what we called the mathematical and physical game (§4.5 to §4.7). In addition there is the development of new theories, a development which progresses against the direction of the arrows in §3. The last section shall be devoted to one aspect of this dynamic process, the development of the "real domain" of a theory.

## §4.9 The Real Domain of a $\mathscr{P}\mathscr{T}$

We want to describe "all" what is real in the context of a $\mathscr{P}\mathscr{T}$ which we presume as g.$\mathscr{G}$-closed. Initially the fundamental domain $\mathscr{G}$ is real, or better: The observational reports are physically real in the sense that they report facts. "All" observational reports describe all facts of the fundamental domain $\mathscr{G}$. The word "all" in this context has nothing to do with the logical sign $\forall$ in $\mathscr{M}\mathscr{T}$. These "all" observational reports are not given. The observational reports develop more and more, and we influence this development. Not $\mathscr{P}\mathscr{T}$ in itself determines the development. Our interests and our free decisions to realize these and not other free possibilities in the physical game are as essential.

The reality domain $\mathscr{W}$ shall extend the domain of "all" observational reports which we have identified with $\mathscr{G}$.

We try to define $\mathscr{W}$ as the domain of "all" certain and determined (i.e. of all physically real) hypotheses.

To make this meaningful, we must consider the two processes

(1) the composition of two hypotheses,
(2) the extension of the observational report.

These two processes are extensively described in §4.3 and §4.4. From there we conclude that the properties "certain" and "determined" are not changed by these processes. (It can be that a certain but not determined hypothesis becomes determined by process 2).

$\mathscr{H}$ may be a certain and determined hypothesis. If we have the theorem that one of the invented elements $x_i$ equals an $a_k$ of the observational report, we replace $x_i$ by $a_k$. If we have the theorem that one of the invented elements $x_i$ equals another one $x_k$, we replace $x_i$ and $x_k$ by a single letter. After the two processes (1) and (2) it may be necessary to make such new replacements; but older replacements cannot be changed since all hypotheses are determined. We say that a certain and determined hypothesis has its normal form if there are no unnecessary letters $x_i$.

The concept "more comprehensive" (defined in §4.4) provides a kind of ordering in the domain of certain and determined hypotheses. We will show that this ordering is directed, i.e. that to any two hypotheses $\mathscr{H}_1$, $\mathscr{H}_2$ of this kind there is a third one $\mathscr{H}_3$ more comprehensive than $\mathscr{H}_1$ and $\mathscr{H}_2$.

To show this we combine the observational reports belonging to $\mathscr{H}_1$ and $\mathscr{H}_2$ to a new observational report which extends the observational reports of $\mathscr{H}_1$ and $\mathscr{H}_2$. Let $\mathscr{H}_1'$ resp. $\mathscr{H}_2'$ be the hypothesis $\mathscr{H}_1$ and $\mathscr{H}_2$ with this extended observational report ($\mathscr{H}_1'$ and $\mathscr{H}_2'$ are again certain and determined). We can go over to the normal form, which does not change the relations that $\mathscr{H}_1'$ is more comprehensive than $\mathscr{H}_1$ and $\mathscr{H}_2'$ more comprehensive than $\mathscr{H}_2$.

Let $\mathscr{H}_3'$ be the composition of $\mathscr{H}_1'$ and $\mathscr{H}_2'$. If we go over to the normal form $\mathscr{H}_3$ of $\mathscr{H}_3'$, we have $\mathscr{H}_3$ more comprehensive than $\mathscr{H}_1'$ and $\mathscr{H}_2'$ and thus more comprehensive than $\mathscr{H}_1$ and $\mathscr{H}_2$.

That the kind of ordering in the domain of certain and directed hypotheses (in normal form) is directed is essential for a meaningful "definition" of $\mathscr{W}$. The domain $\mathscr{G}$ of observational reports is never complete, it just grows. No older parts of the observational report need be dropped. Because of the directness of the ordering of the certain and determined hypotheses, the domain of these hypotheses can only grow if the observational reports are growing and new hypotheses are added. No hypothesis stated as physically real need be dropped later. In this sense we have an evolution of the observational reports and of the physically real hypotheses. This evolution is shortly called the domain $\mathscr{G}$ of the observational reports, resp. the physically real domain $\mathscr{W}$.

Not all physically possible hypotheses can be realized since there are not enough material, not enough time and last not least not enough human beings to do it. Thus the evolution of the physically real domain $\mathscr{W}$ is essentially determined by us.

We have defined $\mathscr{W}$ for one g.$\mathscr{G}$-closed theory. But what about the various $\mathscr{W}$ in the network of physical theories (defined in §3.3)? The intertheory relations "restriction" and "embedding" make it possible to transport parts of $\mathscr{W}$ to less comprehensive theories. This shows that in a chain of the form (3.3.5) it is allowed to form $\mathscr{W}_\nu$ as the domain of certain and determined hypotheses separately for every $\mathscr{PT}_\nu$. Thus $\mathscr{W}_{\nu+1}$ comprises $\mathscr{W}_\nu$. It is only necessary that $\mathscr{W}_n$ is g.$\mathscr{G}$-closed in order to know what can be realized. The $\mathscr{PT}_\nu$ for $\nu<n$ can perhaps suggest something as physically possible what is not so.

We claim that the network of theories is only growing, i.e. no theories are abandoned. One can make mistakes during the development of a new theory, overestimating the fundamental domain or imagining something as physically real what is not so. Such errors are easily possible if one has no axiomatic basis and imagines some theoretical auxiliary terms as "realities".

But it also is an error that there are revolutions in the development of physical theories. In the contrary, we have an evolution of the network of theories and a corresponding evolution (not only of the physically real domain of one theory but also) of the collection of the physically real domains of all theories. For many parts of this evolution of the "total" physically real domain, we as human beings are responsible.

No $\mathscr{M}\mathscr{T}_\Sigma$ together with the corresponding correspondence rules can be evil. But moves in the physical games can indeed be evil, e.g. to make physical experiments with human beings which harm these. In many cases it is not simple to decide what move we ought to do, since many circumstances must be taken into account. Thus different persons can reach different conclusions. It would be bad to suspect all who do not make the same decisions as we ourselves.

# Bibliography


1. G. Ludwig: *Einführung in die Grundlagen der theoretischen Physik*, 4 Vols. (Vieweg, Braunschweig 1974–1979)
2. G. Ludwig: *Foundations of Quantum Mechanics I*, Texts and Monographs in Phys. (Springer, Berlin, Heidelberg, New York, Tokyo 1983); Foundations of Quantum Mechanics II, Texts and Monographs in Phys. (Springer, Berlin, Heidelberg, New York, Tokyo 1985)
3. G. Ludwig: *Grundstrukturen einer physikalischen Theorie* (Springer, Berlin, Heidelberg, New York 1978)
4. G. Ludwig: Makroskopische Systeme und Quantenmechanik. Notes Math. Phys. *5* (Marburg 1972)
5. Bourbaki: *Topology generale* (Herrmann, Paris 1961)
6. G. Ludwig: "A Theoretical Description of Single Microsystems", in: *The Uncertainty Principle and Foundations of Quantum Mechanics*, ed. by W.C. Price, S.S. Chissik (Wiley, New York 1977)
7. H.H. Schaefer: *Topological Vector Spaces* (Macmillan, New York 1966)
8. N. Dunford, J.T. Schwartz: *Linear Operators* (Interscience, New York 1958)
9. V.S. Varadarajan: *Geometry of Quantum Theory*, Vol. 1 (Van Nostrand, Princeton, NJ 1968)
10. G. Köthe: *Topologische lineare Räume* (Springer, Berlin, Göttingen, Heidelberg 1960)
11. A Hartkämper, H. Neumann: *Foundations of Quantum Mechanics and Ordered Linear Spaces*, Lecture Notes, Vol. 29 (Springer, Berlin, Heidelberg, New York 1974)
12. A. Kolmogoroff: *Grundbegriffe der Wahrscheinlichkeitsrechnung* (Springer, Berlin 1933)
13. N. Zierler: Axioms for non-relativistic quantum mechanics. Pac. J. Math. *11*, 1151 (1961)
14. N. Zierler: On the lattice of closed subspaces of Hilbert-space. Pac. J. Math. *19*, 583 (1966)
15. M.D. MacLaren: Atomic orthocomplemented lattices. Pac. J. Math. *14*, 597 (1964)
16. M.D. MacLaren: Notes on axioms for quantum mechanics. ANL-7065 (1965)
17. E. Weiss, N. Zierler: Locally compact division rings. Pac. J. Math. *8*, 369 (1958)
18. L.S. Pontrjagin: *Topologische Gruppen*, Teil 1 (Teubner, Leipzig 1957)
19. G. Birkhoff, J. v. Neumann: The logic of quantum mechanics. Ann. Math. *37*, 823 (1936)
20. C. Piron: "Axiomatic Quantic", Ph. D. Thesis, Université de Lausanne, Faculté des Sciences (Birkhäuser, 1964)
21. M. Jammer: *The Conceptual Development of Quantum Mechanics* (McGraw-Hill, New York 1966)
    M. Jammer: *The Philosophy of Quantum Mechanics* (Wiley, New York 1977)
    E. Scheibe: *The Logical Analysis of Quantum Mechanics* (Pergamon, New York 1973)
22. L. Kanthack: in Vorbereitung
23. K. Drühl: A theory of classical limit for quantum theories which are defined by real Lie Algebras. J. Math. Phys. *19*, 1600 (1978)
24. G. Dähn: Attempt of an axiomatic foundation of quantum mechanics IV. Commun. Math. Phys. *9*, 192 (1968)
25. J. v. Neumann: *Mathematische Grundlagen der Quantenmechanik* (Springer, 1932) [English transl.: *Mathematical Foundation of Quantum Mechanics*, translated by R.T. Beyer (Princeton NJ, 1955)]
26. G. Ludwig: Meß- und Präparierprozesse. Notes Math. Phys. *6* (Marburg 1972)
27. H. Neumann: *A Mathematical model for a set of microsystems.* Int. J. Theor. Phys. *17*, 3 (1978)

H. Neumann: „Zur Verdeutlichung der statistischen Interpretation der Quantenmechanik durch ein mathematisches Modell für eine Menge von Mikrosystemen", in: *Grundlagen der Quantentheorie*, ed. by P. Mittelstaedt, J. Pfarr (B.I. Wissenschaftsverlag, Mannheim 1980)

28. E. Bishop, R.R. Phelps: The support functionals of a convex set. Proc. Symp. Pure Math. 7, 27 (1963)
29. A.J. Ellis: Minimal decompositions in partially ordered normed vector spaces. Proc. Cambridge Philos. Soc. *64*, 989 (1968)
30. G. Ludwig: Axiomatische Basis einer physikalischen Theorie und Theoretische Begriffe. Z. allg. Wissenschaftstheorie *12* (1), 55 (1981)
31. F. Riesz, B. Sz. Nagy: *Vorlesungen über Funktionalanalysis* (Deutscher Verlag der Wissenschaften, Berlin 1956)
32. H. Neumann: A new physical characterization of classical systems in quantum mechanics. Int. J. Theor. Phys. *9*, 225 (1974)
33. H. Neumann: Classical systems and observables in quantum mechanics. Commun. Math. Phys. *23*, 100 (1971)
34. P.R. Halmos: *Measure Theory* (Van Nostrand, Princeton, NJ 1950)
35. G. Ludwig: Der Meßprozeß. Z. Phys. *135*, 483 (1953)

G. Ludwig: Zur Deutung der Beobachtung in der Quantenmechanik. Phys. Bl. *11*, 489 (1955)

G. Ludwig: Zum Ergodensatz und zum Begriff der makroskopischen Observablen. Z. f. Naturforsch. *A 12*, 662 (1957)

G. Ludwig: Zum Ergodensatz und zum Begriff der makroskopischen Observablen I. Z. Phys. *150*, 346 (1958)

G. Ludwig: Zum Ergodensatz und zum Begriff der makroskopischen Observablen II, Z. Phys. *152*, 98 (1958)

G. Ludwig: "Axiomatic Quantum Statistics of Macroscopic Systems", in *Ergodic Theories*, ed. by P. Caldirola (Academic, New York 1960), p. 57

G. Ludwig: „Gelöste und ungelöste Probleme des Meßprozesses in der Quantenmechanik", in: *Werner Heisenberg und die Physik unserer Zeit* (Vieweg, Braunschweig 1961), p. 150

G. Ludwig: Zur Begründung der Thermodynamik auf Grund der Quantenmechanik. Z. Phys. *171*, 476 (1963)

G. Ludwig: Zur Begründung der Thermodynamik auf Grund der Quantenmechanik II, Masterequation. Z. Phys. *173*, 232 (1963)

G. Ludwig: Versuch einer axiomatischen Grundlegung der Quantenmechanik und allgemeinerer physikalischer Theorien. Z. Phys. *181*, 233 (1964)

G. Ludwig: "An Axiomatic Foundation of Quantum Mechanics on a Nonsubjective Basis", *Quantum Theory and Reality*, ed. by M. Bunge (Springer, Berlin, Heidelberg, New York 1967), p. 98

G. Ludwig: Attempt of an axiomatic foundation of quantum mechanics and more general theories II. Commun. Math. Phys. *4*, 331 (1967)

G. Ludwig: Hauptsätze über das Messen als Grundlage der Hilbert-Raum-Struktur der Quantenmechanik, Z. Naturforsch. *A 22*, 1303 (1967)

G. Ludwig: Ein weiterer Hauptsatz über das Messen als Grundlage der Hilbert-Raum-Struktur der Quantenmechanik. Z. Naturforsch. *A 22*, 1324 (1967)

G. Ludwig: Attempt of an axiomatic foundation of quantum mechanics and more general theories III. Commun. Math. Phys. *9* (1968)

G. Dähn: [24]

P. Stolz: Attempt of an axiomatic foundation of quantum mechanics and more general theories V. Commun. Math. Phys. *11*, 303 (1969)

G. Ludwig: Deutung des Begriffs „physikalische Theorie" und axiomatische Grundlegung der Hilbert-Raum-Struktur der Quantenmechanik durch Hauptsätze des Messens. Lecture Notes Phys., Vol. 4 (Springer, Berlin, Heidelberg, New York 1970)

P. Stolz: Attempt of an axiomatic foundation of quantum mechanics and more general theories VI. Commun. math. Phys. *23*, 117 (1971)

G. Ludwig: The Measuring Process and an Axiomatic Foundation of Quantum Mechanics, in *Foundations of Quantum Mechanics*, ed. by B. d'Espagnat (Academic, New York 1971), p. 287

G. Ludwig: A physical interpretation of an axiom within an axiomatic approach to quantum mechanics and a new formulation of this axiom as a general covering condition. Notes in Math. Phys. 1 (Marburg 1971)
G. Ludwig: Transformationen von Gesamtheiten und Effekten. Notes Math. Phys. *4* (Marburg 1971)
G. Ludwig: [4]
G. Ludwig: [26]
G. Ludwig: An improved formulation of some theorems and axioms in the axiomatic foundation of the Hilbert space structure of quantum mechanics. Commun. Math. Phys. *26*, 78 (1972)
G. Ludwig: "Why a New Approach to Found Quantum Theory?" in *The Physicist's Conception of Nature*, ed. by J. Mehra (Reidel, Dordrecht, 1973), p. 702
G. Ludwig: "Measuring and Preparing Processes", in *Foundation of Quantum Mechanics and Ordered Linear Spaces*, ed. by A. Hartkämper, H. Neumann. Lecture Notes Phys. Vol. 29 (Springer, Berlin, Heidelberg, New York 1974), p. 122
G. Ludwig: Measurement as a process of interaction between macroscopic systems. Notes Math. Phys. *14* (Marburg 1974)
G. Ludwig: [6]
G. Ludwig: [47]
G. Ludwig: "An Axiomatic Basis of Quantum Mechanics", in: *Interpretations and Foundations of Quantum Theory*, ed. by H. Neumann (B.I. Wissenschaftsverlag, Mannheim 1981), p. 49
G. Ludwig: Quantum theory as a theory of interactions between macroscopic systems which can be described objectively. Erkenntnis *16*, 359 (1981)
G. Ludwig: "The Connection Between the Objective Description of Macrosystems and Quantum Mechanics of 'Many Particles'", in: *Old and New Questions in Physics, Cosmology, Philosophy and Theoretical Biology*, ed. by A. van der Merwe (Plenum, New York 1983), p. 243

36. B. Mielnik: Geometry of quantum states. Commun. Math. Phys. *9*, 55 (1968)
B. Mielnik: Theory of filters. Commun. Math. Phys. *15*, 1 (1969)
B. Mielnik: Generalized quantum mechanics. Commun. Math. Phys. *31*, 221 (1974)
B. Mielnik: "Quantum Logic: Is It Necessarily Orthocomplemented?" in *Quantum Mechanics, Determinism, Causality and Particles*, ed. by M. Flato, Z. Maric, A. Milojevic, D. Sternheimer, J.P. Vigier (Reidel, Dordrecht 1976)
37. A. Lande: *Foundation of Quantum Theory* (Yale University Press, New Haven 1955)
A. Lande: *New Foundations of Quantum Mechanics* (Cambridge University Press, Cambridge 1965)
38. J.S. Bell: "Introduction to the Hidden-Variable Question", in *Foundations of Quantum Mechanics*, ed. by B. d'Espagnat (Academic, New York 1971), p. 171
39. O.M. Nikodym: Sur l'existance d'une mesure parfaitement additive et non separable. Mém. Acac. Roy. Belg. *17* (1939)
O.M. Nikodym: *The Mathematical Apparatures for Quantum Theories* (Springer, Berlin, Heidelberg, New York 1966), Chapt. A.
40. G. Ludwig: "Imprecision in Physics", in *Structure and Approximation in Physical Theories*, ed. by A. Hartkämper, H.-J. Schmidt (Plenum, New York 1980)
41. H. Neumann: The description of preparation and registration of physical systems and conventional probability theory. Found. Phys. *13*, 761 (1983)
42. P. Janich: *Die Protophysik der Zeit* (Suhrkamp, Frankfurt/Main 1980)
43. R. Werner: "Quantum Harmonic-Analysis on Phase Space", Preprint, Universität Osnabrück (1983)
A. Barchielli, L. Lanz, G.M. Prosperi: Statistics of continuous trajectories in quantum mechanics: operation-valued stochastic-processes. Found. Phys. *13*, 779 (1983)
44. R. Werner: "The Concept of Embeddings in Statistical Mechanics", Ph. D. Thesis, Marburg (1982)
45. R. Werner: Physical uniformities and the state space of nonrelativistic quantum mechanics. Found. Phys. *13*, 859 1983)
46. H. Gerstberger, H. Neumann, R. Werner: „Makroskopische Kausalität und relativistische

Quantenmechanik", in *Grundprobleme der modernen Physik*, ed. by J. Nitsch, J. Pfarr, E.W. Stachow (B.I. Wissenschaftsverlag, Mannheim 1981), p. 205
H. Neumann, R. Werner: Causality between preparation and registration processes in relativistic quantum theory. Int. J. Theor. Phys. *22*, 781 (1983)
47. G. Ludwig: Axiomatische Basis der Quantenmechanik, Notes Math. Phys. *16*, *17*, *18*. Marburg (190)
48. G. Ludwig: "Restriction and Embedding", in *Reduction in Science, Structure, Examples, Philosophical Problems*, ed. by W. Balzer, D. Pearce, H.-J. Schmidt (Reidel, Dordrecht 1984), p. 17
49. A. Nitsch: „Die quantenmechanische Beschreibung eines Meßprozesses; der Stern-Gerlach-Versuch als Beispiel einer Meßumwandlung", Diplomarbeit am Fachbereich Physik der Universität Osnabrück (1984)
50. N. Bourbaki: *Theory of Sets* (Addison-Wesley. Massachusetts-London-DonMills, 1968)
51. C.A. Rogers: *Packing and Covering* (Cambridge University Press, Cambridge 1964)
52. M. Reed, B. Simon: Methods of Modern Mathematical Physics (4 volumes) (Academic, New York 1972, 1972, 1978, 1979)
53. R.A. De Vore: *The Approximation of Continuous Functions by Positive Linear Operators*, Lecture Notes in Math. Vol. 293 (Springer, Berlin, Heidelberg, New York 1972)
54. H. Neumann: "The Representation of Classical Systems in Quantum Mechanics", in *Foundations of Quantum Mechanics and Ordered Linear Spaces*, ed. by A. Hartkämper, H. Neumann, Lecture Notes Phys., Vol. 29 (Springer, Berlin, Heidelberg, New York 1974) p. 316
55. A.S. Holevo: *Probabilistic and Statistical Aspects of Quantum Theory* (North-Holland, Amsterdam, 1983)
56. S. Großmann: Occupation number representation with localized one-particle Functions. Physica *29*, 1373–1392 (1963)
57. S. Großmann: Macroscopic time evolution and masterequations. Physica *30*, 779–807 (1964)
58. W. Balzer, J.D. Sneed: Generalized net structures of empirical theories, I and II. Studia Logica *36* (3), 195–212 (1977) and *37* (2), 168–194 (1978)
59. K.-E. Hellwig: "Measuring Processes and Additive Conservation Laws", in *Foundations of Quantum Mechanics*, ed. by B. d'Espagnat (Academic, New York 1971) p. 338
M.M. Yanase: "Optimal Measuring Apparatus", in *Foundations of Quantum Mechanics*, ed. by B. d'Espagnat (Academic, New York 1971) p. 77
60. P. Mittelstaedt: "The Concepts of Truth, Possibility and Probability in the Language of Quantum Physics", in *Interpretations and Foundations of Quantum Theory*, ed. by H. Neumann (Bibliographisches Institut Mannheim, Wien, Zürich 1981); see also the literature given there
P. Mittelstaedt, E.-W. Stachow (eds.): *Recent Developments in Quantum Logic* (Bibliographisches Institut Mannheim, Wien, Zürich 1985)
61. L. Lanz, O. Melsheimer, E. Wacker: Introduction of a Boltzmann observable and Boltzmann equation. Physica *131 A*, 520 (1985)
L. Lanz, O. Melsheimer, S. Penati: A model for an objective description of macroscopic state variables in mechanics. Physica (to be published)
62. K.R. Popper: *The Logic of Scientific Discovery* (Basic Books, New York 1959)
J. Lakatos: "Falsification and the Methodology of Scientific Research Programmes", in *Criticism and the Growth of Knowledge*, ed. by J. Lakatos, A. Musgrave (Cambridge University Press, Cambridge 1970)

# List of Frequently Used Symbols (1)

(page numbers from the first volume)

# List of Frequently Used Symbols (2)

(page numbers from the second volume)

# List of Axioms

(page numbers of the first volume)

# Index

**G. Ludwig**

# An Axiomatic Basis for Quantum Mechanics

## Volume 1
## Derivation of Hilbert Space Structure

1985. 6 figures. X, 243 pages. ISBN 3-540-13773-4

**Contents:** The Problem of Formulating an Axiomatics for Quantum Mechanics. - Pretheories for Quantum Mechanics. - Base Sets and Fundamental Structure Terms for a Theory of Microsystems. - Embedding of Ensembles and Effect Sets in Topological Vector Spaces. - Observables and Preparators. - Main Laws of Preparation and Registration. - Decision Observables and the Center. - Representation of *B*, *B'* by Banach Spaces of Operators in a Hilbert Space. - Appendices. - Bibliography. - List of Frequently Used Symbols. - List of Axioms. - Index.

This is the first volume of a work on fundamental concepts of quantum mechanics. G. Ludwig's aim is to deduce the description of microscopic objects solely from a macroscopic one of the devices used for their detection. The description of a two-part macrosystem where the microsystem is discovered as the system transmitting the interaction is the main topic of this volume. Empirically founded axioms then give raise to the Hilbert-space structure of quantum mechanics.
In this book the author demonstrates how the now famous approach to the foundations of physics works for quantum mechanics and how it also serves to find a solution to the measuring problem. This monograph will not only be an important source of inspiration for future research but should also appeal to all interested in the fundamental structure of nature and of what we may know about it.

**Springer-Verlag**
Berlin Heidelberg New York
London Paris Tokyo

Springer